Palgrave Series in Indian Ocean World Studies

Series Editor

Gwyn Campbell, Indian Ocean World Centre, McGill University, Montreal, QC, Canada

This is the first scholarly series devoted to the study of the Indian Ocean world from early times to the present day. Encouraging interdisciplinarity, it incorporates and contributes to key debates in a number of areas including history, environmental studies, anthropology, sociology, political science, geography, economics, law, and labor and gender studies. Because it breaks from the restrictions imposed by country/regional studies and Eurocentric periodization, the series provides new frameworks through which to interpret past events, and new insights for present-day policymakers in key areas from labor relations and migration to diplomacy and trade.

More information about this series at
https://link.springer.com/bookseries/14661

Philip Gooding
Editor

Droughts, Floods, and Global Climatic Anomalies in the Indian Ocean World

palgrave
macmillan

Editor
Philip Gooding
Indian Ocean World Centre
McGill University
Montreal, QC, Canada

ISSN 2730-9703 ISSN 2730-9711 (electronic)
Palgrave Series in Indian Ocean World Studies
ISBN 978-3-030-98197-6 ISBN 978-3-030-98198-3 (eBook)
https://doi.org/10.1007/978-3-030-98198-3

Cover illustration: NASA/Dembinsky Photo Associates/Alamy Stock Photo

This Palgrave Macmillan imprint is published by the registered company Springer Nature Switzerland AG
The registered company address is: Gewerbestrasse 11, 6330 Cham, Switzerland

For Émilie, Adèle, and Mathis

Acknowledgements

It is somewhat unusual to write an acknowledgements section for an edited volume, but on this occasion, I wish to break with convention.

As readers of the introductory chapter will discover, this book's origins lie in mid-2019. A lot has changed since then. Originally, the plan was to develop a project leading to a publication through an international conference at the Indian Ocean World Centre, McGill University, entitled 'Drought, floods, and rainfall anomalies in the Indian Ocean World.' That conference was scheduled to be held in May 2020. For obvious reasons, it did not go ahead. With this in mind, I wish to state on record how thankful I am to all of the contributors for seeing this project through. This includes all those who responded to the original conference call and those who joined at a later date. Thank you for your patience, your resilience, and your hard work, especially given the challenging circumstances.

I also wish to thank Peter Hynd for his instrumental role in developing the conference theme and for his encouragement thereafter. Thanks also to Sam Stocker, Supraja Ganesh, Chitra Gopalraj and their colleagues at Palgrave and Springer for turning the manuscript into this book. Thanks furthermore to the Social Sciences and Humanities Research Council (SSHRC) of Canada whose funds via a Partnership Grant held by Prof. Gwyn Campbell and an Insight Development Grant held by myself were essential to the editorship of this book. And thank you to my family, especially to Émilie, Adèle, and Mathis, for all their love, particularly in the last two years.

Philip Gooding

Contents

Notes on Contributors

Gwyn Campbell is founding Director of the Indian Ocean World Centre, McGill University, General Editor of the Palgrave Series in Indian Ocean World Studies, and Editor-in-Chief of the *Journal of Indian Ocean World Studies*. He holds degrees in Economic History from the Universities of Birmingham and Wales, and he has held a Canada Research Chair in Indian Ocean World History (2005–2019) and a Humboldt Award (2017–2019). He is the director of a SSHRC Partnership Project entitled, 'Appraising Risk, Past and Present: Interrogating Historical Data to Enhance Understanding of Environmental Crises in the Indian Ocean World.' Among his many publications are: *Africa and the Indian Ocean World from Early Times to circa 1900* (2019) and *An Economic History of Imperial Madagascar* (2005).

Archisman Chaudhuri received his doctorate in history at Leiden University (2019). Primarily trained as a historian specialising in reading the VOC (Dutch East India Company) archives, his doctoral dissertation, 'From Camp to Port: Mughal warfare and the economy of Coromandel, 1682–1710,' researched the impact of the southern military campaigns (1682–1707) of Mughal emperor Aurangzeb (r. 1658–1707) on the economy of the Coromandel Coast—a major trading hub of the erstwhile Indian Ocean. He was a postdoctoral researcher at the Indian Ocean World Centre (IOWC), McGill University (2019–2021), where he studied coeval climatic anomalies in early modern South Asia and Southeast Asia, and where still serves as a research affiliate. He is a co-editor on the international collaborative book project on the embassy of Sir William Norris to Mughal India (1699–1702), which will result in a scholarly edition of Norris' embassy diaries published by the Hakluyt Society.

William Gervase Clarence-Smith is Emeritus Professor of History at SOAS, University of London, and former chief editor of the *Journal of Global History*. He has taken part in collaborative environmental research projects on the Indian Ocean World, based at McGill University and at the University of Sussex. With Ed Emery, he coordinates the Interdisciplinary Animal Studies Initiative at SOAS. He has published on the history of various animals (including diseases of animals), agricultural and marine commodities, manufacturing, labour, diasporas, sexuality, science and technology, and religion. Much of his research has focused on the Portuguese empire, Central Africa, and Southeast Asia. He taught at the University of Zambia from 1975 to 1977, where the research for his contribution to this volume was largely carried out.

Philip Gooding is a Postdoctoral Fellow at the Indian Ocean World Centre, McGill University. He holds a Ph.D. in History from SOAS, University of London (2017). He is the author of *On the Frontiers of the Indian Ocean World: A History of Lake Tanganyika, c.1830–1890* (forthcoming), co-editor of *Animal Trade Histories in the Indian Ocean World* (2020), and holder of a SSHRC Insight Development Grant for a project entitled: 'Climate History and Human-Environment Interaction in Equatorial Eastern Africa, c.1780–1900.' He has published in several academic journals, including *The Journal of African History* and *Slavery and Abolition*. His research specialisations are the cultural and environmental histories of eastern Africa and the wider Indian Ocean World.

Stephen J. Rockel is Associate Professor of African History in the Department of Historical and Cultural Studies, University of Toronto Scarborough. He is a specialist in Tanzania and East Africa, with interests in African and Indian Ocean labour, slavery, urban, and environmental history. His book, *Carriers of Culture: Labor on the Road in Nineteenth-Century East Africa*, was published in the Heinemann Social History of Africa series (2006) and was awarded the Joel Gregory Prize by the Canadian Association of African Studies. Current projects include a history of slavery in Tanzania and the history of Tabora, a nineteenth-century commercial town.

Angela Schottenhammer (蕭婷) is Professor of the Chinese Middle Period & Early Modern World History at KU Leuven, Belgium, Selected Senior Researcher at the School of Economics at Shanghai University (经济学院, 上海大学), and research affiliate at the Geography Department, UGent. She obtained her Ph.D. in 1993 from Würzburg University with a thesis on 'Song Period Tomb inscriptions' and her Habilitation degree in 2000 from LMU Munich University, with a thesis on the port city of Quanzhou during the Song period (960–1279). She is director of the Crossroads Research Centre, and chief editor of the academic journal *Crossroads* and of two book series (*Crossroads—History of Interactions across the Silk Routes*; *East Asian Maritime History*). Her research focuses on Chinese history, archaeology, science and technology, and culture, and on China's and Asia's increasing worldwide integration, through both maritime and overland routes, with the main focus on the period between 650–1800.

Theresa Ventura is Associate Professor at Concordia University, Montreal where she teaches United States history. Her research on the American colonial Philippines places contests over environmental management and natural resource exploitation at the centre of US overseas state-building and power. More generally, she is concerned with the production of agricultural knowledge and labour regimes in plantation settings and the environmental politics and impact of the Plantationocene. Her articles on beriberi and medicalisation of food scarcity, land surveying and enclosure, and the fashioning of colonial expertise can be found in *Philippine Studies*, *Agricultural History*, *History and Technology*, and the *Journal of the Gilded Age and Progressive Era*.

Silvia Freiin Ebner von Eschenbach is Adjunct Professor for Sinology at the University of Wuerzburg and is currently working on a project at the University of Muenchen. Her project explores the ecology and economy of urban water supply on a micro-historical level under macro-historical conditions. She also has a special interest in Buddhism. Her recent publications include: 'Managing Floods and Droughts by Invocating the Water Spirits: Analyzing Prayers for Rain (*daoyu* 禱雨) and Prayers for a Clear Sky (*qiqing* 祈晴). With Some Examples from Local Source Material of the Song 宋Dynasty (960–1279),' *Zeitschrift der Deutschen Morgenlaendischen Gesellschaft* 169 (2019), 205–229; 'The Dilemma of Ecological and Nutritional Policies in view of Buddhist Campaigning: The Use of Hangzhou's Xihu 西湖 as a Pool for the Release of Living

Beings during the Northern and Southern Song Dynasties (960–1279),' *Monumenta Serica* 68:1 (2020), 69–106.

James Francis Warren is Emeritus Professor of Southeast Asian Modern History at Murdoch University, Perth, Western Australia. He is an award-winning historian who has published numerous monographs, journal articles, and book chapters. His books include *The Sulu Zone 1768–1898: The Dynamics of External Trade, Slavery and Ethnicity in the Transformation of a Southeast Asian Maritime State (1981, 2007, 2021)*; *Iranun and Balangingi: Globalisation, Maritime Raiding and the Birth of Ethnicity* (2002); *Pirates, Prostitutes and Pullers: Explorations in the Ethno-and Social History of Southeast Asia* (2008); *Rickshaw Coolie A Peoples History of Singapore* (1986, 2003); and, *Ah Ku and Karayuki San Prostitution and Singapore Society, 1870–1940* (1993, 2003).

Fiona Williamson is Associate Professor of Science, Technology and Society at Singapore Management University. She specialises in the history of climate and environment and the history of science in colonial Singapore, Malaysia, and Hong Kong. She has worked extensively on the history of meteorological science in these three countries, alongside recovering their historical climate data. Her latest research rests at the nexus of human-environmental-climatic interaction, exploring the dynamics of extreme weather and weather events, society, health, infrastructure, and politics.

List of Figures

List of Tables

CHAPTER 1

Introduction: Droughts, Floods, and Global Climatic Anomalies in the Indian Ocean World

Philip Gooding

The origins of this volume lie in a year of rainfall extremes in the Indian Ocean World (IOW). Between mid-2019 and mid-2020, drought contributed to some of the largest bushfires in Australia's known history, leading to the deaths of around 1 billion animals, among other disasters; drought contributed to haze and an air pollution crisis across much of Southeast Asia, as well as to low stands in the Mekong River and to food insecurity in its basin; drought in and around Chennai, India contributed

Thank you to William Gervase Clarence-Smith, Stephen Rockel, Fiona Williamson, and an anonymous reviewer for their comments on earlier versions of this chapter.

P. Gooding (✉)
Indian Ocean World Centre, McGill University, Montreal, QC, Canada
e-mail: philip.gooding@mcgill.ca

P. Gooding (ed.), *Droughts, Floods, and Global Climatic Anomalies in the Indian Ocean World*, Palgrave Series in Indian Ocean World Studies, https://doi.org/10.1007/978-3-030-98198-3_1

to the drying up of all four of its reservoirs, leading to both governmental and private intervention in water transportation from less-affected areas; above-average rainfall contributed to floods being recorded in different parts of Iran during every month between October 2019 and April 2020; above-average rainfall contributed to floods throughout much of north-eastern and eastern Africa, which further contributed to the formation of the largest swarm of locusts in the region for over 70 years; and persistent drought dating from the end of 2018 led the South African government to declare a state of emergency in March 2020.

On a micro level, all of these droughts and floods and their associated effects may appear distinct. How people living in the IOW experienced them (and continue to experience their after-effects, especially in the context of the Covid 19 pandemic) depended on a range of phenomena, including the significance of the rainfall extremes in each region, the nature of the physical environment, and the mitigative strategies (if any) put in place by governments, private organisations, and individuals. Thinking more broadly, however, they were all linked to broader global climatic teleconnections. These teleconnections were: A small positive El Niño Southern Oscillation (ENSO) anomaly in September 2018–June 2019, and a large positive Indian Ocean Dipole (IOD) anomaly in May 2019–December 2019. ENSO is an anomaly of sea surface temperature (SST) in the East Central Pacific, and IOD refers to SSTs in the Indian Ocean. Anomalies of both ENSO and IOD have significant teleconnections with the Indian Ocean monsoon system, on which all regions of the IOW rely for rainfall. In 2019–20, ENSO and IOD anomalies disrupted the Indian Ocean monsoon system, contributing to droughts and floods across large swathes of the IOW.

Thinking about global climatic anomalies in these terms has several consequences for studying the IOW, which this volume illuminates. Firstly, it has consequences for how the regions around the Indian Ocean are collectively conceived as a distinct 'world.' Since the 1980s, the Indian Ocean monsoon system has frequently been cited as a 'deep structure' underpinning connections across the IOW.[1] How global climatic anomalies affecting this 'deep structure' have then gone on to affect the nature of such connections is one of this volume's key themes. Secondly, it has consequences for understandings the role of global climate in the IOW, past, present, and future. Climatologists and other natural scientists have been concerned in recent years with these understandings in attempts to project the potential effects of global warming. This volume adds historical perspectives, showing the multifaceted effects of global climatic anomalies and associated droughts and floods in the IOW in the past. Finally, it has consequences for interdisciplinary humanities and social sciences. Since the 1990s especially, climatologists have used natural and human proxy records to advance our understandings of past global climate, and global climatic teleconnections therein. This volume is one of the first to integrate their data and models into an interdisciplinary history of the IOW.

'Deep Structure' and the Indian Ocean 'World'

The IOW is defined as the environmental macro-region that is reliant on rainfall from the Indian Ocean monsoon system for its agricultural systems. This includes southeastern, eastern, and northeastern Africa; the Middle East; southern, southeastern, and eastern Asia; and parts of Australasia. The Indian Ocean monsoon system is underpinned by convection. During the Eurasian Summer, the Asian landmass heats up, pushing hot air upwards, and sucking in wet air from the Indian Ocean. This is the southwestern monsoon, and it brings seasonal rainfall to much of the northern and eastern IOW. By contrast, during the Eurasian winter, dry air from the Eurasian landmass is expelled over the Indian Ocean. This is the northeastern monsoon. In theory, then, the Indian Ocean monsoon system is an annual back-and-forth of wet and dry winds, bringing seasonal rainfall to the IOW. The processes underpinning this back-and-forth rhythm are the IOW's 'deep structure.'

[1] See, for example: Michael Pearson, *The Indian Ocean* (London: Routledge, 2003), 13–26; K.N. Chaudhuri, *Trade and Civilisation in the Indian Ocean: An Economic History from the Rise of Islam to 1750* (Cambridge: Cambridge University Press, 1985), 3, 21; Gwyn Campbell, *Africa and the Indian Ocean World from Early Times to circa 1900* (Cambridge: Cambridge University Press, 2019), 1–21.

French historian and leader of the second generation of the so-called *Annales School*, Fernand Braudel (1902–85), was the first to promote the idea of 'deep structures' in history. He challenged the perspectives of his contemporaries, who stressed the importance of events, such as battles, and human structures, such as nation states, in shaping human history. He argued that events and human structures were underpinned by a 'deep structure,' namely the environmental context. Changes in this environmental context, he argued further, may have been barely perceptible to those who witnessed them. As such it was the 'slowest' form of history, and often much harder to comprehend than faster, more 'popular' histories of 'big men,' politics, and economics. Nevertheless, he argued, human actions and structures were constrained by the physical environment within which they occurred. For Braudel, understanding the environmental context as 'deep structure' was integral to the study of history. It was also the key context underpinning his conception of the Mediterranean 'world,' whose late sixteenth-century history was the subject of his most famous scholarly work.[2]

Although his ideas serve as inspiration for perspectives taken in this volume, Braudel can be criticised on several grounds. Many historians and social scientists remain uncomfortable with his apparently 'deterministic' viewpoints, even if few among them offer alternatives for how the environment should be placed within the context of human history.[3] He also underestimated the dynamism of environmental change, caused both by human actions and global climatic anomalies. His famous case study on the Mediterranean world, for example, fails to account for the deleterious effects of deforestation on the quality of soils, agricultural production, and the wider economy in the Spanish Empire.[4] It also does not deeply consider the effects of specific droughts and floods, especially during the 1590s–1610s, which he generally attributed to be symptomatic of broader

[2] See especially: Fernand Braudel, *The Mediterranean and Mediterranean World in the Age of Philip II*, 3 Vols., trans. Sian Reynolds (New York: Harper & Row, 1972–3); Fernand Braudel, *On History*, trans. Sarah Matthews (Chicago: University of Chicago Press, 1980).

[3] See, for example: David Abulafia, *The Great Sea: A Human History of the Mediterranean* (New York: Oxford University press, 2011), xxv–xxvii. For a summary of the critique, see: Peter Burke, *The French Historical Revolution: The Annales School, 1929–2014*, 2nd ed. (Cambridge: Polity, 2015), 99–108.

[4] Burke, *French Historical Revolution*, 104.

(and normal) general cycles. More recent research has shown that these droughts and floods were probably caused by a series of sulphur-rich volcanic eruptions in the context of a protracted cool period during the Little Ice Age (LIA).[5] Their effects on climate contributed to epidemics, mass-starvation, the retreat of the Spanish Empire in North America, and with a series of peasant rebellions in Ottoman Empire.[6] On closer analysis, history in the context of 'deep structure' might not have been so 'slow' after all.[7]

An appreciation of the rapid potentialities of 'deep structure' is at the core of this volume's study of droughts, floods, and global climatic anomalies in the IOW. According to research published in 2016, the IOW contains five of the ten countries most vulnerable to the effects of global warming.[8] Several factors go into making this assessment, including analysis of current mitigation strategies, the physical environment, and population density. Yet, much of the IOW's vulnerability is driven by the effects that global warming will have on the Indian Ocean monsoon system, and the subsequent impacts this will have on regional rainfall.[9] This contrasts with other 'worlds,' such as Braudel's Mediterranean world, where the climate is more temperate and rainfall less seasonal. While Braudel could make the claim that changes to the environmental context might not be visible to people living around the Mediterranean (with the qualifications outlined above), similar claims could not be made about the IOW. In the IOW, global climatic anomalies, through their teleconnections with the Indian Ocean monsoon system, can have significant and lasting consequences on regional environments, societies, and structures.

This volume shows the rapidity of the IOW's 'deep structure' in action. In some ways, it builds on recent developments in disaster history.

[5] Sam White, *A Cold Welcome: The Little Ice Age and Europe's Encounter with North America* (Cambridge, MA: Harvard University Press, 2018), 76–7.

[6] Ibid., 70–87; Sam White, *The Climate of Rebellion in the Early Modern Ottoman Empire* (New York: Cambridge University Press, 2011), 140–86.

[7] Sunil Amrith, *Unruly Waters: How Rains, Rivers, Coasts and Seas have Shaped Asia's History* (New York: Basic Books, 2018), Ch. 1.

[8] Sönke Kreft, David Eckstein, and Inga Melchoir, *Global Climate Risk Index 2017: Who Suffers Most from Extreme Weather Events? Weather-related Loss Events in 2015 and 1996 to 2015* (Bonn: Germanwatch, 2016).

[9] Ibid.

This burgeoning field has in recent years sought to examine the human and environmental contexts that affect how different people experience adverse natural phenomena, including droughts and floods, but also earthquakes, volcanic eruptions, and tsunamis.[10] Recent work challenges the distinctions between 'natural' and 'unnatural' disasters, examining how a 'nexus' between the two—adverse climatic factors and structural vulnerabilities—have combined.[11] But instead of taking disasters as individual, often localised events, this volume seeks to put different 'disasters' in conversation with each other by assessing their respective relationships to global climatic teleconnections that affect wider regions.[12] This approach is also partly inspired by recent understandings about the origins of the current global climate crisis, which scholars such as Jason W. Moore attribute to the birth and subsequent spread of capitalism from the end of the fifteenth century.[13] Climatic extremes, their causes, and societies' vulnerabilities or resilience to them are global phenomena. In this volume, therefore, droughts and floods, which are projected to increase in frequency under global warming, are understood to be a function of both climatic and structural factors, the origins and consequences of which operate at a 'world' scale.

Nevertheless, when thinking about droughts and floods specifically, concentration on 'disasters' has some limitations. Regular rainfall years/seasons can be as historically significant as years/seasons of deficient or overly abundant rainfall. Thus, although a great deal of the historiography on droughts and floods focuses on the challenges associated with

[10] For a fairly recent summary of European historiography, see: Gerrit Jasper Schenk, 'Historical Disaster Research: State of Research, Concepts, Methods, and Case Studies,' *Historical Social Research*, 32, 3 (2007), 9–31. For a more recent related study on the IOW, see: Greg Bankoff and Joseph Christensen, 'Bordering on Danger: An Introduction,' in *Natural Hazards and Peoples in the Indian Ocean World: Bordering on Danger*, eds. Greg Bankoff and Joseph Christensen (New York: Palgrave Macmillan, 2016), 1–30.

[11] Chapter by Williamson, this volume.

[12] See also: Sugata Bose, *A Hundred Horizons: The Indian Ocean in the Age of Global Empire* (Cambridge, MA: Harvard University Press, 2009), 1–2. Here, Bose argues for the unity of the IOW on the basis of collective experience of the tsunami of 26 Dec. 2004.

[13] Jason W. Moore, 'The Capitalocene, Part I: On the Nature and Origins of Our Ecological Crisis,' *Journal of Peasant Studies*, 44, 3 (2017), 594–630; Jason W. Moore, 'The Capitalocene Part II: Accumulation by Appropriation and the Centrality of Unpaid Work/Energy,' *Journal of Peasant Studies*, 45, 2 (2018), 237–79.

these phenomena, preceding periods of regular rainfall and how people acted in those periods, provide important contexts for how they were experienced. Questions need to be continually asked about structural factors that affect levels of vulnerability and resilience to adverse global climatic factors, and how such levels developed over long periods.[14] Additionally, not all droughts and floods can be tied to global climatic anomalies. Floods in flood plains and droughts in drought-prone regions are hardly anomalous conditions. Scholars have provided several examples of communities that have developed robust cultural responses to these environments, divorcing them also from histories of disasters.[15] There are thus nuances to be unpacked about what constitutes a drought or flood in the context of disaster history and global climatic anomalies. In this volume, understanding them is rooted in the idea of the 'deep structure' of the IOW, described above. It examines times during which the Indian Ocean monsoon system has been significantly disrupted, contributing to extreme levels of rainfall in diverse regions. It then analyses the extent to which such anomalies contributed to droughts and floods, whose severity was sometimes such that they constituted 'disasters' that had significant cascading effects on IOW societies.

Global Climate and the Indian Ocean Monsoon System

Understandings of global climatic anomalies and their effects on the Indian Ocean monsoon system have developed significantly in the last three decades. The reasons for this are directly tied to growing concerns with the effects of global warming. Climatologists and other natural scientists have used natural and human proxies to gain a clearer picture of past climatic changes in order to improve models for future climate

[14] For examples of long-term developments in vulnerability and resilience, see respectively: White, *The Climate of Rebellion*, 15–122; Dagomar Degroot, *The Frigid Golden Age: Climate Change, the Little Ice Age, and the Dutch Republic, 1560–1720* (Cambridge: Cambridge University Press, 2018), 18.

[15] For a celebrated example from northeastern Africa, see several chapters in: Douglas H. Johnson and David Anderson, eds. *The Ecology of Survival: Case Studies from Northeast African History* (London: L. Crook Academic Pub., 1988).

scenarios.[16] They have also used this data to model more precisely the effects of global oscillations and anomalies, such as ENSO, IOD, volcanism, sunspot activity, cyclones, migrations of the Intertropical Convergence Zone (ITCZ), and humans' activities on global climate. This section of the introduction represents a summary of the current scientific knowledge of these factors, and how they affect rainfall in different IOW regions, contributing to droughts and floods therein. As such, these factors represent the global climatic contexts that underpin many of the droughts and floods addressed in the chapters that follow.

Before detailing how these factors affect climate in the IOW, it is important to first make some caveats. Firstly, some of these factors are better understood than others, as are some of the links between them. What is written in the following is a summary of 'the best of our knowledge.' Ongoing research may add precision in the relatively near future. Secondly, the list of anomalies addressed is not exhaustive, rather it reflects those that appear more frequently in the remainder of this volume. Scholars may also be interested to research the effects of Madden–Julian Oscillation, the Pacific Decadal Oscillation, and the North Atlantic Oscillation, the latter of which plays a significant role in the climate of the northwestern IOW, for example. Finally, these summaries are descriptions of models and only that. There are several 'exceptions to the rule.'[17] It is partly for this reason that contributors to this volume frequently use the terms 'associated with' or 'contributed to,' rather than 'caused,' to describe the relationships between global climatic anomalies and above/below-average rainfall, and thus also to drought/flood events, in the IOW. These models are useful for drawing global linkages, such as across the regions affected by the Indian Ocean monsoon system, but they should not always be used to imply causality.

Fluctuating SSTs, especially those related to ENSO and the IOD, significantly affect rainfall in the IOW. ENSO refers to anomalies of SST in the equatorial east-central Pacific Ocean. Positive anomalies, when SSTs

[16] For a summary, see: White, *A Cold Welcome*, 6; Christian Pfister, Sam White, and Franz Mauelshagen, 'General Introduction: Weather, Climate, and Human History,' in *The Palgrave Handbook of Climate History*, eds. Sam White, Christian Pfister, and Franz Mauelshagen (London: Palgrave Macmillan, 2018), 3–6.

[17] The most famous recent example of such 'rule breaking' is probably of normal rainfall in India in 1997–8 during the strongest positive ENSO event of the twentieth century. Richard Grove and George Adamson, *El Niño in World History* (London: Palgrave Macmillan, 2018), 2.

are higher than normal, are known as El Niño; negative anomalies as La Nina. IOD refers to a SSTs in the Indian Ocean. A positive IOD refers to when SSTs are warmer in the western half of the ocean than in the eastern half; a negative IOD is the opposite phenomenon. The significance of both ENSO and IOD anomalies and their effects on global climate vary between and during events, and they usually occur every 3–5 years. ENSO anomalies may last a few months or a couple of years; positive IOD anomalies occur between May and December. Both cause changes in atmospheric pressure, which destabilises in the Indian Ocean monsoon system. Broadly speaking, both positive ENSO anomalies and positive IOD anomalies are associated with drought in island southeastern Asia, northern China, mainland South Asia, and northeastern and southeastern Africa, and with above-average rainfall in Sri Lanka and equatorial East Africa—although the extent of the correlations is not consistent. Negative anomalies are generally associated with the opposite phenomena.[18] ENSO and IOD anomalies frequently occur together, suggesting that one (ENSO) can trigger the other (IOD), although the evidence for this is far from conclusive. Some of the largest IOD anomalies on record, such as in 1961 and 2019, occurred independently of ENSO.[19]

Fluctuating global temperatures also affect rainfall in the IOW. The causes of changing global temperatures over the *longue durée* are diverse. For example, sulphur-rich volcanic eruptions obscure both light and radiation from the sun, contributing to global cooling for up to five years[20]; heightened/decreased sunspot activity, which may last more than a century, contributes to global warming/cooling; and increased greenhouse gas emissions, especially in the last c.250 years, contribute to global warming.[21] Both global cooling and warming destabilise the forces of convection that underpin the Indian Ocean monsoon, contributing to erratic rainfall in the IOW. They also affect ENSO: sulphur-rich volcanic

[18] Ibid., 5.

[19] N.H. Saji, B.N. Goswami, P.N. Vinayachandran, and T. Yamagata, 'A Dipole Mode in the Tropical Indian Ocean,' *Nature*, 401 (1999), 360–3.

[20] Campbell, *Africa and the IOW*, 17.

[21] Stefan Brönnimann, 'Global Warming (1970-Present),' in *The Palgrave Handbook*, eds. White, Pfister, and Mauelshagen, 321–8.

eruptions are broadly associated with El Niño events[22]; some past solar minimums (when sunspot activity is low) have been associated with La Nina-like conditions[23]; and current models predict that global warming will lead to ENSO anomalies becoming more extreme.[24]

Thinking more long term, rainfall in the IOW is modulated by migrations of the ITCZ, a belt of low pressure that circles the globe near the equator. In the IOW, the ITCZ moves seasonally with the location where northwestern and southeastern monsoon winds meet, bringing rainfall to the regions it is located over. Therefore, long-term displacements of the ITCZ, in which its annual position is further north or south than normal, significantly affect how much monsoon rainfall different regions in the IOW receive. Southern displacements of the ITCZ are associated with decreased levels of rainfall over southern Asia; northern displacements with the opposite.[25] Such displacements can last centuries. Additionally, global warming is associated with a narrowing of the ITCZ, which contributes to increased levels of rainfall intensity, thus increasing the likelihood of floods in tropical regions of the IOW.[26]

Tropical cyclones also contribute to floods in parts of the IOW. The IOW has three cyclone zones: In the Southwest around Madagascar, the Mascarenes, and Mozambique; in the North on either side of India; and

[22] Shayne McGregor, Myriam Khodri, Nicola Maher, Masamichi Ohba, Francesco S.R. Pausata, and Samantha Stevenson, 'The Effect of Strong Volcanic Eruptions on ENSO,' in *El Niño Southern Oscillation in a Changing Climate*, eds. Michael J. McPhaden, Agus Santoso, and Wenju Cai (Hoboken, NJ: Wiley, 2021), 267–87; Chapter by Ventura, this volume.

[23] Kam-biu Liu, Caiming Shen, and Kin-sheun Louie, 'A 1,000-year History of Typhoon Landfalls in Guangdong, Southern China, Reconstructed from Chinese Historical Documentary Records,' *Annals of the Association of American Geographers*, 91, 3 (2001), 461.

[24] Y.G. Ham, 'El Niño Events will Intensify under Global Warming,' *Nature*, 564, 7735 (2018), 192–3.

[25] Franziska A. Lechleitner, Sebastian F.M. Breitenbach, Kira Rehfeld, Harriet E. Ridley, Yemene Asmerom, Keith M. Prufer, Norbert Marwan, Bedartha Goswami, Douglas J. Kennett, Valorie V. Aquino, Victor Polyak, Gerald H. Haug, Timothy I. Eglinton, and James U.L. Baldini, 'Tropical Rainfall over the Last Two Millennia: Evidence for a Low-latitude Hydraulic Seesaw,' *Scientific Reports*, 7, 45, 809 (2017), 1–9; Campbell, *Africa and the IOW*, 70, 135.

[26] Michael P. Byrne, Angeline G. Pendergrass, Anita D. Rapp, and Kyle R. Wodzicki, 'Response of the Intertropical Convergence Zone to Climate Change: Location, Width, and Strength,' *Current Climate Change*, 4, 4 (2018), 355–70.

in the China Seas, around southeastern China and parts of Southeast Asia (where they are referred to as typhoons). Cyclones form in warm waters, thus global warming is generally associated with their increased strength and frequency the world over.[27] Additionally, positive IODs are associated with increased frequency and intensity of cyclones in the northwest Indian Ocean, as occurred in 2019. However, in eastern and southeastern Asia, cooler SSTs—often associated with La Nina—cause their average place of origin to move westwards, and so global cooling is associated with increased likelihood of typhoons making landfall.[28] Similar patterns are observable with cyclones in the southwest Indian Ocean.[29] The location and frequency of tropical cyclones in the IOW are significantly affected by SSTs and global temperatures.

In summary, although listed independently, the global climatic anomalies that contribute to droughts and floods in the IOW are heavily interlinked, and many likely cannot be understood fully in isolation. A change in the size, frequency, or movement of one often has a cascading effect on another. Thus, the global climatic factors contributing to droughts and floods in the IOW are invariably numerous and complex. This illustrates further the need for historians to not always imply causality between what may appear to be a distinct global climatic anomaly and varying levels of rainfall in the IOW. There may be several climatic factors at play, and the linkages between them may be little understood. It is hoped that this volume's examination of past drought and flood events in the context of global climate may aid climatologists in their efforts to further understand teleconnections between ostensibly distinct global climatic anomalies.

[27] Thomas Knutson, Suzana J. Camargo, Johnny C.L. Chan, Kerry Emanuel, Chang-Hoi Ho, James Kossin, Mrutyunjay Mohapatra, Masaki Satoh, Masato Sugi, Kevin Walsh, and Liguang Wu, 'Tropical Cylcones and Climate Change Assessment: Part II: Projected Response to Anthropogenic Warming,' *Bulletin of the American Meteorological Society*, 101, 3 (2020), E303–22.

[28] Liu, 'A 1,000-year History of Typhoon Landfalls,' 453–64; James B. Elsner and Kam-biu Liu, 'Examining the ENSO-Typhoon Hypothesis,' *Climate Research*, 25 (2003), 43–54; Chapters by Schottenhammer and Warren, this volume.

[29] Lan Xia, Hans von Storch, Frauke Feser, and Jian Wu, 'A Study of Quasi-millennial Extratropical Winter Cyclone Activity over the Southern Hemisphere,' *Climate Dynamics*, 47, 7–8 (2016), 2121–38.

Droughts and Floods in IOW History

Incorporation of global climate data and climatic models situates this volume in the growing interdisciplinary field of climate history.[30] As with the science that underpins it, this field's growth in the last two decades is linked to present-day concerns with global warming. If humans' presents and futures are indelibly shaped by global warming, climate historians ask, how have their pasts also been shaped by their changing climatic contexts?[31] Thinking about this very broad question has led to recent reappraisals of several well-known events and processes in world history, ranging from the European colonisation of the Americas, to the rise and fall of dynasties in China, and the transition between the ancient and medieval eras centred on the Indo-Mediterranean world(s).[32] Climate historians have shown that climatic changes and climatic anomalies have indelibly shaped human history over the *longue durée*, as well as during somewhat singular events. Droughts and floods, underpinned by global climatic anomalies, are a key feature of these histories.

The sources are variable—in number and in quality, depending on region—for making climate histories of the IOW. For rainfall, on which this volume focuses, freely available rain gauge data exists from the beginning of the nineteenth century in some regions. Chennai, for example, has consistent monthly data from 1813 to the present. But, coverage for much of the rest of IOW does not start until around the 1870s–80s, and data for Southeast Asia and eastern Africa is particularly sparse until the beginning of the twentieth century.[33] Climate proxy data, such as tree rings, ice cores, pollen records, corals, stalagmites, and lake sediments, analysed and published by climatologists and other natural scientists, is

[30] For an early summary of what this entails, see: Emmanuel Le Roy Ladurie, *Times of Feast, Times of Famine: A History of Climate since 1000*, trans. Barbara Bray (Garden City: Doubleday, 1971), 18–22.

[31] Sverker Sörlin and Melissa Lane, 'Historicizing Climate Change—Engaging New Approaches to Climate History,' *Climatic Change*, 151, 1 (2018), 1–13.

[32] For the European colonisation of the Americas, see: White, *A Cold Welcome*. For the rise and fall of Chinese dynasties, see: Ka-wai Fan, 'Climatic Change and Dynastic Cycles in Chinese History,' *Climatic Change*, 101, 3–4 (2010), 565–73. For end of the ancient world, see: Campbell, *Africa and the IOW*, 71–2.

[33] This data is available using the World Meteorological Organisation's climate explorer, made available by the Royal Netherlands Meteorological Institute. See: https://climexp.knmi.nl/start.cgi?id=someone@somewhere [Accessed: 11 Jan. 2021].

often necessary to gain a picture of rainfall before the use of rain gauges, and where the rain gauge data is thin. Limnological and paleolimnological studies, for example, measure fluctuating levels of lakes over time, and some such studies analyse periods since before the beginning of the common era. Climatologists have then examined the extent to which such fluctuations are representative of depressed or increased levels of rainfall, thus providing in some circumstances a sense of past drought and flood events.[34] Similarly, tree ring, coral, pollen, and ice core data shed light on regional rainfall and global climatic proxies such as ENSO, among other climatic anomalies.[35] Again, analysis of these kinds of sources can point to past periods of above and below-average rainfall.

The addition of sources that are often more familiar to historians and social scientists can add precision to these climatological studies, particularly for histories of droughts and floods. Many archives, for example, contain rain gauge data. In some cases, such data has been added to climatologists' databases, but continual discoveries show that there is still more to be found.[36] Moreover, observers whose documents are now found in archives or are recorded in oral testimonies frequently reported on unusual weather events. These sources have allowed historians and climatologists 'to grasp the full scale of environmental disasters,' such as droughts and floods.[37] The particularity of such events is that they often last only a season or year. The data from proxy records, by contrast, is sometimes more useful for establishing general trends over longer periods.

[34] See, for example: Sharon E. Nicholson, 'Historical and Modern Fluctuations of Lakes Tanganyika and Rukwa and their Relationship to Rainfall Variability,' *Climatic Change*, 41 (1999), 53–71.

[35] For ENSO, see: Joëlle L. Gergis and Anthony M. Fowler, 'A History of ENSO Events since A.D. 1525: Implications for Future Climate Change,' *Climatic Change*, 92 (2009), 343–87.

[36] Sharon E. Nicholson, 'Climatology: Methods,' *Oxford Research Encyclopedia of African History* (2017) [Accessed: 15 Jan. 2021]; Sharon E. Nicholson, 'A Semi-quantitative, Regional Precipitation Data Set for Studying African Climates of the Nineteenth Century, Part I. Overview of the Dataset,' *Climatic Change*, 50, 3 (2001), 317–53. Note, many such data points are not available in the WMO's climate explorer. As far as I am aware, for example, data in the following source have never been included in an eastern African rainfall dataset: Edward C. Hore, *Tanganyika: Eleven Years in central Africa* (London: Edward Stanford, 1892), 145; See also: Chapters, by Clarence-Smith and Campbell in this volume.

[37] Sarah Kate Raphael, *Climate and Political Climate: Environmental Disasters in the Medieval Levant* (Leiden: Brill, 2013), 1.

Lake levels, for example, are more often subject to changes in long-term cycles of rainfall than they are to a singular season of drought or overly abundant rain—and they may also be affected by other phenomena, such as the development or destruction of naturally-forming dams at their outlets.[38] Such and other proxy data often lend themselves more to analysis of megadroughts (droughts that last over a decade) than to short-term events, such as which form the core of this volume.[39] Traditional historical methods are often a necessary addition to those exclusively used by climatologists for reconstructing somewhat singular rainfall events.

The number and type of historical sources available to climate historians and historical climatologists of the IOW varies depending on region and time period. To take two examples from differing ends of the spectrum: Official Chinese records have detailed accounts of abnormal weather that date from the turn of the common era[40]; by contrast, the documentary record for the Great Lakes region of eastern Africa only begins in the mid-nineteenth century and was first written by European outsiders who had significantly distorted understandings of the climates, environments, and peoples they encountered.[41] Thus, it might not be surprising that significantly more has been written on the climate history of China than on eastern Africa, especially for the deeper past.[42] Even so, both these types of sources pose challenges to historians. Successive Chinese dynasties destroyed their predecessors' records, and so drought and flood events may be absent or unevenly reported in the surviving

[38] Nicholson, 'Historical and Modern Fluctuations,' 53–71; Chapters by Gooding and Rockel, this volume.

[39] See, for example: Brendan M. Buckley, Kevin J. Anchukaitis, Daniel Penny, Roland Fletcher, Edward R. Cook, Masaki Sano, Le Canh Nam, Aroonrut Wichienkeeo, Ton That Minh, and Truong Mai Hong, 'Climate as Contributing Factor in the Demise of Angkor Cambodia,' *Proceedings of the National Academy of Sciences*, 107, 15 (2009), 6748–52.

[40] Fan, 'Climatic Change and Dynastic Cycles,' 568; Chapters by Ebner von Eschenbach and Schottenhammer, this volume.

[41] Philip Gooding, 'Tsetse Flies, ENSO, and Murder: The Church Missionary Society's Failed East African Ox-cart experiment of 1876–78,' *Africa: Rivista semestrale di studi e ricerche*, N.S. 1, 2 (2019), 21–36.

[42] Fiona Williamson, 'The "Cultural Turn" of Climate History: An Emerging Field for Studies of China and East Asia,' *Wiley Interdisciplinary Reviews: Climate Change*, 11, 3 (2020), 2; Nicholson, 'Climatology: Methods.'

archive.[43] Sinologists have thus often supported their archival research through incorporation of climatic proxy data.[44] Eastern Africanists, meanwhile, have sometimes made use of oral traditions (despite the difficulty of placing such traditions in time and space), which have often traced the foundation of kingdoms and societies to migrations, which themselves were sometimes provoked by abnormal weather.[45] This, again in conjunction with climate proxy data, has enabled attempts at reconstructing drought and flood events in the region to the fifteenth-sixteenth centuries.[46] Incorporating diverse source types is a necessary precursor to the climate history of the IOW, especially when referring to the deeper past.

'Traditional' historical material is especially important for understanding drought and flood events. This is because no matter how accurate the climatological record, climate alone cannot explain how droughts and floods are experienced, or their severity.[47] Human activity can both exacerbate or moderate the effects of extreme rainfall through changing land-use and through building mitigative strategies. In terms of mitigative strategies, examples include water and food storage; construction of flood barriers, effective drainage systems, and reservoirs; and mobilisation of effective relief infrastructure. Additionally, the extent to which these strategies work or are accessible varies across time and space, and between social and demographic groups. Women, the poor, the young, and the disenfranchised are often disproportionately vulnerable. As examples: Relief efforts frequently fail to reach the poorest areas; inadequate housing increases the risk of floods washing homes away; women tasked with collecting water and fuel during periods of drought are often forced to walk further, and girls are forced to drop out of school to assist

[43] Wenxian Zhang, 'Dang An: A Brief History of the Chinese Imperial Archives and Its Administration,' *Journal of Archival Organization*, 2, 1–2 (2004), 17–38.

[44] Fan, 'Climatic Change and Dynastic Cycles,' 565–73; Chapters by Ebner von Eschenbach and Schottenhammer, this volume.

[45] Philip Gooding, 'History, Politics, and Culture in Central Tanzania,' *Oxford Research Encyclopedia of African History* (2019) [Accessed 15 Jan. 2021], 3; J.B. Webster, 'Noi! Noi! Famines as an Aid to Interlacustrine Chronology,' in *Chronology, Migration, and Drought in Interlacustrine Africa*, ed. J.B. Webster (New York: Africana Pub. Co., 1979), 1–37.

[46] Nicholson, 'A Semi-quantitative, Regional Precipitation Data Set,' 317–53.

[47] Ladurie, *Times of Feast*, 17.

them. Institutions and structures thus shape the effects of climatic anomalies on environments and societies. Details of these features of history are only available in materials that historians are already familiar with.

Once scholars have used climatological and historical methods to identify and reconstruct past drought and/or flood events, a subsequent challenge is to place them in context. In thinking about this process in the IOW specifically, it is useful to build on Greg Bankoff and Joseph Christensen's recent work on 'natural hazards' in the IOW. They argue that natural hazards, including droughts and floods, are regular enough in the IOW that they 'facilitate cultural adaptation,' but they are also occasionally severe enough that they 'constitute a crisis or a "turning point" in the history of those who experienced [them].'[48] This argument is rooted in a wider 'cultural turn' currently taking place in climate history, whose roots lay in Europe, but which is now spreading to analyses of some IOW regions.[49] Collectively, this work shows the centrality of climate and climatic fluctuations to IOW studies. Thus, droughts and floods are here considered central to understanding the IOW, both in terms of everyday life and in relation to seminal moments and events that have shaped and re-shaped its history. This statement does not only apply to the IOW's past, but also to its present and future. Indeed, it especially applies to the latter, given the projections that suggest that global climatic anomalies, contributing to anomalous levels of rainfall in the IOW, will become more frequent and extreme as a result of global warming.

The importance of droughts and floods in IOW studies is established through analysing the cascading effects that stem from them. This is partly apparent from analysis of the droughts and floods associated with the positive ENSO and IOD anomalies in 2018–19, discussed in the opening paragraphs of this introductory chapter. Droughts and floods are, for example, frequently associated with failed harvests and thus also with food insecurity, as was the case in the Mekong River basin in 2019. Thus, many histories of the most serious droughts in history also analyse famine.[50]

[48] Bankoff and Christensen, 'Bordering on Danger,' 6.

[49] Williamson, 'The "Cultural Turn" of Climate History,' 1–10; Sarah Carson, 'Atmospheric Happening and Weather Reasoning: Climate History in South Asia,' *History Compass* (2020), 1–13; Ruth Morgan, 'Climate, Weather, and Water in History,' *WIREs Climate Change*, 10, 1 (2019), 1–13.

[50] Mike Davis, *Late Victorian Holocausts: El Niño famines and the Making of the Third World* (London: Verso, 2002); Deepti Singh, Richard Seager, Benjamin I. Cook, Mark

Such famines have sparked migrations of people to less-affected areas, or in search of wages to purchase food and/or relief supplies.[51] Thinking about drought and famine in these contexts necessarily invokes histories of labour and of institutional responses to natural disasters.[52] The 2019–20 examples from Chennai and South Africa are examples of the latter, in which states used their response to droughts to extend their influence over their citizen's lives. This has precedent in the IOW's deeper past, as recent studies have shown that 'natural' disasters have enabled some states and empires to strengthen their apparatus, which had long-term ramifications for governmental control.[53] By contrast, the inability to mitigate against the effects of droughts and floods has contributed in other instances to the undermining of central authority.[54] Droughts and floods, therefore, can be highly political events.

Organisms other than humans are also, of course, affected by droughts and floods. This is evident from, in 2019–20, the locust swarms in eastern Africa and the mass deaths of fauna in Australia's bushfires. Analysis of these effects necessarily invokes the growing scholarly field of animal studies. Indeed, scholars studying animals in the IOW have analysed animal vulnerabilities in the context of adverse climatic factors, as well as shifts in fragile human–animal relationships. They have argued that humans often increase their exploitation of animals—for their flesh and for trades of their products—in times of drought and flooding, which is disastrous for non-human animal populations.[55] Additionally, apart from

Cane, Mingfang Ting, Edward Cook, and Mike Davis. 'Climate and the Global Famine of 1876–78,' *Journal of Climate*, 31, 23, (2018), 9445–67; Richard Pankhurst and Douglas H. Johnson, 'The Great Drought and Famine of 1888–92 in Northeast Africa,' in *The Ecology of Survival*, eds. Johnson and Anderson, 47–72; Chapter by Rockel, this volume.

[51] Campbell, *Africa and the IOW*, 248–53.

[52] For labour, see: Chapter by Rockel in this volume.

[53] Michael Christopher Low, *Imperial Mecca: Ottoman Arabia and the Indian Ocean Hajj* (New York: Columbia University Press, 2020), 131; Zozan Pehlivan, 'El Niño and the Nomads: Global Climate, Local Environment, and the Crisis of Pastoralism in Late Ottoman Kurdistan,' *Journal of the Economic and Social History of the Orient*, 63, 3 (2020), 318; Davis, *Late Victorian Holocausts*.

[54] Kathryn Dyt, 'Emperor Tự Đức's 'Bad Weather': Interpreting Natural Disasters in Vietnam, 1847–1883,' in *Natural Hazards and Peoples*, eds. Bankoff and Christensen, 169–98; Chapters by Schottenhammer and Ventura, this volume.

[55] Martha Chaiklin and Philip Gooding, 'Introduction: Investigating Animals, Their Products, and Their Trades in the Indian Ocean World,' in *Animal Trade Histories in the*

food insecurity, the locust swarms in eastern Africa point to the theme of disease in IOW studies. Many insects (though not locusts) in the IOW carry diseases, such as malaria, dengue fever, and sleeping sickness, and their populations often grow during the flood and after drought events. Thus, floods have, in the past, contributed to short-term epidemics of insect-borne diseases, in addition to water-borne diseases, such as cholera, whose relationship to flooding is much better known.[56] This is in addition to the diseases spread by humans during times of drought, as hungry people with weakened immune systems migrate, contributing to regional epidemics.[57]

Of course, these examples are not exhaustive. Recent scholarly work and the chapters of this volume implicate a range of additional scholarly fields, including histories of science, medicine, and infrastructure. But the examples given here, inspired by the effects of the 2018–19 positive ENSO and IOD events, are illustrative of a broader point—that droughts and floods are integral features of IOW studies, past, present, and future. Indeed, considering droughts and floods using a somewhat fluid temporality in IOW studies is key. As several scholars have argued, understanding the past—in terms of climate reconstruction, the effects of 'natural' disasters, and the success or failure of mitigative strategies—is crucial for making preparations against the effects of present- and future-day global warming.[58] It is hoped that, by using interdisciplinary methods to reconstruct past drought and flood events in the context of global climatic anomalies and human–environment interaction, this volume sheds light on the wide-ranging effects that droughts and floods have in the IOW, and on the ways in which humans can contribute to their exacerbation or their mitigation. Histories of droughts and floods are

Indian Ocean World, eds. Martha Chaiklin, Philip Gooding, and Gwyn Campbell (Cham, CH: Palgrave, 2020), 15.

[56] Gooding, 'Tsetse Flies, ENSO, and Murder,' 21–36; Campbell, *Africa and the IOW*; Chapters by Schottenhammer, Chaudhuri, and Warren, this volume.

[57] Chapters by Gooding and Williamson, this volume.

[58] Indian Ocean World Centre, 'Appraising Risk': https://www.appraisingrisk.com [Accessed 12 Jan. 2021]; Bankoff and Christensen, 'Bordering on Danger,' 21; Katie Holmes, Andrea Gaynor, and Ruth Morgan, 'Doing Environmental History in Urgent Times,' *History Australia*, 17, 2 (2020), 230–51; George Adamson, ''The Most Horrible of Evils': Social Responses to Drought and Famine in the Bombay Presidency, 1782–1857,' in *Natural Hazards and Peoples*, eds. Bankoff and Christensen, 79.

crucial to understandings the IOW—past, present, and (especially under the predicted effects of global warming) future.

Chapters

The chapters in this volume are organised in a roughly chronological order. There are some overlaps in time, and so a rigid chronological ordering is impossible, but the intention remains the same: Organising by chronology has a two-fold effect. First, it shows the methods available to climate historians referring to different regions at different points in time. Thus, given the relative abundance of archival materials available to scholars of the IOW focusing on China compared to those of some other IOW regions, it should not be surprising that the first two substantive chapters of this volume focus on China. Meanwhile, as other regions are introduced, much of the focus is on climate reconstruction, using proxy data in conjunction with traditional historical materials. By contrast, the availability of rain gauge data for more recent periods means that the focus shifts in later chapters away from historical climatology, and towards methods that most historians and other social scientists will be more familiar with, often centred on archival materials. The effect, then, is an illustration of the various methodologies used by climate historians analysing different periods in IOW history.

The second effect of the chronological ordering is that it places each of the chapters in a *longue durée* context. This builds on Gwyn Campbell's recent monograph, *Africa and the IOW from early times to circa 1900*, which itself built on some of Braudel's core ideas.[59] Campbell argued that the IOW economy has gone through a series of upturns and downturns over the *longue durée*, and that these trends went in parallel with long-term changes in patterns of human–environment interaction in the context of fluctuations in the nature of the Indian Ocean monsoon system. Thus, many of the chapters in this volume refer to events during periods of generally adverse climatic conditions in the IOW. There are four chapters, for example, that refer to events during the long seventeenth century (c.1585–1710), which was a period of global climatic instability, associated with heightened ENSO activity, several sulphur-rich

[59] Campbell, *Africa and the IOW*; Braudel, *The Mediterranean*; Fernand Braudel, 'History and the Social Sciences: The *Longue Durée*,' trans. Immanuel Wallerstein, *Review (Fernand Braudel Center)*, 32, 2 (2009), 171–203.

volcanic eruptions, and the Maunder solar minimum (c.1645–1715).[60] Additionally, two climate reconstructions add to historical knowledge of droughts and floods in the context of a series of global climatic anomalies associated with the Wolf solar minimum (c.1790–1830), extreme ENSO events, and sulphur-rich volcanic eruptions in the period c.1780–1840.[61] Finally, six chapters focus on case studies between the 1860s and 1910s, which was a period of unusually frequent and strong ENSO events in the context of expanding European imperialism, which increased many IOW regions' and people's vulnerability to the effects of droughts and floods.

In the first substantive chapter (Chapter 2), Silvia Ebner von Eschenbach uses archival materials to reconstruct flood events during China's Southern Song Dynasty (1127–1279) in Hangzhou, the Southern Song's capital. Flood events were especially frequent and concerning to the rulers of the Southern Song. Hangzhou's coastal location, a shift in the Yellow River, climatic cooling in East Asia, and an agricultural regime that was reliant on a limited supply of fresh water, made the capital and its economy vulnerable to the effects of tidal and monsoon-related flooding. Ebner von Eschenbach's chapter thus analyses the mitigative strategies that Southern Song rulers sought to put in place to limit floods' possible effects and their successes and failures therein. The most enduring and effective strategy appears to have been tax relief when floods turned into disasters, despite attempts at building infrastructure to prevent floods from occurring as well.

As with Ebner von Eschenbach, in Chapter 3, Angela Schottenhammer uses archival materials to reconstruct flood events in China. Her focus, though, is on floods resulting from typhoons during a protracted cool period (c.1550–1680) of the LIA, during which La Nina conditions prevailed. Schottenhammer then uses this context to draw links between typhoons, floods, and epidemics, before further examining institutional efforts to mitigate against these phenomena, as well as cultural and

[60] Geoffrey Parker, *War, Climate Change, and Catastrophe in the Seventeenth Century* (New Have, CT: Yale University Press, 2013).

[61] Vinita Damodaran, Rob Allan, Astrid E.J. Ogilvie, Gaston R. Demarée, Joëlle Gergis, Takehiko Mikami, Alan Mikhail, Sharon E. Nicholson, Stefan Norrgård, and James Hamilton, 'The 1780s: Global Climate Anomalies, Floods, Droughts, and Famines,' in *The Palgrave Handbook*, eds. White, Pfister, and Mauelshagen, 517–50; Christian Pfister and Sam White, 'A Year without Summer, 1816,' in *The Palgrave Handbook*, eds. White, Pfister, and Mauelshagen, 551–61.

intellectual responses therein. The discussion implicates a history of infrastructure development, labour recruitment, provisioning, and science—the latter through the development of practical measures and recordings to prevent the spread of disease. As with the volume as a whole, the weaving between various historical themes and sub-fields shows the centrality of droughts, floods, and global climate to understandings of the IOW.

Archisman Chaudhuri's chapter (Chapter 4) picks up the history of the IOW temporally where Schottenhammer's leaves off. His case study uses the archives of the Dutch East India Company (VOC) to analyse the history of the Coromandel Coast, in present-day southeastern India, in 1685–7. In these years, the region experienced both drought and floods, associated with successive El Niño and La Nina events. Chaudhuri argues that developments inland, especially earlier in the 1680s, made the Coromandel Coast especially vulnerable to famine, disease, and other associated effects during severe climatic anomalies. He then draws linkages between ENSO-related famines in southeastern India and similar phenomena in Southeast Asia at the same time, including in Batavia (Jakarta), the VOC's regional capital. In so doing, he explores how global climatic teleconnections affected early European networks in the IOW. This approach positions climatic teleconnections as key to understanding human connections across the macro-region.

William G. Clarence-Smith's chapter (Chapter 5) reconstructs climate in the Upper Zambezi River catchment in the 1680s–1910s. It is somewhat a case study into the methodological challenges and opportunities of climate history in the IOW during different epochs. For earlier periods, Clarence-Smith relies on several proxy records. Then, from the mid-nineteenth century, he incorporates first-hand qualitative data from travelers and missionaries. And finally, for the late nineteenth and early twentieth centuries, he brings in primary rain gauge data. Through this methodological versatility, Clarence-Smith is able to incorporate the Upper Zambezi basin into a history of the IOW for the first time. While 'disasters' are not the core focus of the chapter, he shows the centrality of varying annual floods in two floodplains to the region's history. Touching on several themes, including histories of political change and epidemics, anomalous levels of rainfall, which contribute to overly abundant or insufficient annual floods, are considered alongside years of regular rainfall to reconstruct a *longue durée* enviro-climatic history.

Gwyn Campbell's chapter (Chapter 6) inserts climatic and environmental factors into understandings of political instability in 1820s

Imerina, Madagascar. It is set during a period of heightened climatic volatility in the western IOW—including the regions described by Clarence-Smith—associated with the last decades of the LIA. It builds on Campbell's previous work, in which he attributed the changes described in largely political and economic terms. Here, using archival sources, he inserts the environment into the picture. He shows how changing patterns of human–environment interaction and political and military expansion increased vulnerabilities to the effects of drought. He then ties the effects of two severe droughts in the 1820s to significant political change, which then underpinned much about Imerina's domestic and foreign policies during the remainder of the nineteenth century. Thematically defined, Campbell's chapter analyses the potential intersections of global climate and political (in)stability.

James F. Warren's chapter (Chapter 7) is the first of the remaining chapters to focus on droughts and/or floods in the context of a series of strong ENSO anomalies that took place between the 1860s and 1910s. His contribution examines a 'great flood,' which was underpinned by an El Niño-related typhoon in September 1867, whose worst effects occurred in the Abra River Valley in Northern Luzon, the Philippines. He shows how Spanish colonial policies in the preceding decades increased people's vulnerabilities to the effects of flooding. Using official records and newspapers, he then explores how different people of different backgrounds and demographics were affected by the flood event, and how some—notably a 'female first responder'—were able to provide relief. By linking this event to El Niño, Warren provides a historical perspective on the dangers of heightened SSTs for residents of the Philippines in the context of typhoons. This is a topic that has garnered significant scientific attention in recent years owing to a series of disasters since c.2000, and such disasters' links to global warming and rising SSTs.

Fiona Williamson's chapter (Chapter 8) takes up the theme of El Niño in the eastern IOW by focusing on three droughts in Singapore and its surrounding zones in 1877, 1902, and 1911. A key feature of her analysis is that the year with the deepest drought according to rainfall statistics was not the one with the most deleterious consequences for Singapore's inhabitants. Her analysis thus shows the ways in which human structures can exacerbate or mitigate against the effects of global climate anomalies, building vulnerabilities or resilience therein. Key themes in this context include demographic growth, infrastructure, and colonial governance. Subsequently, Williamson goes on to explore the droughts' effects on disease, economic inequality, and social tensions in Singapore, as well as on colonial science in the wider IOW. Again, droughts are positioned as a crucial context for understanding wide-ranging developments in the IOW's history.

Williamson's chapter provides some of the global climatic contexts for Philip Gooding's chapter (Chapter 9), which also analyses the 1877–1878 El Niño, alongside other concurrent climatic anomalies. Gooding's case study is equatorial eastern Africa, and his chapter builds on several works by historians (including Williamson) and climatologists alike who have analysed this global climatic anomaly's effects on human societies in other parts of the IOW. The chapter analyses both a drought and floods, which occurred in 1876 and 1877–8, respectively. Gooding then links these conditions to grain shortages, increased levels of migration, human and animal diseases, and political instability. The evidence, however, based as it is on early missionary records, is somewhat patchy for making sure conclusions in several contexts, especially in terms of regional variation. The chapter may thus show the limits of some archival materials' utility for climatic reconstructions of the past.

Stephen Rockel's extended chapter (Chapter 10) adds further details to the late nineteenth-century climate history of equatorial eastern Africa, a theme that Gooding in Chapter 9 first brought up. Rockel's chapter is situated in the aftermath of the eruption of the Krakatau volcano in 1883, which, he argues, triggered a severe drought across the entire planting season of 1883–1884. He uses proxy records from various IOW regions to make this argument, although with the lack of data for equatorial eastern Africa itself, he also poses questions for future climatological research. Moreover, in assessing the effects of 1883–4 drought, Rockel uses a close reading of several archival and other documentary materials to cover a large area, including parts of present-day eastern Kenya and northeastern, central, and southeastern Tanzania. His chapter shows the widespread and devastating nature of the drought at a particularly volatile time in Africa's history, occurring, as it did, as European imperial powers were becoming intent on carving up the continent. This drought represents a crucial context for understanding the initial colonial encounter.

The volume concludes with a chapter by Theresa Ventura (Chapter 11) on the effects of a volcanic eruption and an El Niño-related drought in the Philippines. As with parts of Chapter 8, the chapter focuses on the year 1911—and Ventura draws on the drought in mainland southeast

Asia that Williamson analyses (as well as droughts and floods in other IOW regions) as a reason for the effects of drought in the Philippines being so severe. Thus, one of her core contributions is to situate the early twentieth-century history of the Philippines within broader climatic and economic linkages across parts of the wider IOW. Furthermore, Ventura shows how analysis of the fall-out of the 1911 drought adds significantly to understandings of the Green Revolution in the Philippines and India during the 1950s–1960s, which has hitherto mostly been understood in terms of Cold War politics. In this instance, the drought and volcano that preceded it are demonstrated as a significant 'turning point' in IOW history.

Taken collectively, the chapters of this volume contribute case studies on the vast expanse of the IOW. The Middle East and, notwithstanding Chaudhuri's chapter, South Asia are perhaps under-represented—which is unusual in a historical volume on the IOW, as, if anything, these regions are often over-represented. Of course, this does not mean that work into the climate histories of droughts and floods in these regions does not exist. Works by scholars such as Sarah Kate Raphael and Zozan Pehlivan on the Middle East, and Vinita Damodaran and George Adamson on South Asia, are crucial to the conception of this book.[62] The focus on other regions here, though, partly represents a response to a call to de-centre India and surrounding regions from understandings of the IOW. While India and, to a lesser extent, the Middle East may be central geographically to the IOW, they remain only two 'areas' within it. This volume, then, is situated in a (slowly) growing body of work that explores eastern Africa's and Southeast and eastern Asia's connections to the wider IOW, climatic, environmental, or otherwise.

Bibliography

Abulafia, David. *The Great Sea: A Human History of the Mediterranean*. New York: Oxford University Press, 2011.

Adamson, George. 'The Most Horrible of Evils': Social Responses to Drought and Famine in the Bombay Presidency, 1782–1857,' in *Natural Hazards and Peoples*, eds. Bankoff and Christensen: 79–104.

62 Raphael, *Climate and Political Climate*; Pehlivan, 'El Niño and the Nomads,' 316–56; Adamson, "The Most Horrible of Evils',' 79–104; Damodaran, et al., 'The 1780s,' 517–50. Many other works by many other scholars could be cited here.

Amrith, Sunil. *Unruly Waters: How Rains, Rivers, Coasts and Seas have Shaped Asia's History*. New York: Basic Books, 2018.

Ashton, Basil, Kenneth Hill, Alan Piazza, and Robin Zeitz. 'Famine in China, 1958–61.' *Population and Development Review*, 10, 4 (1984): 613–45.

Bankoff, Greg and Joseph Christensen, eds. *Natural Hazards and Peoples in the Indian Ocean World: Bordering on Danger*. New York: Palgrave Macmillan, 2016.

Bankoff, Greg, and Joseph Christensen. 'Bordering on Danger: An Introduction,' in *Natural Hazards and Peoples*, eds. Bankoff and Christensen: 1–30.

Bose, Sugata. *A Hundred Horizons: The Indian Ocean in the Age of Global Empire*. Cambridge, MA: Harvard University Press, 2009.

Braudel, Fernand. *The Mediterranean and Mediterranean World in the Age of Philip II*, 3 Vols., trans. Sian Reynolds. New York: Harper & Row, 1972–3.

Braudel, Fernand. *On History*, trans. Sarah Matthews. Chicago: University of Chicago Press, 1980.

Braudel, Fernand. 'History and the Social Sciences: The *Longue Durée*,' trans. Immanuel Wallerstein. *Review (Fernand Braudel Center)*, 32, 2 (2009): 171–203.

Brönnimann, Stefan. 'Global Warming (1970-Present),' in *The Palgrave Handbook*, eds. White, Pfister, and Mauelshagen: 321–8.

Buckley, Brendan M., Kevin J. Anchukaitis, Daniel Penny, Roland Fletcher, Edward R. Cook, Masaki Sano, Le Canh Nam, Aroonrut Wichienkeeo, Ton That Minh, and Truong Mai Hong. 'Climate as Contributing Factor in the Demise of Angkor Cambodia.' *Proceedings of the National Academy of Sciences*, 107, 15 (2009): 6748–52.

Burke, Peter. *The French Historical Revolution: The Annales School, 1929–2014*, 2nd ed. Cambridge: Polity, 2015.

Byrne, Michael P., Angeline G. Pendergrass, Anita D. Rapp, and Kyle R. Wodzicki. 'Response of the Intertropical Convergence Zone to Climate Change: Location, Width, and Strength.' *Current Climate Change*, 4, 4 (2018): 355–70.

Campbell, Gwyn. *Africa and the Indian Ocean World from Early Times to circa 1900*. Cambridge: Cambridge University Press, 2019.

Carson, Sarah. 'Atmospheric Happening and Weather Reasoning: Climate History in South Asia.' *History Compass* (2020): 1–13.

Chaiklin, Martha, and Philip Gooding. 'Introduction: Investigating Animals, their Products, and their Trades in the Indian Ocean World,' in *Animal Trade Histories*, eds. Chaiklin, Gooding, and Campbell: 1–25.

Chaiklin, Martha, Philip Gooding, and Gwyn Campbell, eds. *Animal Trade Histories in the Indian Ocean World*. Cham, CH: Palgrave, 2020.

Chaudhuri, K.N. *Trade and Civilisation in the Indian Ocean: An Economic History from the Rise of Islam to 1750*. Cambridge: Cambridge University Press, 1985.

Damodaran, Vinita, Rob Allan, Astrid E.J. Ogilvie, Gaston R. Demarée, Joëlle Gergis, Takehiko Mikami, Alan Mikhail, Sharon E. Nicholson, Stefan Norrgård, and James Hamilton. 'The 1780s: Global Climate Anomalies, Floods, Droughts, and Famines,' in *The Palgrave Handbook*, eds. White, Pfister, and Mauelshagen: 517–50.

Davis, Mike. *Late Victorian Holocausts: El Niño Famines and the Making of the Third World*. London: Verso, 2002.

Degroot, Dagomar. *The Frigid Golden Age: Climate Change, the Little Ice Age, and the Dutch Republic, 1560–1720*. Cambridge: Cambridge University Press, 2018.

Dyt, Kathryn. 'Emperor Tự Đức's 'Bad Weather': Interpreting Natural Disasters in Vietnam, 1847–1883,' in *Natural Hazards and Peoples*, eds. Bankoff and Christensen: 169–98.

Elsner, James B., and Kam-biu Liu. 'Examining the ENSO-typhoon Hypothesis.' *Climate Research*, 25 (2003): 43–54.

Fan, Ka-wai. 'Climatic Change and Dynastic Cycles in Chinese History.' *Climatic Change*, 101, 3–4 (2010): 565–73.

Gergis, Joëlle, and Anthony M. Fowler. 'A History of ENSO Events since A.D. 1525: Implications for Future Climate Change.' *Climatic Change*, 92 (2009): 343–87.

Gooding, Philip. 'Tsetse Flies, ENSO, and Murder: The Church Missionary Society's Failed East African Ox-cart Experiment of 1876–78.' *Africa: Rivista semestrale di studi e ricerche*, N.S. 1, 2 (2019): 21–36.

Gooding, Philip. 'History, Politics, and Culture in Central Tanzania,' *Oxford Research Encyclopedia of African History* (2019) [Accessed 15 Jan. 2021].

Grove, Richard. 'The Great El Niño of 1789–93 and Its Global Consequences: Reconstructing an Extreme Climate Event in World Environmental History.' *The Medieval History Journal*, 10, 1–2 (2006): 75–98

Grove, Richard, and George Adamson. *El Niño in World History*. London: Palgrave Macmillan, 2018.

Ham, Y.G. 'El Niño Events will Intensify under Global Warming.' *Nature*, 564, 7735 (2018): 192–3.

Hastenrath, Stefan. 'Variations of East African Climate During the Past Two Centuries.' *Climatic Change*, 50, 1–2 (2001): 209–17.

Holmes, Katie, Andrea Gaynor, and Ruth Morgan. 'Doing Environmental History in Urgent Times.' *History Australia*, 17, 2 (2020): 230–51.

Hore, Edward C. *Tanganyika: Eleven Years in Central Africa*. London: Edward Stanford, 1892.

Indian Ocean World Centre. 'Appraising Risk.' https://www.appraisingrisk.com [Accessed 12 Jan. 2021].

Johnson, Douglas H., and David Anderson, eds. *The Ecology of Survival: Case Studies from Northeast African History*. London: L. Crook Academic Pub., 1988.

Knutson, Thomas, Suzana J. Camargo, Johnny C.L. Chan, Kerry Emanuel, Chang-Hoi Ho, James Kossin, Mrutyunjay Mohapatra, Masaki Satoh, Masato Sugi, Kevin Walsh, and Liguang Wu. 'Tropical Cyclones and Climate Change Assessment: Part II: Projected Response to Anthropogenic Warming.' *Bulletin of the American Meteorological Society*, 101, 3 (2020): E303–22.

Koch, Alexander, Chris Brierley, Mark M. Maslin, and Simon L. Lewis. 'Earth System Impacts of the European Arrival and Great Dying in the Americas after 1492.' *Quaternary Science*, 207 (2019): 13–36.

Kreft, Sönke, David Eckstein, and Inga Melchoir. *Global Climate Risk Index 2017: Who Suffers Most from Extreme Weather events? Weather-related Loss Events in 2015 and 1996 to 2015*. Bonn: Germanwatch, 2016.

Ladurie, Emmanuel Le Roy. *Times of Feast, Times of Famine: A History of Climate since 1000*. Translated by Barbara Bray. Garden City: Doubleday, 1971

Lechleitner, Franziska A., Sebastian F.M. Breitenbach, Kira Rehfeld, Harriet E. Ridley, Yemene Asmerom, Keith M. Prufer, Norbert Marwan, Bedartha Goswami, Douglas J. Kennett, Valorie V. Aquino, Victor Polyak, Gerald H. Haug, Timothy I. Eglinton, and James U.L. Baldini. 'Tropical Rainfall over the Last Two Millennia: Evidence for a Low-latitude Hydrolic Seesaw.' *Scientific Reports*, 7, 45809 (2017): 1–9.

Little, Peter D., Hussein A. Mahmoud, and D. Layne Coppock. 'When Deserts Flood: Risk Management and Climatic Processes among East African Pastoralists.' *Climate Research*, 19, 2 (2001): 149–59.

Liu, Kam-biu, Caiming Shen, and Kin-sheun Louie. 'A 1,000-year History of Typhoon Landfalls in Guangdong, Southern China, Reconstructed from Chinese Historical Documentary Records.' *Annals of the Association of American Geographers*, 91, 3 (2001): 453–64.

Low, Michael Christopher. *Imperial Mecca: Ottoman Arabia and the Indian Ocean Hajj*. New York: Columbia University Press, 2020.

McGregor, Shayne, Myriam Khodri, Nicola Maher, Masamichi Ohba, Francesco S.R. Pausata, and Samantha Stevenson. 'The Effect of Strong Volcanic Eruptions on ENSO,' in *El Niño Southern Oscillation in a Changing Climate*, eds. McPhaden, Santoso, and Cai, 267–87.

McPhaden, Michael J., Agus Santoso, and Wenju Cai, eds. *El Niño Southern Oscillation in a Changing Climate*. Hoboken, NJ: Wiley, 2021.

Moore, Jason W. 'The Capitalocene, Part 1: On the Nature and Origins of Our Ecological Crisis.' *Journal of Peasant Studies*, 44, 3 (2017): 594–630.

Moore, Jason W. 'The Capitalocene Part II: Accumulation by Appropriation and the Centrality of Unpaid Work/Energy.' *Journal of Peasant Studies*, 45, 2 (2018): 237–79.

Morgan, Ruth. 'Climate, Weather, and Water in History.' *WIREs Climate Change*, 10, 1 (2019): 1–13.

Nakagawa, Michiko, Tanaka Kenta, Tohru Nakashizuka, and Tatsuhiro Ohkubo. 'Impact of Severe Drought Associated with the 1997–1998 El Niño in a Tropical Forest in Sarawak.' *Journal of Tropical Ecology*, 16, 3 (2000): 355–67.

Nicholson, Sharon E. 'Historical and Modern Fluctuations of Lakes Tanganyika and Rukwa and their Relationship to Rainfall Variability.' *Climatic Change*, 41 (1999): 53–71.

Nicholson, Sharon E. 'A Semi-quantitative, Regional Precipitation Data Set for Studying African Climates of the Nineteenth Century, Part I. Overview of the Dataset.' *Climatic Change*, 50, 3 (2001): 317–53.

Nicholson, Sharon E. 'Climatology: Methods.' *Oxford Research Encyclopedia of African History* (2017) [Accessed: 15 Jan. 2021].

Nicholson, Sharon E., Douglas Klotter, and Amin K Dezfuli. 'Spatial Reconstruction of Semi-quantitative Precipitation Fields over Africa during the Nineteenth Century from Documentary Evidence and Gauge Data.' *Quaternary Research*, 78, 1 (2012): 13–23.

Pankhurst, Richard, and Douglas H. Johnson. 'The Great Drought and Famine of 1888–92 in Northeast Africa,' in *The Ecology of Survival*, eds. Johnson and Anderson: 47–72.

Parker, Geoffrey. *War, Climate Change, and Catastrophe in the Seventeenth Century*. New Have, CT: Yale University Press, 2013.

Pearson, Michael. *The Indian Ocean*. London: Routledge, 2003.

Pehlivan, Zozan. 'El Niño and the Nomads: Global Climate, Local Environment, and the Crisis of Pastoralism in Late Ottoman Kurdistan.' *Journal of the Economic and Social History of the Orient*, 63, 3 (2020): 316–56.

Pfister, Christian, and Sam White. 'A Year Without Summer,' in *The Palgrave Handbook*, eds. White, Pfister and Mauelshagen: 551–61.

Pfister, Christian, Sam White, and Franz Mauelshagen. 'General Introduction: Weather, Climate, and Human History' in *The Palgrave Handbook*, eds. White, Pfister, and Mauelshagen: 1–17.

Raphael, Sarah Kate. *Climate and Political Climate: Environmental Disasters in the Medieval Levant*. Leiden: Brill, 2013.

Saji, N.H., B.N. Goswami, P.N. Vinayachandran, and T. Yamagata. 'A Dipole Mode in the Tropical Indian Ocean.' *Nature*, 401 (1999): 360–3.

Schenk, Gerrit Jasper. 'Historical Disaster Research: State of Research, Concepts, Methods, and Case Studies.' *Historical Social Research*, 32, 3 (2007): 9–31.

Singh, Deepti, Richard Seager, Benjamin I. Cook, Mark Cane, Mingfang Ting, Edward Cook, and Michael Davis. 'Climate and the Global Famine of 1876–78.' *Journal of Climate*, 31, 23 (2018): 9445–67.

Sörlin, Sverker and Melissa Lane. 'Historicizing Climate Change—Engaging New Approaches to Climate History.' *Climatic Change*, 151, 1 (2018): 1–13.

Webster, J.B. 'Noi! Noi! Famines as an aid to Interlacustrine Chronology,' in *Chronology, Migration, and Drought*, ed. Webster: 1–37.

Webster, J.B., ed. *Chronology, Migration, and Drought in Interlacustrine Africa*. New York: Africana Pub. Co., 1979.

White, Sam. *The Climate of Rebellion in the Early Modern Ottoman Empire*. New York: Cambridge University Press, 2011.

White, Sam. *A Cold Welcome: The Little Ice Age and Europe's Encounter with North America*. Cambridge, MA: Harvard University Press, 2018.

White, Sam, Christian Pfister, and Franz Mauelshagen, eds. *The Palgrave Handbook of Climate History*. London: Palgrave Macmillan, 2018.

Williamson, Fiona. 'The "Cultural Turn" of Climate History: An Emerging Field for Studies of China and East Asia.' *Wiley Interdisciplinary Reviews: Climate Change*, 11, 3 (2020): 1–10.

Xia, Lan, Hans von Storch, Frauke Feser, and Jian Wu. 'A Study of Quasi-millennial Extratropical Winter Cyclone Activity over the Southern Hemisphere.' *Climate Dynamics*, 47, 7–8 (2016): 2121–38.

Zhang, Wenxian. 'Dang An: A Brief History of the Chinese Imperial Archives and Its Administration.' *Journal of Archival Organization*, 2, 1–2 (2004): 17–38.

CHAPTER 2

The Economy of Floods and Inundations in the Southern Song Capital Prefecture Lin'an (Hangzhou) on the Shore of the Zhe River Estuary and the Hangzhou Bay in Southeastern China During the Twelfth and Thirteenth Centuries

Silvia Freiin Ebner von Eschenbach

After the Jurchen conquest of the northern part of the Song 宋 Empire and its capital Kaifeng 開封 in 1127, it was the prefectural city of Hangzhou 杭州 that in 1138 became capital of the Southern Song Dynasty (1127–1279), to be renamed Lin'an 臨安, until it was invaded by the Mongol conquerors in 1276. The city of Lin'an was not only capital

S. F. Ebner von Eschenbach (✉)
Department für Asienstudien, Institut für Sinologie,
Ludwig-Maximilians-Universitaet Muenchen, Munich, Germany
e-mail: ebner@ostasien.fak12.uni-muenchen.de

P. Gooding (ed.), *Droughts, Floods, and Global Climatic Anomalies in the Indian Ocean World*, Palgrave Series in Indian Ocean World Studies, https://doi.org/10.1007/978-3-030-98198-3_2

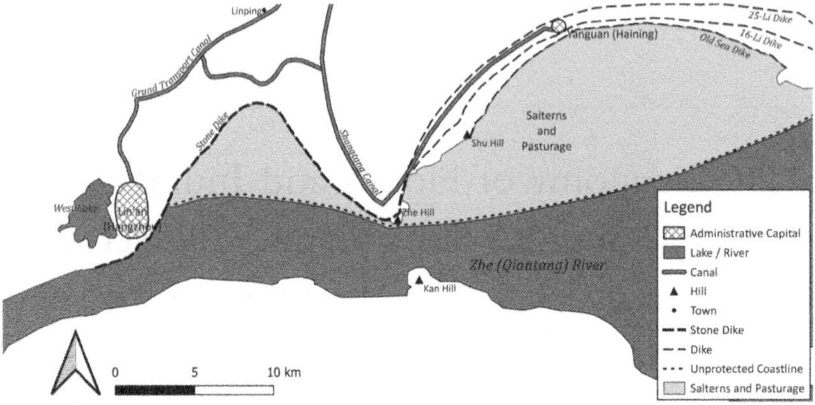

Fig. 2.1 Map of region under review, with places mentioned in-text marked. Loosely based on: Elvin, *The Retreat of the Elephants*, map 3 (p. 142). Drawn by Philip Gooding

of the Southern Song Empire, but also capital of the prefecture with the same name, consisting of nine districts.

The coastal port of Hangzhou in Zhexi 浙西 Province was located in the alluvial land of the northern shore of the Zhe River (Zhejiang 浙江)[1] Estuary debouching into Hangzhou Bay and the sea. It was chosen as a capital as it was difficult for the Jurchen cavalry to traverse the muddy terrain.[2] The city of Lin'an, comprising the districts of Renhe 仁和 and Qiantang 錢塘, and its northeastern agricultural area, situated in the district of Yanguan 鹽官, which produced the city's vegetables, both drew their freshwater from a reservoir west of the city, then called the West Lake (Xihu 西湖), which was a former lagoon. The Zhe River shore delineated the two capital districts on their southern and eastern sides, and the adjacent district of Yanguan on its southern side (Fig. 2.1).

[1] The name Zhe 浙 was written also with variant other characters such as Ci 剌, Zhi 之, Qu 曲 (A.C. Moule, *Qinsai with Other Notes on Marco Polo* [Cambridge: University Press, 1957]), 22. The Zhe River is also known as the Qiantang River (Qiantang jiang 錢塘江) or Grand River (Dajiang 大江).

[2] Jacques Gernet, *La vie quotidienne en Chine à la veille de l'invasion mongole 1250–1276* (*s. l.*: Librairie Hachette, 1959), 22.

Of the huge amounts of sediment washed down into Hangzhou Bay, only a minor part was brought in by the Zhe River, whereas most was discharged from the Yangzi Estuary.[3] After the southern shift of parts of the Yellow River in the years 1166–1194 eventually draining into the Yangzi Estuary, southward ocean currents transported ever more sediment into Hangzhou Bay, where it settled and created a bottle neck with tidal bores piling up.[4] The shift of parts of the Yellow River had occurred as a consequence of the Northern Song defence strategy that opened breaches in the river dikes 80 kilometres north of the capital of Kaifeng in 1128, in the vain hope of stopping the advance of the Jurchen cavalry.[5]

The sedimentation of the Hangzhou Bay made Lin'an Prefecture increasingly vulnerable to tidal inundations. When the Southern Song founding emperor Gaozong 高宗 (r. 1127–1162) chose Hangzhou as his new capital, he probably was not aware of the imminent vulnerability of the region by inundations. Additionally, the tidal waves may also have

[3] Mark Elvin, *The Retreat of the Elephants*: *An Environmental History of China* (New Haven: Yale University Press, 2004), 145.

[4] Mark Elvin and Su Ninghu, 'Action at a Distance: The Influence of the Yellow River on Hangzhou Bay Since A. D. 1000,' in *Sediments of Time: Environment and Society in Chinese History*, eds. Mark Elvin and Liu Ts'ui-jung (Cambridge: Cambridge University Press, 1998), 346, 348. For a more detailed analysis of how the increasing siltation in Hangzhou Bay affected intra-urban traffic on the channels of the Lin'an capital city, see: Silvia Freiin Ebner von Eschenbach, 'Innerstädtische Verkehrsinfrastruktur und Mobilität und ihre Einflussfaktoren. Dargestellt am Beispiel einer südstchinesischen Stadt unter der Qian 錢-Dynastie (907–978) und der Nördlichen und Südlichen Song 宋-Dynastie (960–1127, 1127–1279),' in *Mobilität in China* (forthcoming).

[5] Christian Lamouroux, 'From the Yellow River to the Huai: New Representations of a River Network and the Hydraulic Crisis of 1128,' in *Sediments of Time*, eds. Elvin and Ts-ui-jung, 545, citing: *Song shi* 宋史 (Beijing: Zhonghua shuju, 1977; hereafter: *SS*), *j*. 25, 459; Li Xinchuan 李心傳 (1166–1243), *Jianyan yilai xinian yaolu* 建炎以來繫年要錄 (ed. Qinding Sikuquanshu, Shanghai: Shanghai guji chubanshe), *j*. 18, 19a. The new course of the Yellow River took varying routes, including via the bed of the Qing 清 River, a tributary stream of the Huai 淮, thus taking the Huai River to flow into the Yangzi Estuary (Joseph Needham With the Collaboration of Wang Ling and Lu Gwei-Djen, *Science and Civilisation in China*. Vol. 4, *Physics and Physical Technology*. Part 3: *Civil Engineering and Nautics* (Cambridge University Press, 1971), 209 (Fig. 859), 24–23 (Table 69)). Through the mouth of the Huai River, the loess sediment (*huangtu* 黃土) washed off from the northwest China plane, and was eventually discharged into the Yellow Sea east of Huaiyin 淮陰 (today's Huaian 淮安), thus creating a large new delta. The sediment from the Yellow River's new delta added to the sediment already emptying into the sea through the Yangzi Estuary (Elvin and Su, 'Action at a Distance,' 345, 347 (map 10.1), 361, 362–364, 394).

intensified during this period due to climatic cooling, which contributed to extreme fluctuations in precipitation and increased frequency of flood events in the northern and western foothills and on the Zhe River shore in Lin'an Prefecture. Contrary to other global regions, which generally experienced warmth, data from the Song Empire and from Zhexi Province and Lin'an Prefecture show that climatic cooling was under way already in the twelfth century. Evidence from texts, including poems and paintings, makes clear, for example, that the climate cooling of the twelfth century was a major factor to the eventual collapse of the Northern Song Dynasty (960–1127).[6]

This study shows how floods and inundations wreaked havoc on the alluvial land on the Zhe River's northern shore. Although the region had been protected from salination by dike building from an early stage, flooding and tidal inundations increased at alarming scales during the period under review. The study also explores how the construction, maintenance, and repair of dikes on the Zhe River were precautionary but ultimately futile measures to protect coastal lands, in particular the Yanguan agricultural area, against inundation and ensuing salination. It reveals how the Yanguan agricultural area depended on the preservation of the West Lake freshwater reservoir for irrigation, and how fiscal considerations were involved in these processes. The study will also point out how financial resources were invested not only in dike building and the preservation of the West Lake reservoir, but also in relief measures that were implemented to cope with the damages caused by floods and inundations.

This is a micro-historical study that shows how man-made changes in the context of a long-term climatic shift increased vulnerabilities to

[6] Huiping Pang, 'Strange Weather: Art, Politics, and Climate Change at the Court of Northern Song Emperor Huizong,' *Journal of Song-Yuan Studies*, 39 (2009), 13 (Figs. 9–10), 39–41. For further reading on paleoclimate research, see his Note 12 on pp. 11–12. For the drop of temperature in the beginning of the twelfth century also see: Manfred Domrös and Peng Gongbing, *The Climate of China* (Berlin: Springer, 1988), 133. For the climatic variations in the tenth–thirteenth centuries, see: Silvia Freiin Ebner von Eschenbach, 'Managing Floods and Droughts by Invocating the Water Spirits: Analyzing Prayers for Rain (*daoyu* 禱雨) and Prayers for a Clear Sky (*qiqing* 祈晴). With some examples from local source material of the Song 宋 Dynasty (960–1279),' *Zeitschrift der Deutschen Morgenländischen Gesellschaft*, 169, 1 (2019), 205–29. This pattern of climate cooling is also supported by records of typhoons from around this time. See: Chapter by Schottenhammer, this volume.

the effects of flooding. These vulnerabilities and their exposition therein from abnormal levels of rainfall and tidal inundations contributed to hydrological, economic, and social disasters, and the development of diverse and concurrent mitigative strategies. The relevant source material was published in the Southern Song editions of the Lin'an prefectural gazetteer.[7] Prefects and their staff were required to edit a local gazetter, and to compile data and texts in an encyclopaedically arranged compendium of their prefecture. The prefectural gazetteer of Lin'an contains chapters on inundations and dike building, water management, and local finance. Information on local issues was also edited in the Song documentary *Song huiyao jigao* 宋會要輯稿, an archival compilation in which petitions on local matters and responding imperial orders were included.[8] As an official compilation at the upper level of data aggregation from the local level, the Song dynastic history (*Song shi* 宋史),[9] edited under the surveillance of Toqtogha (Tuotuo 脱脱, 1313–1355) during the Yuan 元 dynasty (1280–1368) in 1343–1345, contains two chapters on 'water' (*shui* 水). They form part of the chapters on the Five Elements (*wuxing* 五行) and include references to flooding and tidal inundations. The study takes the research work on the sedimentation of the Hangzhou Bay by Mark Elvin and Su Ninghu (1998), Mark Elvin (2004), and Christian Lamouroux (1998) as a starting point for research (Figs. 2.2 and 2.3).

[7] Christine Moll-Murata, *Die Chinesische Regionalbeschreibung: Entwicklung und Funktion einer Quellengattung, dargestellt am Beispiel der Präfekturbeschreibungen am Hangzhou* (Wiesbaden: Verlag Harrassowitz, 2001), 59–92. See also: James Hargett, 'Song Dynasty Local Gazetteers and Their Place in the History of *Difangzhi* Writing,' *Harvard Journal of Asiatic Studies*, 56, 2 (1996), 405–42.

[8] It was edited in 1936 as a draft recovered edition under the title *Song huiyao jigao* 宋會要輯稿 (ed. Shanghai: Shanghai guji chubanshe, 2014, hereafter: *SHYJG*). See: Endymion Wilkinson, *Chinese History: A New Manual*, 4th ed. (Cambridge, MA: Harvard University Press, 2015), 757a; Yves Hervouet, ed., *A Sung Bibliography (Bibliographie des Sung)* (Hong Kong: The Chinese University Press, 1978), 154.

[9] Wilkinson, *Chinese History*, 756a–757a.

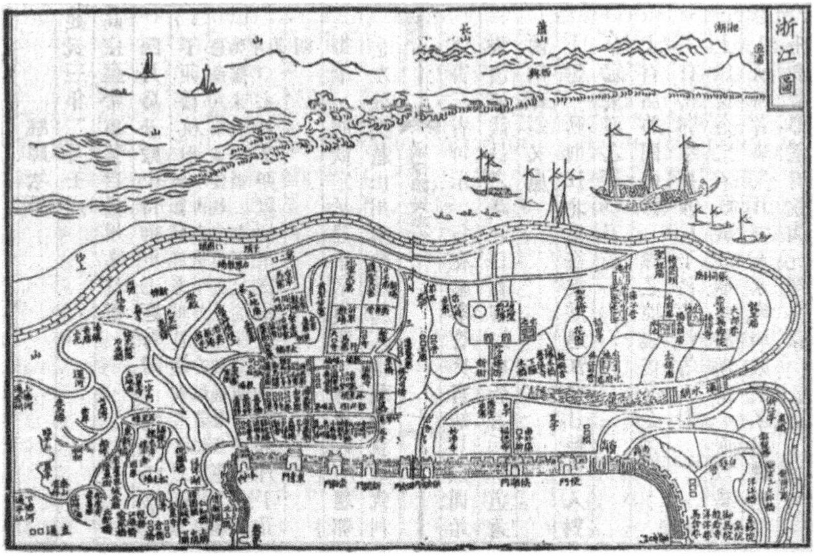

Fig. 2.2 Archival map of Hangzhou area and the Zhe River. *Zhejiang tu* 浙江圖, *XCLAZ*, j.1, p. 7 (north: left side)

Land Use and Damage Due to Flooding and Tidal Inundations

Initial Dike Building for the Use of Coastal Lands

In 910, King Wusu 武肅, that is Liu 鏐 of the Qian錢 Dynasty (852–932, r. 907–931) and founder of the Kingdom of Wu and Yue 吳越, started the construction of a rammed-earth wall, held in place by planks (*banzhu* 版築), which was then called the 'Dike to Ward off the Sea' (Hanhai tang 捍海塘), or simply the 'Sea Dike' (Haitang 海塘). But at that time, the 'strokes' (*chongtu* 衝突) of tidal water did not occur frequently. During the years 1034–1038 of the Northern Song Dynasty (960–1127), the Fiscal Commissioner (*zhuanyun shi* 轉運使)[10] of the Liang Zhe 兩浙 Provinces, Zhang Xia 張夏 (dates unknown), built the first dike made

[10] Charles O. Hucker, *A Dictionary of Official Titles in Imperial China* (Stanford, CA: Stanford University Press, 1985), no. 1490, 2.

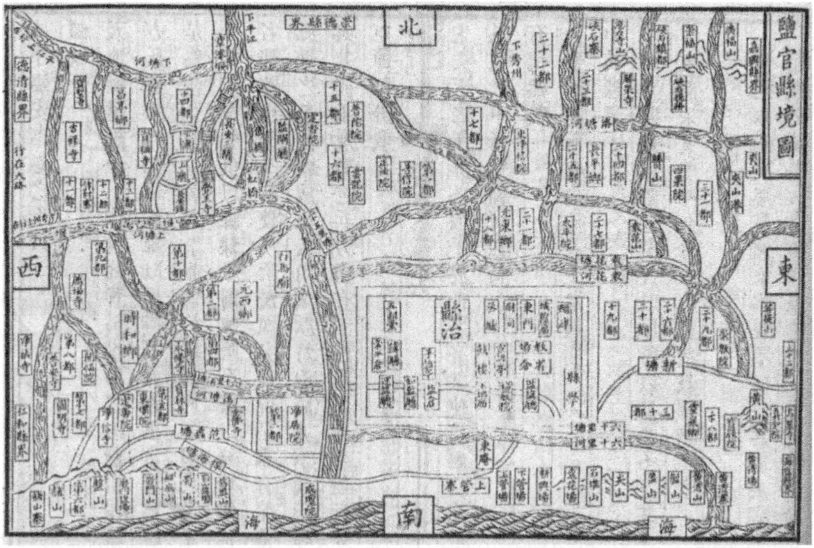

Fig. 2.3 Archival map of Yanguan District. *Yanguan xian jingtu* 鹽官縣境圖, *XCLAZ*, j.16, p. 8 (north: on top)[11]

of stones on the river, known as 'Stone Dike' (Shidi 石隄).[12] With a length of 12 *li* 里 (approximately 6.7 kilometres) it protected the city of Hangzhou against flooding, stretching alongside the Zhe River from

[11] The maps in Figs. 2.2 and 2.3 were supplemented in the 1867 edition of the The Lin'an Gazetteer of the Xianchun (1265–1274) era (*Xianchun Lin'an zhi* 咸淳臨安志; hereafter: *XCLAZ*). See: Moll-Murata, *Die Chinesische Regionalbeschreibung*, 43. They were copied from maps made by Lu Wenchao 盧文弨 (1717–1795) in the second half of the eighteenth century from the originals of ca. 1274 that were printed at Hangzhou. The maps drawn by Lu Wenchao show squares where he could no longer read the original characters (Moule, *Qinsai with other Notes*, 12). For a reconstruction of the original maps of the Southern Song dynasty, see: Jiang Qingqing 姜青青, *≪Xianchun Lin'an zhi≫ Song pan 'jingcheng si tu' fuyuan yanjiu* ≪咸淳临安志≫ 宋版'京城四图'复原研究 (Shanghai: Shanghai guji chubanshe, 2015), 350, maps 1–4.

[12] Probably identical with the Stone Dike (Shitang石塘), situated on the shore of the Zhe River south and southeast of the city of Hangzhou (see: the *Zhejiang tu* 浙江圖, in: *XCLAZ*, *j*.1, p. 7).

the Liuhe Stupa (Liuhe ta 六和塔)[13] to the northeastern Dongqing City Gate (Dongqing men 東青門) of Hangzhou.[14]

By the Southern Song, Yanguan District was protected by the '25-*li*-Dike' (Ershi wu li tang 二十五里塘), which was approximately 14 kilometres long, and ran along the inside of the coastline that connected the three-step Chang'an Lock (Chang'an zha 長安閘)[15] in the west of Yanguan District to Linping 臨平 Garrison[16] and Chongde 崇德 District,[17] bordering the north of Yanguan District.[18]

While the land north of the dike was used for growing vegetables, the alluvial land south of it was used for salterns (*yanchang* 鹽場 or *tingchang* 亭場).[19] The salt furnaces (*yanzao* 鹽竈) were reported to have been

[13] Situated southwest of the city on the bank of the Zhe River (see: the *Xihu tu* 西湖圖, in *XCLAZ*, *j*.1, p. 8.

[14] 'Hanhai tang Tiezhuang pu,' in Lin'an Gazetteer of the Chunyou (1241–1252) era (*Chunyou Lin'an zhi* 淳祐臨安志; hereafter: *CYLAZ*), *j*.10, p. 7a, l.7–8, p. 7b, l.5–6; 'Hanhai tang 捍海塘,' in *XCLAZ*, *j*.31, p. 7a, 5–7, p. 8a, l.5; Ye Shaoweng 葉紹翁 (c. 1175–1230), 'Zhang si feng miao 張司封廟,' in *Sichao wenjian lu* 四朝聞見錄 (ed. Beijing: Zhonghua shuju, 2006), *jiaji* 甲集, p. 32, l.14–15, p. 33, l.4–6. By contrast, Shiba says that the Wu Yue sea wall was built of stone (Shiba Yoshinobu, 'Environment Versus Water Control: The Case of the Southern Hangzhou Bay Area from the Mid-Tang Through the Qing,' in *Sediments of Time*, eds. Elvin and Ts-ui-jung, 138).

[15] The Three-Step Chang'an Lock (Chang'an sanzha 長安三閘) is located in the western part of Yanguan 鹽官 District (*Yanguan xian jingtu* 鹽官縣境圖, in *XCLAZ*, *j*.16, p. 8).

[16] A garrison (*zhen* 鎮) in Renhe District (Hope Wright, *Alphabetical List of Geographical Names in Sung China* [Paris: École Pratique des Hautes Études, Centre de Recherches Historiques, 1956] [Sung Project], 79), situated north of the capital city (see: *Xihu tu* 西湖圖, in *XCLAZ*, *j*.1, p. 8).

[17] A district belonging to neighbouring Xiuzhou 秀州 (Wright, *Alphabetical List*, 68).

[18] 'Yanguan haishui,' in *SS*, *j*.97, p. 2401, l.11–12.

[19] During the Shaoxing 紹興 era (1131–1162) there were 10 salterns in Lin'an Prefecture. See also: Edmund H. Worthy, 'Regional Control in the Southern Sung Salt Administration,' in *Crisis and Prosperity in Sung China*, ed. John Winthrop Haeger (Tuscon: The University of Arizona Press, 1975), 106; Shiba Yoshinobu, 'Environment Versus Water Control: The Case of the Southern Hangzhou Bay Area from the Mid-Tang Through the Qing Period,' *Sediments of Time*, eds. Elvin and Ts-ui-jung, 154; Cecilia Lee-fang Chien, *Salt and State: An Annotated Translation of the Songshi Salt Monopoly Treatise* (Ann Arbor: Center for Chinese Studies, The University of Michigan, 2004), 170–71.

most prosperous, and that they yielded steadily rising tax revenues (*keli* 課利).[20] Although the furnaces were mainly fueled with ashes from reeds, their demand for firewood may have led to deforestation of the nearby hills, followed by erosion in cases of heavy rainfall.[21]

Damage Wrought on Land by Flooding and Tidal Inundations

In Lin'an Prefecture, flooding due to rainwater run-off from eroded hills and the Zhe River, as well as tidal inundations were continuously reported at short intervals during the twelfth and thirteenth centuries.[22] Although various dikes had been built, floods and inundations wrought havoc on Lin'an Prefecture. In the following discussion, some examples of severe incidents illustrate the whole range of damage caused to fields, farmhouses, and people.

A major incident occurred, for example, in the fifth month of 1160. After heavy monsoon rains in Yuqian 於潛 and Lin'an 臨安 Districts, water from the hills suddenly poured down, leaving houses, fields, and mulberry trees destroyed and numerous people drowned.[23] In the seventh month of 1167, after heavy monsoon rains, again large water masses (*hongshui* 洪水) suddenly streamed down from Tianmu Hill

[20] 'Shuili zalu 水利雜錄', 1219, in *SHYJG*, vol. 12, *j.Shihuo* 食貨 61, p. 7546b, l.8–9; cp. the almost identical text of 'Yanguan haishui', in: *SS*, *j.*97, p. 2401, l.8–9.

[21] Chien, *Salt and State*, 170; Shiba, 'Environment Versus Water Control,' 161.

[22] Notable years include: 1124, 1129, 1131, 1132, 1135, 1158, 1160, 1162, 1163, 1164, 1167, 1176, 1194, 1195, 1207, 1210, 1213, 1214, 1216, 1218, 1222, 1223, 1229, 1238, 1251, 1274 (see: *SS*, *j.*61, p. 1329, l.11–12; p. 1330, l.6, 8, 10–11; p. 1331, l.1, 12–15; p. 1332, l.2, 5–6, 9; p. 1336, l.4, 8–10, 13–15; p. 1337, l.5–7, 11, 13–15; p. 1338, l.6; *j.*62, p. 1347, l.8; p. 1423, l.10, 13–15; *j.*65, p. 1423, l.10, 13–15; p. 1426, l.6–9; Chen Shan 陳善 (comp. 1579), ***Wanli Hangzhou fuzhi*** 萬曆杭州府志 (repr. ***Zhongguo fangzhi congshu***, no. 524, Taibei: chengwen chubanshe, 1983), *j.*4 'Junshi ji 郡事記,' *xia* 下, p. 27a, l.3; 'Hanhai tang Tiezhuang pu,' in *CYLAZ*, *j.*10, p. 8a, l.6–8, 9; 'Shuizai 水災,' 1167, in *SHYJG*, vol. 5, *j. Ruiyi* 瑞異 3, p. 2653a, l.7–9, p. 2659a, l.19–23; 'Zhenhuo 賑貨,' 1164, 1195, 1214, 1223, in *SHYJG*, vol. 13, *j. Shihuo* 66, p. 7983a, l.9, p. 8007a, l.20, 22, p. 8011b, l.3, p. 8013a, l.1). For the crop failures with famine caused by the climate change in Zhexi Province, see also: Joseph P. McDermott and Shiba Yoshinobu, 'Economic Change in China, 960–1279,' in *The Cambridge History of China*. Vol. 5, Part 2, *Sung China, 960–1279*, eds. John W. Chaffee and Denis Twitchett (Cambridge: Cambridge University Press, 2015), 416–17.

[23] *SS*, *j.*61, p. 1330, l.8.

(Tianmu shan 天目山),[24] situated on the border between Yuqian and Lin'an Districts. The water crushed 285 houses in five suburbs (*xiang* 鄉) of Lin'an District, causing the death by drowning of many people.[25]

In the seventh month of the year 1174 a heavy storm surge (*dafeng tao* 大風濤) broke off (*jue* 決) the Stone Dike (*di* 隄) on the Zhe River at a length of 1660 *zhang* 丈 (approximately 5.1 kilometres), and washed away 630 houses. In two suburbs of Renhe District, situated close to the Zhe River, the storm surge destroyed fields and gardens.[26] In the fifth month of 1177, billows (*tao* 濤) from the Zhe River spilled over and ruined the Stone Dike (*di*) in Lin'an Prefecture at a length of 80 *zhang* (approximately 250 metres) and, after that, another 100 *zhang* (approximately 309 metres).[27]

In the fifth month of 1207, a clash between water masses (*hongshui*) from the upper reaches of the Zhe River, probably due to monsoon rainfall, and tidal floods that entered the river bank (*pu* 浦) occurred. The water burst into (*yongru* 湧入) Qiantang District, submerging (*jinmo* 浸汲) fields, main roads, and dwellings. Places by the post road (*yilu* 驛路) were covered in water 8 *chi* 尺 (approximately 2.5 metres) deep.[28] Outside the Genshan City Gate (Genshan men 艮山門)[29] in the northeastern part of the Lin'an city, the tides dashed (*chongdang* 衝蕩) people's houses at the Stone Dike (Shitang石塘)[30] alongside the Zhe River to pieces.[31]

[24] Of the twin peaks of Tianmu Hill, the eastern one is situated in the northwest of Lin'an 臨安 District and the western one in the northeast of Yuqian District (*XCLAZ*, *j.*25, p. 1a, l.5, 8a-9b; *j.*26, p. 1b, l.4–5, maps 4–5, inserted between *j.*15/16), west of the West Lake.

[25] 'Shuizai,' 1167, in *SHYJG*, vol. 5, *j. Ruiyi* 3, p. 2653a, l.7–9; *SS*, *j.*61, p. 1331, l.1.

[26] *SS*, *j.*61, p. 1331, l.12–13.

[27] *SS*, *j.*61, p. 1332, l.2.

[28] 'Shuizai,' 1207, in *SHYJG*, vol. 5, *j. Ruiyi* 3, p. 2659a, l.19–23; cp. *SS*, *j.*61, p. 1336, l.4.

[29] Situated in the northeastern part of the Lin'an city wall.

[30] Probably identical with the Stone Dike (Shidi 石隄), situated on the shore of the Zhe River south and southeast of the city of Hangzhou (see the *Zhejiang tu*, in *XCLAZ*, *j.*1, p. 7).

[31] 'Hanhai tang Tiezhuang pu,' in *CYLAZ*, *j.*10, p. 8a, l.9.

In the fifth month of 1210 there was flooding after heavy rains (*dayu shui*) in Fuyang 富陽, Yuhang 餘杭, Yanguan, and Xincheng 新城 Districts. Numerous people drowned; the first seed in the fields rotted, and houses, markets, and suburbs were destroyed. Even the West Lake spilled over, inundating 5300 houses in the capital city.[32]

Higher Risks and Repair Measures on the Dike System

Warnings and Inundation Disasters in 1219 and 1222

The continuous series of flood and inundation disasters was alarming, but officials gave warnings and made repair suggestions late, and only at a time when the situation was grave. It was not until 1219 that the prefectural officials warned of an impending inundation if the 25-*li* Dike was not repaired in time:

> In the past years the sea caused inundations (*fanzhang* 泛漲) when torrential floods (*tuanji* 湍激) hit [the sand embankments] at right angles (*hengchong* 橫衝) [so that] they constantly broke off at several tens of *zhang* (1 *zhang* = approximately 3 metres) with every surge. Since this happened repeatedly day by day, [the brine] immersed the salty soil (*yandi* 鹽地) [so that] the canals and drains (*gangdu* 港瀆) in the alluvial lands with reeds (*luzhou* 蘆洲) were washed away turning into one single gully.
>
> Today we heard that through the pressure of the tides (*chaoshi* 潮勢) [the water] penetrated deeply into [the land] close to the border of the inhabited [area].
>
> If in spring the water rises rapidly, the raging billows (*nutao* 怒濤) will surge and the wind from the sea will contribute to it, then through thrust and suction (*huxi* 呼吸)[33] [of the tides, the soil] will be washed out. Would not people [living] at 100 *li* (approximately 55.6 kilometres) [from the sea] be buried altogether in the stomach of the fish (*yufu* 魚腹)?[34,35]

[32] *SS*, *j*.61, p. 1336, l.8–10.

[33] Literally: Exhaling and inhaling.

[34] Phrase taken from 'Yufu 漁父,' in *Chuci* 楚辭, *j*.7.

[35] 'Yanguan haishui,' in *SS j*.97, p. 2401, l.9–11. See also the translation in: Elvin, *Retreat of the Elephants*, 148. The translation comes from a similar text in: Gu Yanwu 顧炎武 (1613–1682), *Tianxia junguo libing shu* 天下郡國利病書 (repr. Shanghai: Shangwu

If the 25-*li* Dike were to be breached, the capital city with its two districts Qiantang and Renhe and the Yanguan agricultural area would be submerged by the brine, and the diked embankments (*di'an* 堤岸) of the inner canals (*lihe* 裏河) would burst.[36] Thus, this year, the prefectural officials petitioned for measures to repair (*zhuna* 築捺) the 25-*li* Dike and withhold sufficient money and rice for the expenditure of labour and material. The dike urgently needed to be filled in with soil for reinforcement.[37]

It was probably shortly after that that tidal floods broke nearly 16.7 kilometres into the flat open land of Yanguan District, stretching as far as the district capital:

> In the twelfth year [of the Jiading 嘉定 era (1219)] the sea lost its former course in Yanguan District and the tidal floods (*chaoxi* 潮汐) broke through (*chong* 衝) into over 30 *li* (approximately 16.7 kilometres) of the flat open land. When they encroached on the district capital, the open space of the alluvial land with its reeds (*luzhou* 臚州),[38] the canals and drains (*gangdu*), and its upper and lower pipelines (*guan* 管) as well as the Huangwan Mound (Huangwan gang 黃灣岡)[39] were all destroyed. The Shu Hill (Shushan 蜀山)[40] sank into the sea. Half of the villages and fields were lost.[41]

In the fall of 1222, discussions arose among the District Magistrates (*yizhang* 邑長)[42] about taking precautionary measures against flooding from the Zhe River (*fang* Jiang 防江), seeing as, after repairs, the dike was likely to be destroyed again.[43]

yinshuguan, 1936, Siku shanben facsimile repr., reissued Tabei, s. a., hereafter: *TXJGLBS*), 'Zhejiang浙江,' part *xia* 下, p. 42a–b.

36 'Yanguan haishui,' in *SS*, *j*.97, p. 2401, l.11–13.

37 'Shuili zalu,' 1219, in *SHYJG*, vol. 12, *j*. *Shihuo* 61, p. 7546b, l.14–18. See also: The partly identical text of 'Yanguan haishui,' in *SS*, *j*.97, p. 2401, l.13–14.

38 Probably a misreading of *luzhou* 蘆洲.

39 Location unknown.

40 An islet in the Zhe River (see the reconstructed *Zhejiang tu* [Fig. 2.2] and fn. 10.).

41 *SS*, *j*.61, p. 1337, l.2–3.

42 Hucker, *A Dictionary of Official Titles*, no. 2925,2 and no. 84,2.

43 'Hanhai tang Tiezhuang pu,' in *CYLAZ*, *j*.10, p. 8b, l.1–2.

In 1222, the Supervisor (*tiju* 提舉)[44] of Zhexi Province Liu Hou 劉垕 (dates unknown) specified in his petition that originally the coastline was 40 *li* (approximately 22.3 kilometres) away from the district capital of Yanguan. But in the last couple of years, the morning and evening tides had steadily moved further northward so that land 10 *li* (approximately 5.6 kilometres) south of the district capital of Yanguan was lost to the sea. Even the old Sea Dike with its original length of 20 *li* (approximately 11.2 kilometres) situated at just 10 *li* south of the district capital was endangered, as the tides were encroaching on its eastern and western endpieces for another 3–4 *li* (approximately 1.68–2.24 kilometres), moving further toward the district capital and leaving a stretch of only 10 *li* of the old Sea Dike—half of its original length.[45]

In the same year, 1222, Liu Hou 劉垕 (dates unknown) warned that the whole region bordering the north of Yanguan District might additionally be inundated by brine, leaving fields unfarmable due to salination[46]:

> Speaking about the present calamities in more detail, there are in general two: The one is that the dry land (*ludi* 陸地) will be destroyed, the other is that the brackish tidal water (*xianchao* 鹹潮) will spill over (*fanyi* 泛溢). When the dry land (*ludi*) is destroyed, there will be no effort that can be put forth [to save it]. If the brackish tidal water (*xianchao*) spills over and consequently the Old Dike to Ward off the Sea (Hanhai tang 捍海古塘) is damaged, it may happen that the bore (*dachao* 大潮) coils up and gushes to the north.[47]

In the fall of 1222, the tides broke through (*chongtu* 衝突) northeast of the city wall of the Lin'an capital city, closing in on the boundary of the district capital of Yanguan to a distance of only 3 *li* (approximately 1.7 kilometres).[48] When in the same year, 1222, the old Sea Dike burst

[44] Hucker, *A Dictionary of Official Titles*, no. 6395.

[45] 'Yanguan haishui,' in *SS*, *j*.97, p. 2402, l.1–4.

[46] Ibid., l. 4–5. See also: Elvin, *Retreat of the Elephants*, 148, which cites: *TXJGLBS*, 'Zhejiang,' part *xia*, pp. 42b–43a, and refers to *Haining xianzhi* 海寧縣誌 (ed. 1765, repr. Taibei: Chengwen, 1983, Zhongguo fangzhi congshu, no. 516, hereafter: *HNXZ*), 463.

[47] Ibid.

[48] 'Hanhai tang Tiezhuang pu,' in *CYLAZ*, *j*.10, p. 8a, l.10; See also: Elvin, *Retreat of the Elephants*, 148, cites *HNXZ*, p. 1663.

because of horizontal waves (*hengjue* 横決), Liu Hou pleaded for its maintenance as a protection measure of the Yanguan district capital against the sea, seeing as it was situated only about 1 *li* (approximately 0.56 kilometres) south of the district capital. The old Sea Dike needed a check-up of its stakes (*zhuang* 樁) and stones, and some earth had to be added to fortify it.[49]

According to Liu Hou, the old Sea Dike was to be supplemented by two rows of earth dikes (*tutang* 土塘) to protect the fields from brackish tidal water (*xianchao*). Following this suggestion, the '16-*li* Dike Against the Brine' (Shiliu li xiantang 十六里鹹塘)[50] and the 'Dike Against the Brine' (Xiantang 鹹塘) were built south of the Yanguan district capital, and two more called the 'Yuanhua Dike' (Yuanhua tang 袁花塘)[51] and the 'Shallow Water Dike' (Qiantang 淺塘) were erected north of it. The latter protruded into the surroundings west of the district capital at a total length of 50 *li* (approximately 28 kilometres).[52]

Inundations in 1238 and Ensuing Repair Measures in 1239

And yet, in the fall of 1238, the endpieces (*tou* 頭) of all these crescent-shaped dikes (*yuetang* 月塘), arranged in three rows behind one another, were steadily scraped off, as the Zhe River spilled over,[53] and the tides from the strait of Haimen 海門[54] devastated the coastal stretch at a width of 40 *li* (approximately 22.3 kilometres).[55] Therefore, in 1239 Emperor Lizong 理宗 (r. 1225–1265) ordered Prefect Zhao Yuguan 趙與灌[56] (*jinshi* 1214) to carry out the repair measures (*xiuzhu* 修築) he

[49] 'Yanguan haishui,' in *SS*, *j*.97, p. 2402, l.12–13.

[50] Situated in the southern part of Yanguan District southeast of the district capital (*Yanguan xian jingtu* 鹽官縣境圖, in *XCLAZ*, *j*.16, p. 8).

[51] Situated in the northern part of Yanguan District north of the district capital (ibid.).

[52] 'Yanguan haishui,' in *SS*, *j*.97, p. 2401, l.15—p. 2402, l.8–9, 12.

[53] *SS*, *j*.61, p. 1337, l.13.

[54] The strait between the two islets called Zhe Hill (Zheshan 赭山) and Kan Hill (Kanshan 龕山), situated 65 *li* (approximately 36 kilometres) northeast of Renhe District ('Hai 海,' in *XCLAZ*, *j*.31, p. 12a, l.8–10; see the reconstructed *Zhejiang tu*, in Jiang, ≪*Xianchun Lin'an zhi*≫ *Song pan 'jingcheng si tu' fuyuan yanjiu*, 350/4.

[55] 'Hanhai tang,' in *XCLAZ*, *j*.31, p. 8b, l.1–2.

[56] Chang Bide 昌彼德, *Song ren zhuanji ziliao suoyin* 宋人傳記資料索引 (Taibei: Dingwen shuju, 1988), IV, 3593–94.

had suggested: that is, to first repair (*zhu* 築) the earth dikes (*tutang*) in front and to the side of the existing dikes, and only then repair (*zhu*) the Stone Dike (Shitang) behind the earth dikes (*tutang*)—or at least such a procedure can be assumed.[57] He also petitioned for the deployment of 5500 government soldiers and the hiring of 3000 labourers and soldiers from the Office for Repair on the Zhe River (*xiu Jiang si* 修江司)[58] to have them stuff baskets filled with stones (*shicang* 石倉) between the bamboo fences (*zhuangba* 樁笆) and wooden planks (*banmu* 板木) as a foundation, and to fill in earth. Within three months, they repaired the embankment (*ba* 壩) with a south-to-north length of 150 *zhang* (approximately 464 metres) stretching from the Buddhist Shuilu Monastery (Shuilu si 水陸寺)[59] southwest of the Yanguan district capital up to the Jiang jia Bridge (Jiang jia qiao 江家橋).[60] Starting from the Surrounding Endpiece (Tuanwei tou 團圍頭)[61] of the Stone Dike (Shitang) southeast of the Lin'an capital city and reaching up to near the [Zhe] River (Jiang 江), the soldiers repaired (*zhuna* 築捺)[62] the dikes against flooding (*shuitang* 水塘), each at a length of 600 *zhang* (approximately 1.9 kilometres). Starting from the Liuhe Stupa (Liuhe ta 六和塔) to the east, they added a belt of new elements to the Stone Dike (Shidi) and repaired what had fallen into ruin at a length of 400 *zhang* (approximately 1.2 kilometres). After three months, they had finished their work and the water returned to its former course.[63]

57 'Hanhai tang,' in *XCLAZ*, *j*.31, p. 8b, l.3–5.

58 Not indicated in: Hucker, *A Dictionary of Official Titles*.

59 Situated in the southwestern part of Yanguan District (*Yanguan xian jingtu* 鹽官縣境圖, in *XCLAZ*, *j*.16, p. 8).

60 Location unknown.

61 Situated on the shore of the Zhe River east or southeast of the city of Lin'an (see the *Zhejiang tu*, in *XCLAZ*, *j*.1, p. 7 and the reconstructed *Zhejiang tu*, in Jiang, *≪Xianchun Lin'an zhi≫ Song pan 'jingcheng si tu' fuyuan yanjiu*, 350/4).

62 The character in the original obviously is a misprint for *na* 捺: *CYLAZ*, j.10, p. 9a, l.8.

63 'Hanhai tang,' in *XCLAZ*, *j*.31, p. 8b, l.7—p. 9a, l.3.

Vulnerability of Coastal Agriculture

The Irrigation of the Coastal Agricultural Area with Water from the West Lake

Concurrently, the supply of Lin'an city with agricultural products also became precarious. This was not only because the Yanguan agricultural area, situated on the coast northeast of Lin'an city, was endangered by flooding and tidal inundation. It was also vulnerable because the only freshwater supply for the irrigation of its fields came from the West Lake.

Since the eighth century, freshwater from the West Lake had been conducted through the Shangtang Canal (Shangtang he 上塘河[64]) to irrigate the fields of Yanguan District covering an area of—allegedly—up to 1000 *qing* 頃 (approximately 5800 hectares).[65]

With reference to former Prefect Su Shi's 蘇軾 (1037–1101) agenda detailing five reasons for the preservation of the West Lake that he presented in one of his petitions of 1090, the *Xihu youlan zhi* 西湖遊覽志 makes it clear that the area northeast of Hangzhou was the only area where vegetables for the supply of the capital city were grown, and that the irrigation of its fields depended on the freshwater that was conducted from the West Lake via the Shangtang Canal:

> Southwest of Hang 杭 city the hills are many and the fields are few. The requirements for rice (*gumi* 穀米) and vegetables (*shusu* 蔬蔌)[66] entirely depend [on the fields] in its northeast. The fields (*tiandi* 田地) on the canal banks of the Shangtang [Canal] (Shantang上塘) reach from Renhe 仁和 to Haining 海寧 (i.e., former Yanguan District). — How can [fields of] only 1000 *qing* 頃 (approximately 5,800 hectares) rely on water from the

[64] The Shangtang Canal (Shangtang he) joins the Grand Transport Canal (Da yunhe 大運河) in the north and the sea in the south via other branch canals (Moule, *Qinsai with Other Notes*, 21).

[65] Bai Juyi 白居易 (722–846), 'Jiwen shiji 記文石記,' in *XCLAZ*, *j*.33, p. 6a, l.3a–8a; 'Xihu 西湖,' in *XCLAZ*, *j*.32, p. 1a, l.10—p. 1b, l.1; Su Shi 蘇軾 (1037–1101), 'Hangzhou qi dudie kai Xihu zhuang 杭州乞度牒開西湖狀,' in *Su Shi wenji* 蘇軾文集 (Beijing: Zhonghua Shuju, 1992, hereafter: *SSWJ*), *j*.30, p. 864, l.10–11; Su Shi, 'Shen Sansheng qiqing kaihu liu tiao zhuang 申三省起請開湖六條狀,' in *SSWJ*, *j*.30, p. 867, l.12–14; *SS*, *j*.96, p. 2382, l.13–14; Moule, *Qinsai with Other Notes*, 30. In the twelfth century, for comparison, 5800 ha of irrigated fields in the Yuhang 餘杭 basin were cultivated by 7000 households (Shiba, 'Environment Versus Water Control,' 241).

[66] The original has a character variant.

> lake for relief during hot dry weather? When the West Lake is encroached upon and obstructed, then the people [living] on the Shangtang [Canal] have nothing to rely upon in case of an emergency (*huanji* 緩急).[67]

The Shangtang Canal branched off from the extra-urban canal system north of the Lin'an capital city to flow to the northeast, where it joined the canal system of the Yanguan agricultural area.[68]

Irrigation Problems in the Coastal Agricultural Area Caused by the Cultivation of the West Lake

Although irrigation of the Yanguan agricultural area was wholly dependent on the supply of freshwater from the West Lake reservoir, the capacity of the reservoir to hold water was slowly reduced by farming of edible plants in the lake and ensuing sedimentation.

Contrary to the necessity of supplying the Yanguan agricultural area with freshwater for irrigation, there was a tendency for powerful families and eunuchs to encroach on parts of the West Lake in order to lease the water surface to tenants for the cultivation of edible water plants, such as Indian lotus (*he* 荷), water chestnut (*ling* 菱), and wild rice (*jiao* 茭, *feng* 葑).[69] Tenants might have come from the Yanguan agricultural area after

[67] Tian Rucheng 田汝成 (c. 1500–1563), *Xihu youlan zhi* 西湖遊覽志 (Shanghai: Shanghai guji chubanshe, 1980), *j*.1, p. 6, l. 8–9.

[68] 'Chengnei si he,' in *CYLAZ*, *j*.10, p. 24a, l. 7—p. 24b, l.3; 'He 河,' 'Chengnei 城内,' in *XCLAZ, j*.35, p. 8b, l.8—p. 9a, l.4; Su Shi, 'Shen Sansheng qiqing kaihu liu tiao zhuang,' in *SSWJ, j*.30, p. 868, l.6–7; Shiba Yoshinobu, 'The Business Nucleus of the Southern Song Capital of Hangzhou,' in *The Diversity of the Socio-economy*, eds. Shiba [Originally published as: 'Sôtô Kôshû no shôgyô kaku 宋都杭州の商業核,' in *Sôdai Kônan keizaishi no kenkyû* 宋代江南經濟史の研究 [Studies in the Economy of the Lower Yangzi in the Song] (Tôkyô: Tôkyô Daigaku Tôyô Bunka Kenkyûjo, 1988), 112. See also: *Zhejiang tu*, in *XCLAZ*, *j*.1, p. 7.

[69] *He* 荷: 'Indian lotus' (*Nelumbo nucifera, Gaertn.*); *ling* 菱, also written *ling* 蔆: 'water chestnut,' also translated as: 'water caltrop' (*Trapa bispinosa* Roxb., *Trapa bicornis*, Osbeck, *Trapa bisponosa* or *Trapa natans*); *jiao* 茭: 'wild rice,' also translated as 'water bamboo' or 'Indian rice' (*Hydropyrum latifolium* or *Zizania aquatica*, *Zizania latifolia* Turcz. or *Zizania caduciflora* (Turcz.) Hand.–*Mazz.*). The root of the wild rice is called *feng* 葑 (Bernard E. Read, *Chinese Medicinal Plants from the Pen Ts'ao Kang Mu* 本草綱目 *A. D. 1596* (Shanghai: Peking Natural History Bulletin, 1936, repr. Taipei: Southern Materials Center, 1982) [Chinese Materia Medica 5], nos. 243, 542; G.A. Stuart, *Vegetable Kingdom* (Shanghai 1911, repr. Taipei: Southern Materials Center, 1979) [Chinese Materia Medica 1], 278–81, 440; E.N. Anderson, Jr. and Marja L. Anderson,

inundation disasters had devastated their fields, looking for other farming opportunities.

Since the cultivation of the water plants led to the sedimentation of the lake, thus reducing its water storage capacity, various attempts were made by the prefectural administration to curb the production of water plants and to restore the lake to its former size. In 1149, for example, Prefect Tang Pengju 湯鵬擧 (*jinshi* 1118) repeatedly dredged the West Lake and petitioned for the prohibition of leasing its surface to tenants for the purpose of farming aquatic plants.[70] Additionally, in 1169, Prefect Zhou Cong 周淙 (c. 1115–1175) pleaded for an enlargement of the lake's surface and the prohibition to fill the lake up. He criticized that people encroached on the lake in order to grow water plants.[71] Also, in 1173, the Prefect—probably Shen Du 沈度 (dates unknown)—complained about the many people who leased the water surface to tenants. Since a whole belt in the southwestern part of the lake was farmed using enclosures (*weili* 圍裏) to turn it into dry ground, he requested the clearance of the lake.[72] Again, in 1185, Prefect Zhang Biao 張杓 (dates unknown) petitioned to inspect for the dredging of water plants. He furthermore observed that whenever heavy rainfall occurred, loose soil that had been eroded from the surrounding hills would spill down, so that the West Lake silted up even more.[73]

'Modern China: South,' in *Food in Chinese Culture: Anthropological and Historical Perspectives*, ed. K.C. Chang (New Haven: Yale University Press, 1977), 330; H.T. Huang, *Science and Civilisation in China*. Vol. 6, *Biology and Biological Technology*. Part 5, *Biological Technology, Fermentation and Food Sciences* (Cambridge: Cambridge University Press, 2000), 36, 40, 330; Francine Fèvre and Georges Métailié, *Dictionnaire Ricci des plantes de Chine: Chinois-français, latin, anglais* (Paris: Association Ricci—Les Éditions du Cerf, 2005), 230, 272b–273a, 278b–279a; Francesca Bray, *Science and Civilisation in China*. Vol. 6, *Biology and Biological Technology*. Part 2, *Agriculture* (Cambridge: Cambridge University Press, 1984), 119. For the nutritional dilemma in the use of the West Lake, see: Silvia Freiin Ebner von Eschenbach, 'The Dilemma of Ecological and Nutritional Policies in View of Buddhist Campaigning: The Use of Hangzhou's Xihu 西湖 as a Pool for the Release of Living Beings During the Northern and Southern Song Dynasties (960–1279),' *Monumenta Serica*, 68, 1 (2020), 69–106.

[70] 'Xihu 西湖,' in *CYLAZ*, *j*.10, p. 13b, l.2; 'Xihu,' in *XCLAZ*, *j*.32, p. 5b, l.6a–b.

[71] 'Xihu,' in *CYLAZ*, *j*.10, p. 13b, l.3–4; *SS*, *j*.97, p. 2398, l. 8–10.

[72] Ibid., l.12–14.

[73] 'Shuili zalu,' 1185, in *SHYJG*, vol. 12, *j*. *Shihuo* 61, p. 7538a, l.8–11, 15, citing from: Su Shi, 'Shen Sansheng qiqing kaihu liu tiao zhuang,' in *SSWJ*, *j*.30, p. 871, l.13–15.

When Military Commissioner (*anfu* 安撫)[74] Zhao Yuchou 趙與籌[75] (1179–1260), who served as Prefect of Lin'an from 1241 to 1252, submitted a petition, he complained about a belt of marshes with aquatic plants extending alongside the eastern shore from north of the Yongjin City Gate (Yongjin men 湧金門) up to the Qiantang City Gate (Qiantang men 錢塘門). In answer to his petition, an order was passed that all water plants be dug up.[76] When eventually in 1247, after a severe drought, the West Lake was drying up, Emperor Lizong commissioned Prefect Zhao Yuchou to dredge the Lake on all sides, and to weed out the marshes of water plants.[77]

Reinvestment of Tax Revenue for Securing Irrigation of the Coastal Agriculture

Financial Means for Securing the Water Supply of the Coastal Agricultural Area

Dredging the West Lake to preserve it as a reservoir for the irrigation of the Yanguan agricultural area was a very costly undertaking. Investments were required for it, but dike repair for the protection of the Yanguan agricultural area from salination was expensive, too. But, in order to secure the vulnerable agricultural area of Yanguan District, it was necessary to repair the dikes on the Zhe River shore and maintain the West Lake as a reservoir concurrently. The money needed for such investments probably came from the tax revenue collected from agricultural and salt yields in Yanguan District and—curiously enough—probably also from the levy (*zuke* 租課) of the land yield tax (*guanqian* 官錢) on the farmed sections of the West Lake.[78]

Although Prefect Tang Pengju pleaded against leasing the surface of the West Lake to tenants in 1147, he—in an inconsistent and contradictory political move—recommended investing the land yield tax revenue

[74] See also: Hucker, *A Dictionary of Official Titles*, no. 17.

[75] The character in the original is a variant.

[76] 'Liujing,' in *XCLAZ*, *j*.33, p. 4b, l.1, 6.

[77] 'Xihu,' in *CYLAZ*, *j*.10, p. 13b, l.5–6; 'Xihu,' in *XCLAZ*, *j*.32, p. 6a, l.10—p. 6b, l.1. See also: 'Liujing,' in *XCLAZ*, *j*.33, p. 4a, l.9.

[78] 'Xihu,' in *XCLAZ*, *j*.32, p. 5b, l.6b–7a.

from the cultivation of water plants in the dredging of these same plants in order to prevent any further crops growing on the lake.[79]

In 1185, Prefect Zhang Biao, too, petitioned for the reinvestment of the land yield tax revenue from the farming of the West Lake's water surface in the dredging of the lake:

> According to the one reason as it is contained in the uniform [series] of petitions on dredging the West Lake that Prefect (*shouchen* 守臣)[80] Su Shi submitted in the fifth year of the Yuanyou 元祐 [era] (1090), I ask to care for the cash tax revenue (*keli* 課利) from recent and former marshes (*dang* 蕩) of water chestnut in the West Lake and fully send [the tax revenue] to the office (*si* 司)[81] of the [District] Defender (*xianwei* 縣尉)[82] of Qiantang to receive and look after it so as to provide it for clearing the roots of the wild rice and dredging [the lake] year by year.[83]

In 1214, officials raised accusations of what they considered treacherous abuse of power while leasing the enclosed (*weili*) parts of the surface of the West Lake to tenants for the cultivation of aquatic plants. Yet they called the taxes collected from farming the lake surface only 'trifling':

> We have already outlined in our budget that from the approximately 400 *mu* 畝 of the former marshes (*dang*) of Lin'an Prefecture the yearly income from the land yield tax (*zuqian* 租錢) increased by around 1000 strings (*guan* 貫). In view of the many different kinds of financial transactions (*caiji* 財計) of the Imperial Treasury (*tianfu* 天府)[84] these thousand or hundred strings (*min* 緡) are really only trifling (*suosuo* 瑣瑣)![85]

[79] Ibid., l.3a–b.

[80] See also: Hucker, *A Dictionary of Official Titles*, no. 5355,3.

[81] Ibid., no. 5533.

[82] Ibid., no. 2549.

[83] 'Shuili zalu,' 1185, in *SHYJG*, vol. 12, *j. Shihuo* 61, p. 7538a, l.12–14, citing from: Su Shi, 'Shen Sansheng qiqing kaihu liu tiao zhuang,' *SSWJ*, *j*.30, p. 871, l.11–12.

[84] See also: Hucker, *A Dictionary of Official Titles*, no. 6696.

[85] 'Shuili zalu,' 1214, in *SHYJG*, vol. 12, *j. Shihuo* 61, p. 7546b, l.1–2, also cp. p. 7546a, l.18–23.

The alleged triviality may have been an argument for the tenant households to be exempt from taxation in order to alleviate their burden.[86] This strategy might have turned out contradictorily as well, since it supported the lease of the lake's surface for cultivation.

Financing the Dike Repairs

For the maintenance of the dike system, construction material had to be delivered and workers had to be hired, but financial resources from tax revenue paid in strings of copper coins, rice, and bolts of various silk qualities were limited. This is why Prefect Zhang Biao petitioned in 1185 for the reinvestment of the land yield tax revenue from crops grown on the West Lake in the fortification of the Stone Dike on the Zhe River, setting up a total calculation of approximately 2900 strings of cash for every year.[87] Additionally, because in 1222 the budget of Yanguan District was already strained, the tax-registered households of the district were required to fortify the 16-*li* Dike Against the Brine (Shiliu li xiantang) in the southeast of the district capital themselves. For the repair of the Yuanhua Dike (Yuanhua tang) northeast of the capital, timbers had to be used as stakes (*zhuang*) because they were less expensive than stones.[88]

When during the Jiaxi 嘉熙 era (1237–1241) five suburbs in Renhe District that were situated close to the Zhe River were affected by tidal inundations, Military Commissioner Zhao Yuchou handed in a request to reinvest (*bo* 撥) the income based on the tax quotas (*shui'e* 税額) in the repair of the diked embankment (*tang'an* 塘岸) of the Stone Dike, amounting to roughly 25,000 strings as well as 332 bolts (*pi* 匹) of silk taffeta (*juan* 絹), 2000 *liang* 兩 (approximately 75.6 kilograms) of silk floss (*mian* 緜) and 2500 *shi* 石 (approximately 144 metric tons) of rice.[89]

86 Ibid., p. 7546b, l.4–5.

87 'Shuili zalu,' 1185, in *SHYJG*, vol. 12, *j. Shihuo* 61, p. 7538a, l.17–19.

88 'Yanguan haishui,' in *SS*, *j.*97, p. 2402, l.12—p. 2403, l.1.

89 'Renhe xian 仁和縣,' in 'Jiu xian suijie zhi e 九縣歲解之額,' in 'Gongfu 貢賦,' in *XCLAZ*, *j.*59, p. 8a, l.9—p. 8b, l.3.

Financing of Relief Measures for Inundated or Flooded Areas

First-Aid Provisioning with Rice and Money

Apart from dredging work and dike repairs, there was another area that needed to be financed by tax revenue: The provisioning of aid to the victims of flood and inundation disasters with rice and money to prevent famine.[90]

In the seventh month of 1167, after heavy rains, water masses poured down from Tianmu Hill. Thus, Prefect Zhou Cong specially commissioned the Associate District Magistrate (*tongling* 同令)[91] to personally go to the flood-stricken 285 tax-registered households and provide them with rice and money as an aid measure of famine relief (*zhenji* 賑濟).[92] Similarly, in the eighth month of 1194 after severe flooding (*dashui* 大水) in the six districts of Yuhang, Lin'an, Xincheng, Fuyang, Qiantang, and Yuqian, residents did not have enough to eat. Thus, the prefectural administration allotted (*zhibo* 支撥) rice from the land yield tax (*guanqian*) revenue, as it had been stored in the Price Regulation Granary (*changping* 常平).[93] The prefectural administration commissioned the district officials to go from house to house to distribute the relief supplies (*zhenji* 賑給): Rice rations for ten days were distributed (*jisan* 給散) to the taxable persons from the registered households of the fourth and fifth ranks who had suffered as a result of the floods; adults received 1 *dou* (approximately 5.84 kilograms of rice), that is, roughly 580 grams of rice

[90] Other kinds of famine relief measures included 'substituting labor [with rice or money] as a relief measure' (*yigong daizhen* 以工代賑) or simply 'relief through labor' (*gongzhen* 工賑), that is, hiring starving workers and paying them with rice or money or selling them rice at a reduced price from the Price Regulation Granary (*changping cang* 常平倉) (Yang Lien-sheng, *Les aspects économiques des travaux publics dans la Chine impériale – quatre conférences* (Paris: Collège de France, 1964), 76; Peter J. Golas, 'The Sung Fiscal Administration,' in *The Cambridge History of China*. Vol. 5, Part 2, 205), are not mentioned in this context.

[91] Hucker, *A Dictionary of Official Titles*, nos. 3733, 7464.

[92] 'Shuizai,' 1167, in *SHYJG*, vol. 5, *j*. *Ruiyi* 3, p. 2653a, l.7–9.

[93] That is, *changping cang* 常平倉 (Hucker, *A Dictionary of Official Titles*, no. 257). Price Regulation Granaries bought rice after good harvests thus raising its price, and sold it after bad harvests thus lowering its price. This stabilized the price over longer periods. They also served for direct famine relief by selling, lending, or distributing rice for free (Golas, 'The Sung Fiscal Administration,' 204–5).

per day, and children half that amount. Even Vice Prefect Huang Han 黃瀚 (dates unknown) himself went to all districts and handed out relief supplies (*zhenji*). When he came to the district of Yuhang, the families whose homes were flooded—a total of 20,000 tax-registered households with 3–5 persons each, thus totaling 80,000 taxable persons—were to be allotted 7000 *shi* (approximately 585 metric tons) of rice. Overall, it was instructed by imperial order to allot (*jiebo* 借撥) 10,000 *shi* (approximately 835 metric tons) in relief supplies (*zhenji*).[94] When in 1195 Emperor Ningzong 寧宗 (r. 1195–1225) ordered Prefect Xu Yi 徐誼 (1144–1208) to sell rice for relief (*zhentiao* 賑糶) at a reduced price, the Prefect replied that the hungry people had no money to purchase rice (*shoudi* 收糴) and suggested granting them famine relief (*zhenji* 賑濟) for free for five days instead.[95]

In 1207, the prefectural officials commissioned the Magistrate (*zhixian* 知縣)[96] of Qiantang District to go to the suburbs and villages and record the tax-registered households affected by the flooding and to take care of the money and rice of the prefecture as it was put aside for relief supplies (*zhenji*) so that the people had enough to eat, everybody resettled in their work, and nobody lost their home.[97]

After the flooding of 1214, Emperor Ningzong ordered to allot (*zhibo*) approximately 70,000 strings in paper exchange bills (*huizi* 會子)[98] from the Emergency Reserves Storehouses (Fengzhuang ku 封樁庫)[99] as relief

94 'Shuizai,' 1194, in *SHYJG*, vol. 5, *j. Ruiyi* 3, p. 2658b, l.17–22, p. 2659a, l.6–7.

95 'Zhenhuo,' 1195, in *SHYJG*, vol. 13, *j. Shihuo* 68, p. 8007a, l.20—p. 8007b, l.2.

96 Hucker, *A Dictionary of Official Titles*, no. 993.

97 'Shuizai,' 1207, in *SHYJG*, vol. 5, *j. Ruiyi* 3, p. 2659a, l.23—p. 2659b, l.3.

98 In 1160, after the war with the Jurchen was resumed, the Southern Song introduced new paper exchange bills, also termed 'paper monies' (*huizi*). During another war with the Jurchen 1206–1208, they issued paper monies in even greater quantities, leading to inflation and depreciation in value. Officially, 1 string was shortened from its original countervalue of 1000 coins of cash to 770 cash. At market rates, 1 string was shortend from 750 cash in 1186 to 620 cash in 1195, then to 300 cash by 1210, and to only 50 cash in the 1230s (Richard von Glahn, *Fountain of Fortune: Money and Monetary Policy in China, 1000–1700* [Berkeley: University of California Press, 1996], 51–53).

99 Hucker, *A Dictionary of Official Titles*, no. 1975; situated north of the imperial palace in Lin'an capital city (see: *Jingcheng tu* 京城圖, in *XCLAZ*, *j.*1, p. 6). There was an Upper and a Lower Emergency Reserves Storehouse (*XCLAZ*, *j.*8, p. 15b, l.3—p. 16a, l.1). For its history under Emperor Xiaozong (r. 1163–1190) and its crucial role in emergency relief towards the end of the dynasty see: Gong Wei Ai 江偉愛, 'Imperial

supplies (*zhenji*) to poor, needy, old, and sick tax-registered persons (*jikou* 計口).[100]

Tax Exemption as a Long-Term Relief Measure

Another measure of flood and inundation relief was the exemption from agricultural taxes. As a long-term relief measure, tax exemption, limited or unlimited, was to be granted at the prefectural and district levels.[101]

In short, the tax-registered households in the districts of Lin'an Prefecture were subject to the twice-a-year land yield tax (*liangshui* 兩稅), calculated as an original quota (*yuan'e* 元額) of 10% of the yield. The tax was payable to the prefecture in summer in silk and in fall in rice.[102] The original tax quota of the prefecture amounted to roughly 96,000 bolts of silk taffeta and other kinds of silk, and of approximately 133,000 *shi* (roughly 11,000 metric tons) of rice, respectively.[103] Additionally, there was a poll tax for adult men (*dingshen qian* 丁身錢) as well as various kinds of surcharges.

When in 1160 fields in Yuqian District were destroyed by flooding caused by rainfall, the district officials requested an exemption (*yige* 倚閣) from the fall tax of 36 *shi* (approximately 2.1 metric tons) of rice, as well as from the poll tax. In 1169, further damage was caused by flooding following heavy rainfall, and an exemption of another 56 *shi* (approximately 3.3 metric tons) of rice from the fall tax was requested. This meant

Policy and National Finance: The Role of Hsiao-tsung in the Consolidation of Southern Sung Finance,' in *Guoji Song shi yantao hui lunwen ji* 國際宋史研討會論文集. *Proceedings of the International Symposium on Sung History* (Taibei: Zhongguo wenhua daxue, 1988), 149–52.

100 'Zhenhuo,' 1214, in *SHYJG*, vol. 13, *j. Shihuo* 68, p. 8011b, l.3–5.

101 'Jiu xian suijie zhi e,' in 'Gongfu,' in *XCLAZ*, *j.*59, p. 5b, l.5–7.

102 Ibid., l.3–4. Certain sections of land in the two capital districts of Qiantang and Renhe that belonged to palaces, official buildings, Buddhist monasteries, Daoist temples, and military barracks were not required to pay taxes (ibid., l.4–5). Their tax-exempt status (*mianshui* 免稅) was granted on grounds of initial repair work necessary after the shift of the capital in 1138 ('Qiantang xian 錢塘縣,' in 'Jiu xian suijie zhi e,' in 'Gongfu,' in *XCLAZ*, *j.*59, p. 6b, l.1–2).

103 'Jiu xian suijie zhi e,' in 'Gongfu,' in *XCLAZ*, *j.*59, p. 3b, l. 2–10, rsp. p. 4a, l.8–9.

that only approximately 1300 *shi* (approximately 75.7 metric tons) were actually collected (*cui* 催).[104]

After water masses had poured down from Tianmu Hill in the seventh month of 1167, Prefect Zhou Cong's petition to Emperor Xiaozong 孝宗 (r. 1163–1190) was quickly approved and a tax exemption for 280 households in Lin'an District was granted, differentiated according to household category, and specified according to damage severity: 235 households were categorized below the fifth rank and 45 households were categorized above it. They were accordingly granted an exemption from the summer tax and partly also the fall tax of the relevant year, and some even from the following years as well.[105]

When during the Jiaxi era (1237–1241) the tides of the Zhe River broke through (*chongtu*) in five suburbs of Renhe District, Prefect Zhao Yuguan petitioned for complete exemption (*juan* 蠲) from the fall tax (*miaoshui* 苗税) in the hope that cultivation and reclamation of land would gradually be resumed as soon as the water had returned to its former course.[106]

Still another means to alleviate the tax burden on flood-stricken households was the 'substitutionary payment' (*daishu* 代輸) of 'irregular' (*jiling* 畸零) taxes, probably a kind of surcharge,[107] both of the summer and fall taxes, in advance (*yu* 預) or else a kind of repayment of loans granted by the government in advance (*yu*), similar to the 'harmonious purchase in advance' (*heyu mai* 和預買), that is, the granting of loans in cash on the expected crops to be repaid in kind later.[108]

In the winter of 1269, Military Commissioner Qian Yueyou 潛說友 (c.1200–1280), who served as Prefect of Lin'an in 1268–1271, carried

[104] 'Yuqian xian 於潛縣,' in 'Jiu xian suijie zhi e,' in 'Gongfu,' in *XCLAZ*, *j*.59, p. 10a, l.7–10; 'Juanfang 蠲放,' 1160, in *SHYJG*, vol. 13, *j. Shihuo* 63, p. 7612a, l.18–20.

[105] 'Shuizai,' 1167, in *SHYJG*, vol. 5, *j. Ruiyi* 3, p. 2653a, l.6–7, 12–22.

[106] 'Renhe xian,' in 'Jiu xian suijie zhi e,' in 'Gongfu,' in *XCLAZ*, *j*.9, p. 8a, l.6–8.

[107] The 'irregular' taxes (*jiling*) may have been a surcharge, similar to the rice surcharge (*haomi* 耗米) that was introduced by the Southern Song government, and amounted to 37% of the fall tax (*qiushui* 秋税). This meant that on top of 1 *shi* of the fall tax an additional 3 *shi* 7 *dou* (3.7 *shi*) had to be paid, thence called the three-seven-surcharge (*sanqi hao* 三七耗) (Golas, 'The Sung Fiscal Administration,' 160). For the large number of supplementary taxes, see: Gong, 'Imperial Policy and National Finance,' 152.

[108] Golas, 'The Sung Fiscal Administration,' 199.

out relief aid (*zhenxu* 賑恤) and substitutionary payments of the irregular summer tax in advance. In 1270, he was granted an exemption (*chu* 除) from the irregular summer tax of more than 1 *chi* (approximately 31 centimetres) of silk taffeta for the lower fourth- and fifth-rank tax-registered households. On behalf of the prefecture, Qian Yueyou thus carried out substitutionary payments in advance for all nine districts amounting to a total of approximately 147,000 *zhang* (approximately 453 kilometres) of silk taffeta—roughly corresponding to 435,000 strings in paper exchange bills with a limit of validity (*jie* 界) of 18 years.[109]

In the summer of 1270, Qian Yueyou petitioned for an exemption from the fall tax (*qiumiao* 秋苗) of less than 1 *sheng* (approximately 580 grams) of rice. He was granted an exemption (*juanfang* 蠲放) of more than 1 *sheng*, but less than 1 *dou* (approximately 5.8 kilograms), and carried out substitutionary payments (*daina* 代納) for all nine districts in advance, amounting to a total of approximately 8800 *shi* (approximately 514 metric tons) of rice at a conversion price in cash of approximately 270,000 strings in paper exchange bills with a validity of 18 years.[110]

In the winter of 1270, Quan Yueyou petitioned for a deadline extension in order to improve the situation for the people, and in the seventh month of 1271 was granted a one-time exemption from the irregular summer tax due and payable by the fourth- and fifth-rank households for all the nine districts amounting to approximately 470,000 strings in cash.[111]

When in 1270 the Magistrate (*ling* 令)[112] of Fuyang District, Wang Jiweng 王積翁 (dates unknown), handed in a request saying that his district could not fulfil its annual quota of the fall tax because people had fled their devastated fields, Qian Yueyou submitted a petition to Emperor Duzong 度宗 (r. 1265–1275) and was granted an unlimited and even retroactive tax exemption (*chuhuo* 除豁) effective from two years before.[113]

[109] 'Benfu daishu jiling 本府代輸畸零,' in 'Jiu xian suijie zhi e,' in 'Gongfu,' in *XCLAZ*, *j.*59, p. 15a, l.1–7, p. 15b, l.6, p. 16a, l.1–2.

[110] Ibid., *j.*59, p. 15b, l.5–9, p. 16a, l.2.

[111] Ibid., *j.*59, p. 16a, l.9—p. 16b, l.2.

[112] Hucker, *A Dictionary of Official Titles*, nos. 3733, 2518.

[113] 'Fuyang xian 富陽縣,' in 'Jiu xian suijie zhi e,' in 'Gongfu,' in *XCLAZ*, *j.*59, p. 11b, l.3–5.

This policy meant a diminution in public (*gong* 公) tax revenue but an increase in private (*si* 私) income.[114] It may be surmised that the substitutionary payment came from the prefectural treasuries and granaries.

Conclusion

Although many incidents may not have been reported by the relevant prefect, perhaps for political reasons, the events recorded in the sources reveal the inner inconsistencies and conflicts of water control and economy arising from an increase in flood and inundation disasters that hit Lin'an Prefecture.

From the beginning of dike building in 910, the goal had been to protect the alluvial land, the agricultural area in Yanguan District, and the city of Hangzhou (Lin'an) in particular, from salination. This was all the more important because the agricultural area supplied the city with food. Already from the eighth century onward, the agricultural area had been dependent on the West Lake reservoir as its exclusive source of freshwater for irrigation.

When in the twelfth century climatic cooling, related to an extraordinary cold period in the northeastern IOW, contributed to increased vulnerability and exposure to flood events, caused by heavy precipitation and tidal inundations, disasters caused great loss of life and significant economic damage. Damage wrought by floods and tidal inundations on the Yanguan agricultural area may have favoured encroachings on the West Lake for cultivating edible plants there. Thus, farming activities on the West Lake may have been intended to counterbalance the loss of fields in Yanguan District. The farming of the West Lake, however, resulted in sedimentation and shrinking of the lake's storage capacity.

In order to improve the lake's storage capacity for the irrigation of the Yanguan agricultural area, the plants cultivated on the lake had to be cleared. The dredging of the lake became necessary to secure the irrigation of the Yanguan agricultural area, but it required public investments. However, public investments were also dearly needed for dike building. There may have been a conflict of reinvesting tax revenue either in the

[114] 'Benfu daishu jiling,' in 'Jiu xian suijie zhi e,' in 'Gongfu,' in *XCLAZ*, *j*.59, p. 16b, l.3.

dredging of the West Lake or in the repair of the Stone Dike and the various earth dikes in Yanguan District.

There may also have been a further conflict between such investments and the financing of relief aid for the victims of floods and inundations. For immediate aid, rice and money were allotted. For long-term relief, exemption from taxes was granted to prevent farmers from fleeing their fields and to keep them to recultivate their land.

Since this was an investigation ceteris paribus, it is, however, possible that expenditure for timely repair work on the dike system, for the dredging of the lake, and for relief aid may have been neglected in favour of other issues, such as military priorities to meet the challenge from the Jurchen Empire in the north.[115]

Bibliography

Primary Sources

CYLAZ: Zhao Yuchou 趙與籌 (1179–1260). ***Chunyou Lin'an zhi*** 淳祐臨安志 (1250). ***Song Yuan difang zhi congshu*** 宋元地方叢書.

HNXZ: ***Haining xianzhi*** 海寧縣誌. Edited. 1765. Repr., Taibei: Chengwen, 1983, Zhongguo fangzhi congshu, no. 516.

Jianyan yilai xinian yaolu 建炎以來繫年要錄. ***Qinding Siku quanshu*** 欽定四庫全書. Repr., Shanghai: Shanghai guji chubanshe.

SHYJG: ***Song huiyao jigao*** 宋會要輯稿. Edited by Xu Song 徐松. Shanghai: Shanghai guji chubanshe, 2014.

SS: Su Shi's 蘇軾 (1037–1101). ***Song shi*** 宋史. Compiled by Toqtogha (Tuotuo 脱脱, 1313–1355) et al. (1345). Edited Beijing: Zhonghua shuju, 1977.

SSWJ: ***Su Shi wenji*** 蘇軾文集. Edited Beijing: Zhonghua shuju, 1992.

TXJGLBS: Gu Yanwu 顧炎武 (1613–1682). ***Tianxia junguo libing shu*** 天下郡國利病書. ***Siku shanben*** 四庫善本. Repr., Shanghai: Shangwu yinshuguan, 1936, reissued Tabei, s. a.

XCLAZ: Qian Yueyou 潛說友 (c. 1200–1280). ***Xianchun Lin'an zhi*** 咸淳臨安志 (1273). ***Song Yuan difang zhi congshu***.

Tian Rucheng 田汝成 (c. 1500–1563). ***Xihu youlan zhi*** 西湖遊覽志. Shanghai: Shanghai guji chubanshe, 1980.

Ye Shaoweng 葉紹翁 (c. 1175–1230). ***Sichao wenjian lu*** 四朝聞見錄. Beijing: Zhonghua shuju, 2006.

[115] For the fiscal policies to meet high military costs, see: Gong, 'Imperial Policy and National Finance,' 142, 147–48, 152; Golas, 'The Sung Fiscal Administration,' 156–57.

Secondary Sources

Anderson, E.N., Jr., and Marja L. Anderson. 'Modern China: South,' in *Food in Chinese Culture: Anthropological and Historical Perspectives*, ed. K.C. Chang. New Haven: Yale University Press, 1977: 317–82.

Bray, Francesca. *Science and Civilisation in China*. Vol. 6: *Biology and Biological Technology*. Part 2, *Agriculture*. Cambridge: Cambridge University Press, 1984.

Chaffee, John W., and Denis Twitchett, eds. *The Cambridge History of China*. Vol. 5, Part 2, *Sung China, 960–1279*. Cambridge: Cambridge University Press, 2015.

Chang, Bide 昌彼德. *Song ren zhuanji ziliao suoyin* 宋人傳記資料索引. 6 Vols. Taibei: Dingwen shuju, 1988.

Chang, K.C., ed. *Food in Chinese Culture: Anthropological and Historical Perspectives*. New Haven: Yale University Press, 1977.

Chien, Cecilia Lee-fang. *Salt and State: An Annotated Translation of the Songshi Salt Monopoly Treatise*. Ann Arbor: Center for Chinese Studies, The University of Michigan, 2004.

Domrös, Manfred, and Peng Gongbing. *The Climate of China*. Berlin: Springer, 1988.

Ebner von Eschenbach, Silvia Freiin. 'Managing Floods and Droughts by Invocating the Water Spirits: Analyzing Prayers for Rain (*daoyu* 禱雨) and Prayers for a Clear Sky (*qiqing* 祈晴). With Some Examples from Local Source Material of the Song 宋 Dynasty (960–1279).' *Zeitschrift der Deutschen Morgenländischen Gesellschaft*, 169, 1 (2019): 205–29.

Ebner von Eschenbach, Silvia Freiin. 'The Dilemma of Ecological and Nutritional Policies in View of Buddhist Campaigning: The Use of Hangzhou's Xihu 西湖 as a Pool for the Release of Living Beings During the Northern and Southern Song Dynasties (960–1279).' *Monumenta Serica*, 68, 1 (2020): 69–106.

Ebner von Eschenbach, Silvia Freiin. 'Innerstädtische Verkehrsinfrastruktur und Mobilität und ihre Einflussfaktoren. Dargestellt am Beispiel einer südostchinesischen Stadt unter der Qian 錢-Dynastie (907–978) und der Nördlichen und Südlichen Song 宋-Dynastie (960–1127, 1127–1279).' *Mobilität in China*, 2021 [Jahrbuch der Deutschen Vereinigung für Chinastudien, 15] [20 pages + 2 maps. Print confirmed].

Elvin, Mark. *The Retreat of the Elephants: An Environmental History of China*. New Haven: Yale University Press, 2004.

Elvin, Mark, and Liu Ts'ui-jung, eds. *Sediments of Time: Environment and Society in Chinese History*. Cambridge: Cambridge University Press, 1998.

Elvin, Mark, and Su Ninghu. 'Action at a Distance: The Influence of the Yellow River on Hangzhou Bay Since A. D. 1000,' in *Sediments of Time: Environment and Society in Chinese History*, eds. Mark Elvin and Liu Ts'ui-jung. Cambridge: Cambridge University Press, 1998: 344–407.

Fèvre, Francine and Georges Métailié. *Dictionnaire Ricci des plantes de Chine: Chinois –français, latin, anglais*. Paris: Association Ricci – Les Éditions du Cerf, 2005.

Gernet, Jacques. *La vie quotidienne en Chine à la veille de l'invasion mongole 1250–1276*. S.l.: Librairie Hachette, 1959.

Glahn, Richard von. *Fountain of Fortune: Money and Monetary Policy in China, 1000–1700*. Berkeley: University of California Press, 1996.

Golas, Peter J. 'The Sung Fiscal Administration,' in *The Cambridge History of China*. Vol. 5, Part 2, *Sung China, 960–1279*, eds. Chaffee and Twitchett: 139–213.

Gong, Wei Ai 江偉愛. 'Imperial Policy and National Finance: The Role of Hsiao-tsung in the Consolidation of Southern Sung Finance.' *GuojiSong shi yantao hui lunwen ji* 國際宋史研討會論文集. *Proceedings of the International Symposium on Sung History*. Taibei: Zhongguo wenhua daxue, 1988: 141–58.

Haeger, John Winthrop, ed. *Crisis and Prosperity in Sung China*. Tuscon: The University of Arizona Press, 1975.

Hargett, James. 'Song Dynasty Local Gazetteers and Their Place in the History of *Difangzhi* Writing.' *Harvard Journal of Asiatic Studies*, 56, 2 (1996): 405–42.

Hervouet, Yves, ed. *A Sung Bibliography (Bibliographie des Sung)*. Hong Kong: The Chinese University Press, 1978.

Huang, H.T. *Science and Civilisation in China*. Vol. 6: *Biology and Biological Technology*. Part 5, *Biological Technology, Fermentation and Food Sciences*. Cambridge: Cambridge University Press, 2000.

Hucker, Charles O. *A Dictionary of Official Titles in Imperial China*. Stanford, CA: Stanford University Press, 1985.

Jiang, Qingqing 姜青青. <<*Xianchun Lin'an zhi*>> Song pan 'jingcheng si tu' fuyuan yanjiu <<咸淳临安志>>宋版' '京城四图' 复原研究. Shanghai: Shanghai guji chubanshe, 2015.

Lamouroux, Christian. 'From the Yellow River to the Huai: New Representations of a River Network and the Hydraulic Crisis of 1128,' in *Sediments of Time*, eds. Elvin and Ts'ui-jung: 545–84.

McDermott, Joseph P., and Shiba Yoshinobu. 'Economic Change in China, 960–1279,' in *The Cambridge History of China*. Vol. 5, Part 2, *Sung China, 960–1279*, eds. Chaffee and Twitchett: 321–436.

Moll-Murata, Christine. *Die chinesische Regionalbeschreibung*: *Entwicklung und Funktion einer Quellengattung, dargestellt am Beispiel der Präfekturbeschreibungen am Hangzhou*. Wiesbaden: Verlag Harrassowitz, 2001.

Moule, A. C. *Qinsai with Other Notes on Marco Polo*. Cambridge: University Press, 1957.

Needham, Joseph with the Collaboration of Wang Ling and Lu Gwei-Djen. *Science and Civilisation in China*. Vol. 4: *Physics and Physical Technology*. Part

3, *Civil Engineering and Nautics*. Cambridge: Cambridge University Press, 1971.

Pang, Huiping. 'Strange Weather: Art, Politics, and Climate Change at the Court of Northern Song Emperor Huizong.' *Journal of Song-Yuan Studies*, 39 (2009): 1–41.

Read, Bernard E. *Chinese Medicinal Plants from the Pen Ts'ao Kang Mu* 本草綱目 *A. D. 1596*. Shanghai: Peking Natural History Bulletin, 1936, repr. Taipei: Southern Materials Center, 1982 [Chinese Materia Medica, 5].

Shiba, Yoshinobu. 'Environment Versus Water Control: The Case of the Southern Hangzhou Bay Area from the Mid-Tang Through the Qing,' in *Sediments of Time*, eds. Elvin and Ts'ui-jung: 135–64.

Shiba, Yoshinobu, ed. *The Diversity of the Socio-economy in Song China, 960–1279*. Tôkyô: Tôyô Bunko, 2011 [Revised edition, originally published 2002].

Shiba, Yoshinobu. 'The Business Nucleus of the Southern Song Capital of Hangzhou,' in *The Diversity of the Socio-economy in Song China,* ed. Shiba: 89–128 [Originally published as 'Sôtô Kôshû no shôgyô kaku 宋都杭州の商業核,' in *Sôdai Kônan keizaishi no kenkyû* 宋代江南經濟史の研究 [Studies in the Economy of the Lower Yangzi in the Song]. Tôkyô: Tôkyô Daigaku Tôyô Bunka Kenkyûjo, 1988: 312–39].

Stuart, G.A. *Vegetable Kingdom*. Shanghai 1911, Repr. Taipei: Southern Materials Center, 1979 [Chinese Materia Medica, 1].

Wilkinson, Endymion. *Chinese History: A New Manual*. 4th edition. Cambridge, MA: Harvard University Press, 2015.

Worthy, Edmund H. 'Regional Control in the Southern Sung Salt Administration,' in *Crisis and Prosperity in Sung China*, ed. Haeger: 101–41.

Wright, Hope. *Alphabetical List of Geographical Names in Sung China*. Paris: École Pratique des Hautes Études, Centre de Recherches Historiques, 1956 [Sung Project].

Yang, Lien-sheng. *Les aspects économiques des travaux publics dans la Chine impériale – quatre conférences*. Paris: Collège de France, 1964.

CHAPTER 3

Epidemic and Environmental Change in China's Early Modern Maritime World During the 'Little Ice Age' (ca. 1500–1680)

Angela Schottenhammer

Lipotidae
Large ones get more than 6 m long; on their back they have like a sharp blade.

This research was supported by the ERC AdG project TRANSPACIFIC which has received funding from the European Research Council (ERC) under the European Union's Horizon 2020 Research and Innovation Programme (Grant agreement No. 833143). It also contributes to the research project 'Appraising Risk, Past and Present: Interrogating Historical Data to Enhance Understanding of Environmental Crises in the Indian Ocean World,' sponsored by the Social Sciences and Humanities Research Council of Canada (SSHRC).

The original version of this chapter was revised: Incorrect text has been updated for reference "Schottenhammer, Angela" in list. The correction to this chapter is available at https://doi.org/10.1007/978-3-030-98198-3_12

A. Schottenhammer (✉)
KU Leuven, Leuven, Belgium
e-mail: angela.schottenhammer@kuleuven.be

UGent, Gent, Belgium

P. Gooding (ed.), *Droughts, Floods, and Global Climatic Anomalies in the Indian Ocean World*, Palgrave Series in Indian Ocean World Studies, https://doi.org/10.1007/978-3-030-98198-3_3

When they come in front of the South Seas God Temple, this is called 'the arrival'.
Sometimes [this occurs] several times a year, sometimes once every few decades.
When they come frequently, then the people have diseases.

鱀魚
大者長二丈餘脊若鋒刃
嘗至南海廟前謂之來朝
或一年數至或數十年一至
若來數, 則人有疫疾.
[*Guangdong xinyu* 廣東新語, *j*.22 (*linyu* 鱗語)]

The environment and climate have become ever more prominent components of recent research into China's history.[1] A great deal of this scholarship has focused on recurring floods and droughts, which have constituted a perpetual problem in China's past. Such catastrophes were not seldom accompanied by severe outbreaks of disease, often of epidemic proportions. Debates still prevail about the linkages between these phenomena. If, for example, significant climatic changes, such as the rise or drop of temperatures, decisively influenced typhoons, floods, or droughts, how did these broader climatic anomalies contribute to the spread of epidemics and related phenomena? This chapter analyses potential answers to this question through an analysis of the linkages between El Niño Southern Oscillation (ENSO) anomalies, typhoons, and

[1] Mark Elvin was certainly one of the first sinologists to be mentioned in this context. See: Mark Elvin, *The Retreat of the Elephant: An Environmental History of China* (New Haven: Stanford University Press, 2004); Mark Elvin and Liu Ts'ui-jung, eds., *Sediments of Time: Environment and Society in Chinese History* (Cambridge: Cambridge University Press, 1998). Subsequently, I will particularly refer to the research of Timothy Brook. See, for example: Timothy Brook, 'Nine Sloughs: Profiling the Climate History of the Yuan and Ming Dynasties, 1260–1644,' *Journal of Chinese History*, 1 (2017), 27–58; Timothy Brook, 'Differential Effects of Global and Local Climate Data in Assessing Environmental Drivers of Epidemic Outbreaks,' *PNAS*, 114, 49 (2017). See also the results of collaborative work between Chinese historians and climate scientists, such as: Quansheng Ge, Jingyun Zheng, Yanyu Tian, Wenxiang Wu, Xiuqi Fang, and Wei-Chyung Wang, 'Coherence of Climatic Reconstruction from Historical Documents in China by Different Studies,' *International Journal of Climatology*, 28 (2008), 1007–24. Several more examples are referred to below.

epidemics during a protracted period of global cooling, known as the Little Ice Age (LIA), focusing on the period c.1500–1680.[2] It builds upon the knowledge of climate historians but restricts itself to social historic analysis.

Recent climatological analyses suggest that ENSO, an anomaly of sea surface temperatures (SSTs) in the eastern equatorial Pacific Ocean, has significant teleconnections with the frequency, origin points, tracks, and strength of tropical cyclones. During La Niña years, when SSTs in the east-central equatorial Pacific are anomalously low, the chance of typhoons making landfall on much of the Chinese coastline is increased.[3] Similarly, colder temperatures in northern and central China have been reported to increase the intensification of westerlies, thereby further increasing typhoons' chances of making landfall.[4]

It is in this context that the period c.1560–1650—a core period within that which is under review—represents a particularly interesting case study. Widely recognized as a period of global cooling, recent climatic reconstructions using proxy data have suggested prevailing La Niña-like conditions. According to Joëlle Gergis and Anthony Fowler, 47 of these 90 years were La Niña years. Moreover, consecutive years of La Niña anomalies occurred in 1571–1573, 1576–1584 (all but one), 1600–1605, 1622–1632, and 1637–1639.[5] Thus, based on recent models connecting these phenomena, historians looking in the archive might expect to find frequent reports of typhoons making landfall in this period. This is further supported by research by Kam-biu Liu, Caiming Shen, and Kin-sheun Louie, who suggested that the years 1660–1680—a particularly cool and dry period in Chinese history—represent one of the most active periods of local typhoon landfalls in Guangdong.[6]

[2] For more on the LIA, see: Chapter by Chaudhuri, this volume.

[3] James B. Elsner and Kam-biu Liu, 'Examining the ENSO-Typhoon Hypothesis,' *Climate Research*, 25, 1 (2003), 43–54. See also: César Caviedes, *El Niño in History: Storming Through the Ages* (Gainesville: University Press of Florida, 2001).

[4] Kam-biu Liu, Caiming Shen, and Kin-sheun Louie, 'A 1,000-Year History of Typhoon Landfalls in Guangdong, Southern China, Reconstructed from Chinese Historical Documentary Records,' *Annals of the Association of American Geographers*, 91, 3 (2001), 460–61.

[5] Joëlle L. Gergis and Anthony M. Fowler, 'A History of ENSO Events Since A.D. 1525: Implications for Future Climate Change,' *Climatic Change*, 92 (2009), 371.

[6] Kam-biu Liu, Caiming Shen, and Kin-sheun Louie, 'A 1,000-Year History of Typhoon Landfalls,' 460.

Building on such work, this chapter examines the frequency of typhoon landfalls in China, especially Fujian, Guangdong, and Zhejiang, in c.1500–1680, and explores their potential connections to flood events and epidemics. Further correlations are of interest in this context. Recent work by Kong Dongyan, Li Gang, and Wang Huijuan has also suggested a close relationship between increased frequency of La Niña events, solar activity, and locust attacks. The authors also argue that La Niña conditions increase the chances of tidal inundations.[7] Similarly, scholars have elsewhere argued that 'successive natural disasters caused by the climate of the LIA were the main natural factors of the plague epidemic during the late Ming dynasty.'[8] Thus, the current scholarship, although only in a formative stage, suggests cautious correlations between global cooling, negative ENSO anomalies, increased frequency of flood events in China, and the outbreak of epidemics. This chapter seeks to explore these possible correlations further, incorporating the existing research, and providing a wider geographic research area (Fig. 3.1).

Writing as a historian, I have to emphasize that this research necessarily remains speculative in various aspects. The data is often incomplete, fragmentary, or too general. Sometimes the sources provide descriptions in some detail; sometimes they just record that a disaster occurred. Records for many parameters that we use today for weather prediction, temperature measurements, and reconstruction of ocean currents, do not exist for most historical periods. In addition, we have yet to gather more data and information from all Chinese coastal provinces, other East Asian coastal

[7] Kong Dongyan 孔冬艳, Li Gang 李钢, Wang Huijuan 王会娟, 'Ming Qing shiqi Zhongguo yanhai diqu haichao zaihai yanjiu 明清时期中国沿海地区海潮灾害研究,' *Journal of Natural Disasters / Ziran zaihai yanjiu* 自然灾害学报, 25, 5 (2016), 93. The article explains that typhoons need strong convective movements, absorbing lots of heat, to develop. Because in El Niño years, the equatorial water surface temperature in the Eastern Pacific is high, while they are relatively low in the Western Pacific, heat and water vapours decrease so that the tendency of increasing atmospheric energy, of strong convective movements in other words, also decreases. When the surface temperatures in the Western Pacific are relatively low, the atmospheric energy to develop typhoons also decreases.

[8] Qiu-Hua Li, Yue-Hai Ma, Ning Wang, Ying Hu, and Zhao-Zhe Liu, 'Overview of the Plague in the Late Ming Dynasty and Its Prevention and Control Measures,' *TMR Journals*, 5, 3 (2020), 138–39.

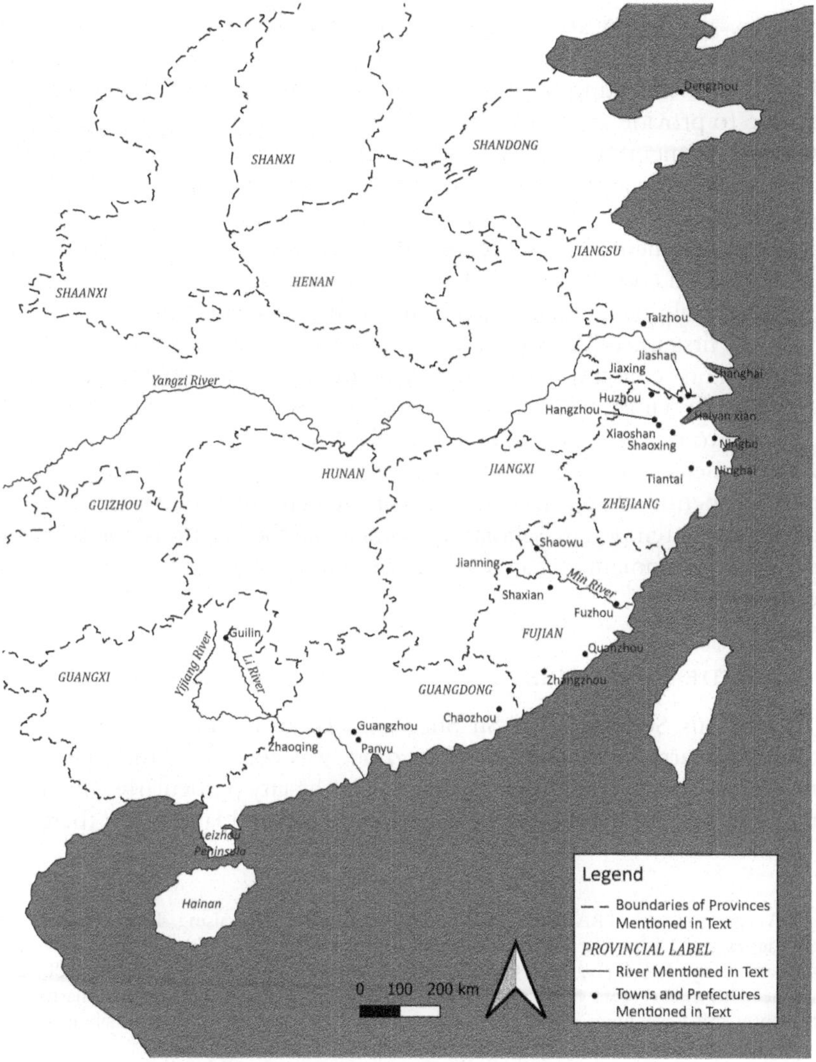

Fig. 3.1 Map of coastal China, including locations of places and features mentioned in-text. Drawn by Philip Gooding

regions, and island archipelagos.[9] Consulting other studies by Chinese historians and environmental and climate specialists, and by analysing data I have been collecting on this topic in cooperation with colleagues, I am unable to provide any generalized tendencies. Being confronted with the restrictions mentioned above and analysing here only a fraction of the body of sources available to us on this topic, I can simply provide some snapshots of micro-environmental events and some particular stories of crisis management. The reader should consequently take this chapter as another small contribution to the question of to what extent climatic and other natural phenomena fostered the outbreak of epidemics.

I will first introduce examples of typhoons, tide disasters, and storm surges, and contemporary outbreaks of epidemics relying on data collected by climate historians and on data we have been gathering as part of two ongoing projects entitled 'TRANSPACIFIC' and 'Appraising Risk.'[10] Before coming to a tentative conclusion, I will introduce forms of Ming-period (1368–1644) crisis management, incorporating the views of some contemporary scholars in China on the relationships between nature, environmental changes, and the outbreak of certain diseases and epidemics.[11]

Tide Disasters, Tsunamis, and Storm Surges

The famous Song scholar, Su Shi 蘇軾 (1036–1101), who established the first pharmacy in 1089 in Hangzhou, was convinced that the dense networks of waterways in Hangzhou made the city particularly vulnerable to epidemics: 'Hangzhou is a place where water and land meet; therefore

[9] As part of my TRANSPACIFIC project and the 'Appraising Risk' project, my colleagues and I have started to collect and organize such data. We already possess year-for-year proxy data for various Chinese coastal provinces for the years 1500–1700. But we need to collect data from many more regions. In addition, our geoinformatician is currently still developing our spatial–temporal database that will enable visualisation and systematic analysis of relationships and correlations between data-points.

[10] See credentials.

[11] The selection of the scholars was partly arbitrary. However, the fact that a scholar like Su Jun dedicated a special chapter to the question of 'climate' attracted my attention. He also directly addressed the question of local climate, environmental change, and epidemics, in this case various forms of malaria. Qu Dajun has been chosen because he speaks especially about the causes and development of typhoons, a major periodic calamity in his home province Guangdong.

there are more illnesses than elsewhere.'[12] He believed in a correlation between plenty of water and the outbreak of disease. In the early centuries CE, most floods occurred in north and northwest China, but from sometime during the Song (960–1279) onwards, the frequency of these catastrophes shifted to the southeastern regions, especially Zhejiang and Jiangsu.[13] This shift may be associated with a process of general cooling in East Asia dating from the twelfth century.[14] As the famous maritime historian Lo Jung-pang has noted:

> The southeast coastal lands of China are a region that feels the greatest effect of the continental cyclonic storms. It was therefore a region that suffered more from floods than any other region of China. Estimated in terms of the number of floods per thousand square kilometres per century, Zhejiang ranked the highest in the nation.[15]

Jiangsu came second. While Lo Jung-pang described the situation during the Song and Yuan (1271–1368), the figures presented in this chapter on the Ming and early Qing (1644–1912) generally confirm the picture.

Based on the statistical data on tropical cyclones gathered by Yen-Chu Liu, Huei-Fen Chen, Xingqi Liu, and Yuan-Pin Chang, it is evident that the number of typhoons increased remarkably after 1400 CE.[16] Referring

[12] Angela Ki Che Leung, 'Organized Medicine in Ming-Qing China: State and Private Medical Institutions in the Lower Yangzi Region,' *Late Imperial China*, 8, 1 (1987), 136: With reference to: *Xu zizhi tongjian changbian* 續資治通鑑長編, by Li Dao 李燾 [1115–1184] (Taibei: Shijie shuju, 1965), 435.20b.

[13] Lo Jung-pang stated that in Yuan times, Zhejiang and Jiangsu provinces had 33.7% of the floods per 1000 square kilometers per century, and 27.5% of all the droughts per 1000 square kilometers per century from 206 BCE to 1911 CE. See: Lo Jung-pang, *China as a Sea Power, 1127–1368: A Preliminary Survey of the Maritime Expansion and Naval Exploits of the Chinese People During the Southern Song and Yuan Periods*, ed. Bruce A. Elleman (Singapore: National University of Singapore Press, 2011), 76: With reference to: Yao Shan-yu, 'The Chronological and Seasonal Distribution of Floods and Droughts in Chinese History, 206 B.C.–A.D. 1911,' *Harvard Journal of Asiatic Studies*, 6, 3–4 (1942), 363.

[14] See also: Chapter by Ebner von Eschenbach, this volume.

[15] Lo Jung-pang, *China as a Sea Power*, 89.

[16] Yen-Chu Liu, Huei-Fen Chen, Xingqi Liu, and Yuan-Pin Chang, 'Insight into Tropical Cyclone Behaviour Through Examining Maritime Disasters Over the Past 1000 Years Based on the Dynastic Histories of China—A Dedication to Ocean Researcher V,' *Quaternary International*, 440, A (2017), 72–81.

to the Ming Period in this context, Zhou Zhiyuan 周致元 has identified the following characteristics[17]: First, most tidal inundations were caused by typhoons and cyclones. Second, the occurrence of these inundations followed certain rules: They primarily occurred during summer and fall, with a peak during the months July to September, and most of the typhoons developed east or southeast of Taiwan and the Philippine Archipelago. Third, whereas floods caused by river inundations were mostly preceded by heavy rainfall, so that people were alarmed in advance and could take precautionary measures, tidal disasters mainly occurred ad hoc, leaving the local population without much time to prepare. The human catastrophes were, consequently, sometimes more drastic in the latter instances.

These patterns are also reflected in the sources investigated for this chapter, which frequently report of thousands, or even tens of thousands, of people drowning, and of water approaching rapidly and rising 'like a mountain.' Often, the water mass arrived with such violence that all dikes or protecting walls were smashed and washed away. The violence of the floods was often significant: Descriptions speak of huge waves flooding rice fields over large areas, and of enormous winds and rains causing additional damages.

To give some examples roughly corresponding to the period under review: In 1472, more than 28,400 people drowned after a severe storm in Zhejiang (浙江大風雨, 海水暴溢,……溺死者二萬八千四百餘人).[18] In 1539, a storm surge had the water rise to more than 6.66 m and more than 29,000 people drowned (海溢, 高二丈餘, 溺死民灶男妇二萬九千餘口).[19] 1568, a severe typhoon hit Taizhou, Zhejiang, and an enormous flood surge covered all districts and towns in Taizhou, only retreating after three days. More than 30,000 people drowned (浙江台州府颶風大作, 海潮汛涨, 天台諸山水驟合, 冲入台州府城三日乃退。溺死人民三萬餘口).[20] In Guangdong, in 1618, lightning, thunder, and a typhoon raged,

[17] Zhou Zhiyuan 周致元, 'Mingdai Dongnan diqu de haichao zaihai 明代东南地区的海潮灾害,' *Shikue jikan* 史学集刊/*Collected Papers of History Studies*, 2 (2005), 83–93.

[18] Zhou Zhiyuan, 'Mingdai Dongnan diqu de haichao zaihai,' 89: With reference to *Ming Xianzong shilu* 明憲宗實錄, *juan*, 106.

[19] Ibid., with reference to the Wanli edition of *Tongzhou zhi* 通州志.

[20] Ibid., with reference to *Ming Muzong shilu* 明穆宗實錄, *juan* 22.

causing 12,530 men and women to drown.[21] In 1628, in Hangzhou 杭州, Jiaxing 嘉興, Huzhou 湖州, and Shaoxing 紹興, Zhejiang, a severe storm caused a tidal surge that drifted away uncountable fields and grain, destroying homes and killing several tens of thousands of people (大風雨, 海溢, 漂没田禾無算, 壞民居數萬間, 殺傷居民數萬人).[22] In 1631, a flood surge in Zhejiang was so high that it buried 70,000 people under the mass of water (海潮狂溢, 漂溺人民七萬).[23] These descriptions provide an insight into the extent of damage caused by such flood surges. The sources are sometimes detailed enough that we do not need to look at the damage of hurricanes and typhoons in more recent times to be able to get an idea of the destruction.

The destruction was sometimes so extensive that it took years, or even decades, for infrastructure to be repaired. Sources repeatedly speak of 'uncountable' (*wusuan* 無算) homes being washed or blown away. Sometimes even concrete figures are mentioned: In 1619, during a tide disaster in Chaozhou 潮州, Guangdong, 31,867 houses were destroyed. In an event during the reign of Emperor Yingzong 英宗 (r. 1435–1449 and 1457–1464), more than 3000 *zhang* of protective dikes were destroyed (that is approximately 10 km) in one district. Sometimes, the entire dike construction, including the base layers, were washed away.[24]

The consequences for agriculture were also often disastrous. Salty water and brine destroyed plants, grain, and the soil.[25] Entire harvests, grain stocks, and fields could be destroyed, as a consequence of which famines followed. In 1422, a flood surge in Guangdong destroyed 1200 homes and more than 25,300 *dan* of grain (i.e. more than approx. 1315 tons).[26] With the shortage of grain and rice, prices increased, and the distress of the people was consequently further exacerbated. Frequently, the number of people who died following such natural disasters still exceeded the quantity of those who drowned or died directly in the disaster. Famine and epidemics were among the most serious of such

[21] Ibid., with reference to *Ming Shenzong shilu* 明神宗實錄, *juan* 583.

[22] Ibid., with reference to Bo Sen 伯森 et al., *Ming shilu leizuan* 明實錄類纂, *juan Ziran zaiyi*·自然災異卷 (Wuhan: Wuhan chubanshe, 1993), *juan* 11.

[23] Ibid., with reference to Bo Sen, *Ming shilu leizuan*, *juan* 45.

[24] Ibid., 90.

[25] See also: Chapter by Ebner von Eschenbach, this volume.

[26] Zhou Zhiyuan, 'Mingdai Dongnan diqu de haichao zaihai,' 90.

consequences. Because of the immediate impacts of climate changes and environmental disasters on agriculture, the basis of the early modern Ming and Qing economies, rulers have been sensitive to such problems ever since.[27]

Outbreaks of Diseases and Epidemics

Corresponding with the climatic effects of the core of the LIA, epidemics were frequent during the late Ming dynasty. Nevertheless, we have to be cautious in drawing simple correlations here. Social, hygienic, and economic factors significantly affected the likelihood that typhoon-related floods would contribute to an epidemic. We also have to take into account that the mid-seventeenth century was a time of war and unrest, characterized by the military conquest of China by the Manchus. Contemporary life was thus also negatively influenced by wars and man-made destruction. Even so, Timothy Brook has stressed that disasters, droughts, and famines were omnipresent in the last years of the Ming. In Shanghai, the corpses of the dead lay in the streets: 'Epidemics followed in the wake of drought and famine.'[28]

China was struck by various infectious diseases and epidemics (*dayi* 大疫; *wenyi* 瘟疫; literally: 'febrile pestilence') from early Ming times. For example, in 1408, more than 78,400 people passed away in Jiangxi and Fujian; in 1411, more than 6000 people died in Dengzhou 鄧州 and Ninghai 寧海; in 1435–1436, 30,000 deaths occurred in Shaoxing 紹興, Ningbo 寧波 and Taizhou 台州; in 1455 more than 20,000 died in Guizhou 貴州; and in 1475, there were innumerable deaths in Fujian and Jiangxi.[29] Bubonic plague and smallpox were likely the most common epidemics in these contexts. In the south, malaria constituted a big

[27] See, for example: Robert B. Marks, '"It Never Used to Snow". Climate Variablility and Harvest Yields in Late-Imperial South China, 1650–1850,' in *Sediments of Time*, eds. Elvin and Ts'ui-jung, 411–12.

[28] Timothy Brook, *The Troubled Empire: China in the Yuan and Ming Dynasties* (Cambridge, MA: Belknap Press of Harvard University Press, 2010), 250.

[29] *Mingshi* 明史 [1739], by Zhang Tingyu 張廷玉 [1672–1755] et al. (Beijing: Zhonghua shuju, 1974), 28.442–442 includes a paragraph on epidemics during Ming times.

problem. Malarial intermittent fevers (*zhangnüe* 瘴虐) are mentioned as early as in an early fourth-century Chinese source.[30]

The frequency of widespread disease outbreaks then reached a peak in the late fifteenth and early sixteenth centuries, and again in the late sixteenth and early seventeenth centuries. During the Ming dynasty, at least 22 major outbreaks of contagious diseases occurred in Zhejiang, 22 in Fujian, 5 in Guangxi, 2 in Guangdong, and 11 in Shandong—in comparison to Huguang with 26, Jiangxi with 16, Shanxi with 24, the metropolitan area with 20, and Shaanxi with 12.[31] Additionally, 137 cases of epidemics have elsewhere been counted for all of Southeast China during the Ming period; 197 cases for the Qing period.[32] Typhoons, unhygienic circumstances, and an extensive river system with frequent floods provided positive conditions for a spreading of contagious diseases in these contexts. Many epidemics, with over 28% of all catastrophes recorded, occurred at the same time as inundations.[33]

Wang Shuanghuai 王雙懷 has already investigated the nature and frequency of natural disasters alongside the outbreak of epidemics (*yizai* 疫災) in Southeast China during the Ming Dynasty. He argues that Fujian 'suffered most,' especially Fuzhou, Quanzhou, and Zhangzhou. Next was Guangdong, with mainly Guangzhou, Chaozhou, Zhaoqing and Qiongzhou being affected.[34] But he does not include Zhejiang or Jiangsu provinces. At least as far as inundations and epidemics are concerned, our research shows that among the Southeast coastal regions, Zhejiang province 'suffered most,' and in terms of floods and 'water calamities' (*shuizai* 水災), Jiangsu was still more heavily affected than Zhejiang (see Fig. 3.2). Temporally speaking, the period between 1522 and 1619, especially the last third of the sixteenth century and the second decade of the seventeenth century, was particularly severe. In terms of epidemics,

[30] *Bao Puzi neipian* 抱朴子内篇 (ca. 320), by Ge Hong 葛洪 (283–363). See: Erhard Rosner, *Miasmen. Studien zur Geschichte der Malaria in Südchina* [Veröffentlichungen des Ostasien-Instituts der Ruhr-Universität Bochum 69] (Wiesbaden: Harrassowitz Verlag 2019), 21.

[31] Mei Li 梅莉 and Yan Changgui 晏昌贵, 'Guanyu Mingdai chuanranbing chubu kao 關於明代传染病的初步考察,' *Hubei daxue xuebao* 湖北大學學報, 5 (1996), 85.

[32] Wen Zongdian 閔宗殿, 'MingQing shiqi Dongnan diqi yiqing yanjiu 明清時期東南地區疫情研究,' *Xueshu yanjiu* 學術研究, 10 (2003), 109.

[33] Ibid., 159.

[34] Wang Shuanghuai, 'Mingdai Hua'nan de ziran zaihai,' *Dili yanjiu*, 18 (1999), 160.

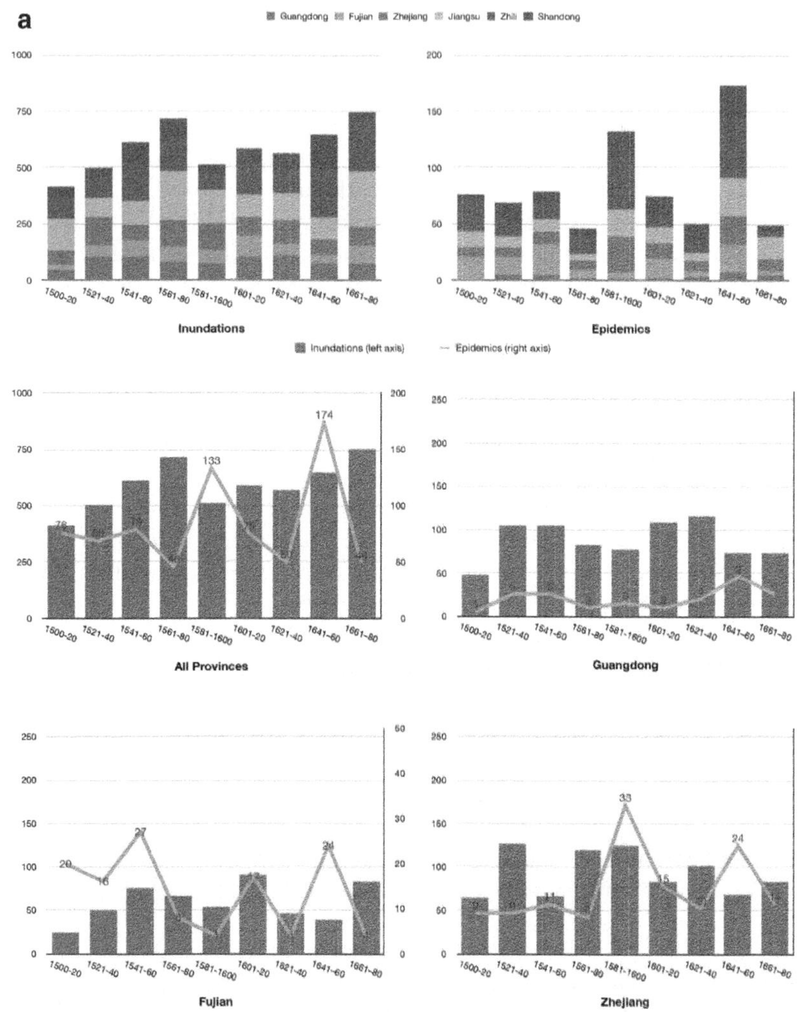

Fig. 3.2 Graphs showing numbers of inundations and epidemics in coastal Chinese provinces in the period c.1500–1680, according to the data collected as part of the ongoing TRANSPACIFIC and 'Appraising Risk' projects

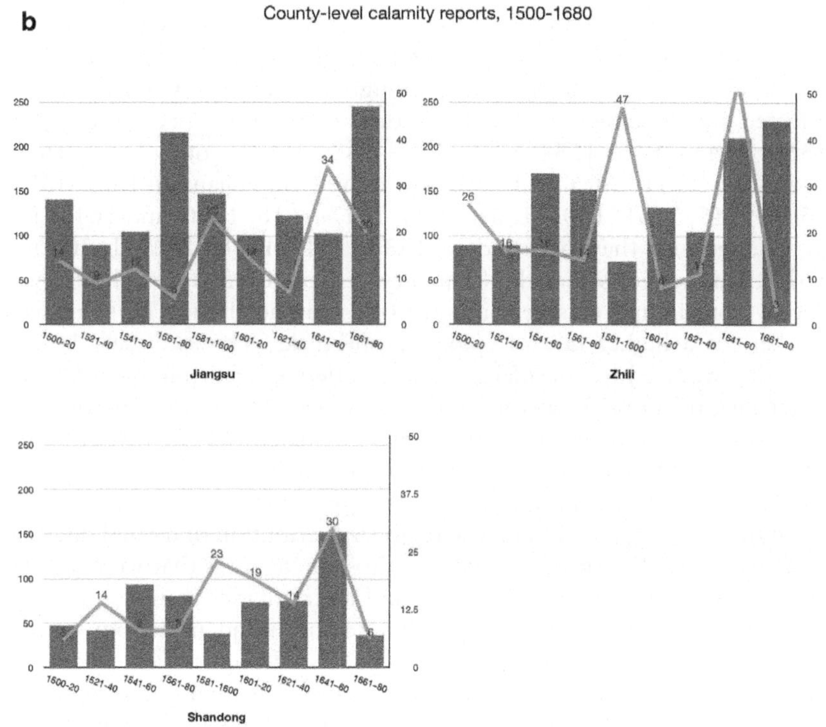

Fig. 3.2 (continued)

most occurred during the Jiajing (1521–1567) and Wanli (1573–1620) periods.[35] The early seventeenth century, especially around 1602, saw an extreme frequency of large epidemics, as Helen Dunstan has also shown—although her study looks at the Ming as a whole and does not specifically focus on China's coastal regions.[36]

[35] Chen Xu 陈旭, *Mingdai yiwen yu Mingdai shehui* 明代瘟疫与明代社会. Chengdu: Xinan caijing daxue chubanshe, 2016, Ch. 2, *Mingdai wenyi de tedian* 明代瘟疫的特點, 25.

[36] Helen Dunstan, 'The Late Ming Epidemics: A Preliminary Survey,' *Ch'ing-shih wen-t'i*, 3, 3 (1975), 1–59.

The data we have so far collected for our TRANSPACIFIC project suggests that Guangdong experienced significantly more typhoon disasters than other coastal provinces, though Zhejiang province was most affected by epidemics. In the latter case, severe outbreaks occurred in 1534, 1545–1546, 1554–1555, 1588–1589, 1601–1603/1604, 1639, and 1640–1641. Meanwhile, they occurred in Fujian in 1522–1523, 1544–1545, 1561, 1601–1603, 1608, 1617–1618, 1640, and 1642; and in Guangdong, they occurred in 1533, 1547–1548, 1629–1630, and 1632. This suggests that the frequency of typhoon disasters alone, even if they contributed to severe inundations, may not hastily be taken as a marker for a higher rate of epidemic outbreaks. This is a clear indication that we need to consider a broader variety of environmental factors. Most investigations, however, agree that water, floods, and inundations, coupled with high temperatures, played a major role in the outbreak of epidemics and constituted ideal circumstances for the breeding of pathogenic microorganisms.[37]

In this context, research on the regional distribution of natural disasters in Ming period Fujian has shown that those prefectures that were located close to water, that is the Min River and its tributaries, namely Jianning, Shaowu, Yanping and Fuzhou, suffered most frequently from natural disasters, especially flooding, typhoons, but also droughts.[38] Moreover, as Xu Zhexin has emphasized, among the recorded natural disasters, water-related hazards played the most direct role in causing widespread diseases, as 79.41% of the epidemic diseases that broke out in Ming Fujian were related to floods and typhoons.[39]

These linkages are borne out by some written records, in which gazetteers discussed epidemics in the context of inundations. For example,

[37] See, for example: Gong Shengsheng 龚胜生, Wang Xiaowei 王晓伟, and Zhang Chou 张涛, 'Mingdai Jiangnan diqu de yizaidili 明代江南地区的疫灾地理,' *Dili yanjiu* 地理研究 *Geographical Research*, 33, 8 (2014), 1569–78.

[38] Wang Shuanghuai 王雙懷, 'Mingdai Huanan de ziran zaihai jiqi shikong tezheng,' 明代華南的自然災害及其時空特征, *Dili yanjiu* 地理研究 18 (1999), 158.

[39] Xu Zhexin, 'The Environment, Perceptions, and Publication of Medical Texts in Fujian During the Ming Period (1368 to 1644), in *Seafaring, Trade, and Knowledge Transfer: Maritime Politics and Commerce in Middle Period and Early Modern China*, eds., Wim De Winter, Angela Schottenhammer, and Mathieu Torck [Crossroads—History of Interactions across the Silk Routes] (Leiden: Brill Publishers, in print, paper held at Ghent University, Ghent in 2017): With reference to: Wang Shuanghuai, 'Mingdai Hua'nan de ziran zaihai jiqi shikong tezheng,' 158.

1544 was a year of various catastrophes in Fujian. Droughts alternated with inundations, and a typhoon struck the coast. Epidemics then followed during the winter. They broke out in Jianning 建寧 (大疫) and Shaxian 沙縣 (瘟疫), as well as in various other districts. In Jianning, the District Magistrate, He Menglun 何孟倫, prohibited private transportation of goods and leaving the district.[40] In addition, medicines were purchased and doctors dispatched to each village to prepare them accordingly (市藥發醫分遣各鄉調製).[41] Similarly, for the 7th month of 1579, a major epidemic (大疫) is mentioned in relation to a typhoon, and many districts suffered from severe inundations.[42]

Meanwhile, 1587–1589 were very bad years for most of Zhejiang and for Guangdong. Floods, storms, and typhoons occurred at the same time as epidemics, low temperatures, famines, and locusts, especially in Zhejiang, although Fujian seems to have been less affected. Timothy Brook has written in this context of an 'environmental collapse on a scale that stunned the regime and established a new benchmark for social disaster.'[43] Furthermore, in Tiantai 天台, Zhejiang, 'another' major epidemic (復大疫) broke out in the 7th month of 1587, accompanied by storms and heavy rains. People consequently ate the bark of trees and the roots of grass and weeds.[44] In 1588, the Regional Inspector (*xun'an yushi* 巡按御史), Cai Xizhou 蔡系周, went through the prefectures and provided medicinals, and so he was able to save tens of thousands of lives. The upright scholars (義士) Du Tan 杜潭 and Ye Shiyuan 葉世源 also provided medicine for help.[45]

Epidemics continued to rage during the years 1588–9, for example in various districts of Shanghai (1588 Fengxian 奉賢: 疫; Baoshan 寶

[40] In addition to controlling the price of rice and securing the local provisions therein, this measure might also have been considered necessary to control the further spread of the epidemics.

[41] Zhang De'er 张德二, *Zhongguo sanqian nian qixiang jilu zongji* 中國三千年氣象紀錄總集, vol. 2 (Nanjing: Jiangsu jiaoyu chubanshe, 2013), 1163. Further source descriptions from Zhang De'er, vol. 2, are abbreviated as 'Description,' followed by the relevant page number.

[42] Description, 1339.

[43] Timothy Brook, *The Troubled Empire*, 242.

[44] Description, 1387.

[45] Description, 1397.

山: 大疫; 1588–89 Qingpu 青浦: 大疫).[46] Most entries for this year in Shanghai speak of terrible droughts and famines. For Baoshan, the records tell us that many people died. A great epidemic is also mentioned for Nanjing and districts of Jiangxi Province. In Hangzhou 杭州, for example, from the 3rd to the 5th month, the rain did not stop, and inundations were widespread. Also, febrile epidemics (*wenyi* 瘟疫) broke out while many places were buried by water.[47] For Xiaoshan 蕭山, Zhejiang, the sources speak of a major epidemic pestilence (大疫癘) raging in both years. Large numbers of dead were left on the street, and the Xiaoshan District Magistrate, Liu Hui 劉會, from Quanzhou, Fujian, was ordered to select physicians to provide medicines to cure the sick. People and monks were hired to cover the skeletons. Meanwhile, a typhoon hit on the 9th day of the 6th month, 1589, overflooding large parts of the region. In Jiashan district one heard the sound of crying people all over the streets.[48]

Moving to examples from the seventeenth century, in early summer 1609, a major epidemic broke out in Fenxian 奉賢, Jiading 嘉定, and Baoshan 寶山 districts in Shanghai, and they occurred at the same time as inundations. Also, many districts in Fujian, for example, Shaowu, were struck by epidemics and severe floods.[49] Tens of thousands of people drowned all over the region. A record for Shaowu explicitly states that 'when the water was gone, the epidemic came' (水過疫作).[50]

In 1639 and 1641, two severe epidemics struck the Yangzi valley. Also, in the years 1641–1643, Zhejiang, Shandong, and almost the entire eastern coast was once again affected by plague and epidemics. Droughts, locust plagues, and epidemics frequently alternated.[51] Timothy Brook speaks of this series of disasters as the 'Chongzhen slough,' the Chongzhen emperor's reign period lasting from 1627 to 1643.[52] No emperor of the Yuan or Ming dynasty before him faced as abnormal and severe climatic conditions as the Chongzhen Emperor. Tempera-

[46] Description, 1396–1398.

[47] Description, 1407.

[48] Ibid.

[49] Description, 1518–1519.

[50] Description, 1519.

[51] Description, 1730–1731.

[52] Brook, *The Troubled Empire*, 429.

tures reached a cold minimum between approximately 1629 and the 1640s.[53] Peak periods of LIA cooling coincided with erratic rainfall, typhoon-related inundations, and epidemics in the sixteenth–seventeenth centuries.

Some Remarks on Ming-Qing China's Crisis Management

An eighteenth-century scholar, Liu Kui 劉奎 (fl. late eighteenth century) listed over seventy types of epidemic diseases, and integrated a discussion of religious and other healing practices in a work entitled *Songfeng shuoyi* 松峰説疫 (Speaking of Epidemics, 1786).[54] Like in most societies all over the world, religion played a major role in crisis management in Ming China. To sacrifice and pray to the gods and spirits of the ocean was an essential part of official ceremonies seeking to prevent damage and disaster. In Haiyan xian 海盐县, a 'Temple to the Sea God,' *Haishen ci* 海神祠, was established in 1405, and the local population called it 'Dragon King Temple,' *Longwang miao* 龍王廟. Every year 'in fall on the 18th day of the 8th month officials proceeded there to pray' (秋及八月十八日有司致祭).[55] In 1569, however, during a disaster, this temple was washed away.

Direct crisis management during floods included burying the dead, except not, of course, the many that drifted out to sea. Official subsidies and subventions, as well as private donations, were also essential. When too many corpses lay around, people and monks were hired to cover them. Government funds were used to reconstruct buildings and dikes, and to provide food, clothes, and medication. Tax exemptions were essential to encourage peasants to resume work, and so the sources repeatedly discuss tax exemptions after serious disasters.[56] Dams and dikes were built

[53] Ibid.

[54] Marta E. Hanson, *Speaking of Epidemics in Chinese Medicine: Disease and the Geographic Imagination in Late Imperial China* (New York: Routledge, 2013), 118. *Songfeng shuoyi* 松峰説疫, by Liu Kui 劉奎, in XXSKQS, 子部-醫家類.

[55] Zhou Zhiyuan, 'Mingdai Dongnan diqu de haichao zaihai,' 91: With reference to: Haiyanxian tujing 海盐縣圖經 (1624), by Hu Zhenheng 胡震亨 and Fan Weicheng 樊维城, *juan* 2.

[56] Parallels can be drawn here with specific cases in Song China. See: Chapter by Ebner von Eschenbach, this volume.

for prevention purposes and had to be repaired and reconstructed expensively when destroyed after disasters. Such dikes and the reclamation of shorelands were also responsible for the growth of cities in Zhejiang and Jiangsu, as early as the Five Dynasties and the Song period (tenth through thirteenth centuries). Shanghai is one of the many cities originally built on reclaimed land.[57]

In 1587, the seawalls in Jiaxing were destroyed after a storm flood and had to be rebuilt. The local military governor (*xunfu* 巡撫), Teng Bolun 滕伯輪 (1528–1589), who had formerly worked in Panyu, Guangdong, where he had been involved both in the construction of the new town of Guangzhou and in repelling pirates, managed the repair of 571 *zhang* (approximately 1.9 km) and the reconstruction of 600 *zhang* (approximately 2 km) of sea dikes. He also added more than 2000 *zhang* (approximately 6.7 km) of a new dike section, in total spending more than 6800 *liang* of silver (roughly 212 kg). The construction works were not completed when he passed away by exhaustion, but he left his 'Ten Point Discussion' (*Shiyi* 十議), a summarized description of requirements when constructing dike and embankments to ward off storm surges.

The ten points were included in the Qing edition of the *Local Gazetteer of Zhejiang* (*Zhejiang tongzhi*). Here we only introduce his first discussion point, which provides insight into the share of responsibilities:

> For the construction works, obligatory *corvée* labour should be used, organized in large transportation units, under the supervision of the Irrigation Circuit (*shuili dao* 水利道). Your humble servant's responsibility lies in going from district to district to supervise the works, (*corvée* labour and the craftsmen), and one associate administrator from the local prefecture has to particularly take care of retracting funds and food supplies. As for the gathering of stones in the two prefectures Su(zhou) and Hu(zhou), two prefecture assistants are to be put in charge to jointly control the works; and for the dyke construction, sixteen officials should be appointed. Four officials shared control of the collecting of stones. Collectively, they commissioned guards to pass through the districts; horse recorders and other officials went through the province to select and recruit staff, so that there are enough people to be sent and to fill the positions.[58]

[57] H. von Heidenstam, 'The Growth the Yangtze Delta,' *Journal of the Royal Asiatic Society, North China Branch*, LIII (1922), 30–31.

[58] *Zhejiang tongzhi* 浙江通志, by Ji Zengjun 嵇曾筠 (1670–1738) and Shen Yiji 沈翼機 (Jinshi 1706), 63.62-13b (海塘二), Siku quanshu-edition, fasc. 519–526: https://www.kanripo.org/text/KR2k0044/062 [Accessed: 7 July 2021]. The original text is as

The (re-)construction of dike was consequently a task that required well-organized supervisory structures and close cooperation between local and central authorities. The sources indicate that it seems to have remained primarily a government task, even if local authorities received private donations.

Private donations played a larger role when we look at the distribution of food, clothes, and medication after a local catastrophe had happened. Local authorities or officials often remained involved in crisis management though. As we have seen above, sometimes local officials were ordered to travel to the various districts and prefectures of a province to distribute medicinals, rice, and clothing. The case of Liu Hui from Quanzhou, Fujian, demonstrates that they sometimes also selected physicians to provide medicines and to cure the sick. Studying such phenomena, Angela Ki Che Leung has argued that healthcare became increasingly commercialized in the Ming and Qing periods, suggesting a gradual retreat of government.[59] But given these examples of governmental roles in crisis management, more research may needed to discuss general tendencies or trends.

It is nevertheless clear that the failure to effectively control epidemics was not necessarily only the result of a lack of knowledge. Already in the early Ming period, infectious and non-infectious diseases were identified, and the symptoms were described in detail—as were treatments. Besides plague, smallpox was one of the great pestilences in this time. The Ming Chinese were already practising variolation (a simple form of inoculation), while the Manchus were not—definitely a major reason why the Manchus paid so much attention to quarantine and segregating people.[60] We possess good documentation of the widespread use of anti-smallpox

follows: 議委官塘工大役總大綱者水利道臣之責移駐該縣督理其董 率官役工匠收放錢糧本府同知一員專理之次/則蘇湖二府採石合委府佐二員分管塘工應用 官十六員分管採石應用官四員俱合委衛經縣丞簿等職於通省選取庶足充任使.

59 Angela Ki Che Leung, 'Organized Medicine,' 134–66.

60 'Interestingly, the Manchu idea of segregation was to quarantine and protect those royalty who had not had smallpox rather than to isolate the sufferers. In case of any smallpox alert, the Manchu emperor and royal family members immediate escaped to their respective shelters,' called *bidou suo* 逼痘所 (shelters for avoiding smallpox). See: Chia-Feng Chang, 'Aspects of Smallpox and Its Significance in Chinese History' (Unpublished PhD diss., SOAS, 1996), 181. But they also banished infected individuals outside the city walls.

inoculation in the seventeenth century. A Ming dynasty physician, Leng Kaitai 冷開泰, wrote a treatise on smallpox during the Wanli period, entitled *Tianhua pushi* 天花譜史, *tianhua* or 'heavenly flowers' being another name for smallpox.[61]

Joseph Needham and Nathan Sivin additionally have drawn our attention to a certain Zhang Lu 張璐 (1695), whose work provides us with further information on how such an inoculation was practically carried out:

> If you are unable to take [literally: 'steal'] lymph from the pustules, you can use scabs to culture the inoculum. If there are no scabs to be taken, you can obtain clothing from a child who has just developed smallpox and give it to another child to wear; it too will develop smallpox. The point is to employ a similar pneuma (*chhi*); inchoate though it be, it can serve to guide out the womb poison.[62]

On the other hand, as Angela Ki Che Leung has observed, the Ming government did at no time 'attempt to follow the example of Cai Jing or Su Shi of the Song, and segregate the ill from the well. ... The only important effort at quarantine undertaken in late imperial China occurred not during the late Ming epidemics but in seventeenth-century Peking under the Manchus, and its circumstances underscore that this was an alien tactic.'[63] She concluded that, while quarantine measures were more common in early modern Europe, the Ming state actually at no time took serious measures to impose a strict quarantine or otherwise try to segregate the infected people. We are, to use her words, rather confronted with 'a slackening of bureaucratic intervention, and a democratization of medical knowledge.'[64]

Nevertheless, 'social distancing' in times of epidemics was not unknown to the Ming. The Ming scholar, Gao Cheng 高澄 (1494–1552),

[61] Available online under: https://www.loc.gov/resource/lcnclscd.2012402208.1A001/?st=gallery [Accessed: 7 July 2021].

[62] Joseph Needham, *Science and Civilisation in China*, vol. 6, *Biology and Biological Technology*, Part 6, *Medicine*, ed. Nathan Sivin (Cambridge: Cambridge University Press, 2000), 123.

[63] Angela Ki Che Leung, 'Organized Medicine,' 142.

[64] Ibid., 154.

described the untenable circumstances on board many ships. He accompanied Chen Kan 陳侃 (1489–1538), who was sent to the Ryūkyū Islands as head of a mission to officially install Shang Qing 尚清 (Jap. Shō Sei, r. 1526–1555) as the new ruler of the Chūzan 中山 kingdom.[65] Gao Cheng was clearly aware of the fact that missing space greatly enhanced the transmission of germs and diseases, which had particularly negative effects on board ships, where hygienic conditions were invariably unsanitary.[66]

Contemporary Theories About Environment, Storms, and Diseases

Chinese thinkers and officials have been concerned about the relationship between Heaven, earth, and mankind since ancient times. The order of the world was supposed, according to traditional Chinese thinking, to depend on a harmonious balance of these three powers. Since the Han Dynasty (206 BCE–220 CE), disasters, including earthquakes, inundations, and locust plagues, were recorded in the chapter 'Records on the Five Elements' (*Wuxing zhi* 五行志) of the official dynastic histories. Correct behaviour of the rulers was believed to be responsible for harmony on earth or for disasters.[67] If an emperor wanted to rule well, he was supposed to correctly consider the so-called 'five elements' (*wuxing* 五行)—wood, fire, earth, metal, and water—and their relation to Heaven, earth, and mankind.[68] This 'moral meteorology' placed a heavy burden

[65] *Shi Liuqiu lu* 使琉球錄, by Chen Kan 陳侃, in ed., Guoli Beiping tushuguan shanben congshu 國立北平圖書館善本 (Shanghai: Commercial Press, 1937).

[66] *Shi Liuqiu lu* 使琉球錄 (1579), by Xiao Chongye 蕭崇業 (*jinshi* 1571) and Xie Jie 謝杰 (*jinshi* 1574), in *Shi Liuqiu lu sanzhong* 使琉球錄三種 [Taiwan wenxian shiliao congkan 台灣文獻史料叢刊, 287] (Taibei: Taiwan datong shuju, 1970), 91 (使琉球錄卷上, 造舟): With reference to: Gao Cheng's *Caozhou ji*操舟記. For a translation, see: Angela Schottenhammer, 'Maritime Disasters and Risk Appraisals in the East Asian Waters,' *Études thématiques* (2022).

[67] See, for example: Angela Schottenhammer, 'Erdbeben in China: Entzug des "Himmlischen Mandats" oder Verlust des Yin-Yang-Equilibriums,' in *Naturkatastrophen. Dramatische Naturereignisse aus kulturwissenschaftlicher Perspektive*, eds. Ilja Steffelbauer and Christa Hammerl (Wien: Mandelbaum Verlag, 2014), 90–129.

[68] The power and force of the five elements corresponded in Heaven to the celestial bodies of Jupiter, Mars, Saturn, Venus, and Mercury; and to virtue, integrity, justice, rationality, and reliability (or trustworthiness) as far as mankind is concerned.

on all emperors,[69] especially during years or periods in which many disasters occurred.[70] These cosmological linkages are exemplified by the works of contemporary scholars, two of whose works are discussed below, with special reference to climate, typhoons, and epidemics.

Yuexi wenzai 粵西文載 (Compendium of documents on the region west of Yue, i.e. Guangxi) includes an interesting discussion by Su Jun 蘇濬 (1541–1599) on the climate (*qihou lun* 氣候論).[71] Su Jun was employed in the southwestern province of Guangxi after 1590, and carefully observed there the relationship between climatic phenomena, the natural environment, and their impacts on the local population:

> Chao Cuo 晁錯 (200–154 BCE)[72] says: "The territory that spreads across Yue [= Guangdong] has little *Yin* and much *Yang*.' Li Daizhi 李待制says: 'The earth in the south is inferior and the soil thin; when the soil is thin, then the *Yang* vapours frequently leak out. When the earth is inferior, the *Yin* vapours flow abundantly. When *Yang* vapours leak out, flowers frequently bloom in the four seasons. In three winters there is no snow, and in one year the hot summer time lasts longer than half a year. When people live there, the vapours rise and are obstructed, the skin produces lots of sweat, the pores do not close; this is caused by the *Yang* that cannot return properly. When *Yin* vapours are abundant, there is a lot of dew during dawns and dusks, in spring and summer, there is excessive rain; within one year, more than half of the time, [the weather of steaming vapours] dominates; in midsummer there is continuous rain, and when it gets extremely cold, the clothes are covered with white mould. The people then frequently [suffer from] dampness, their limbs get heavy and tired, and they frequently get diseases such as *beriberi* (脚氣). It is generally so because the *Yin* is constantly abundant. When *Yin* and *Yang* vapours are extreme and mutually shifting, then, within a day, the climate

[69] Brook, *The Troubled Empire*, 73.

[70] Paolo Santangelo discusses ecologism versus moralism in Ming-Qing times, while Helen Dunstan has provided an overview of official thinking on environmental issues in the eighteenth century. See: Paolo Santangelo, 'Ecologism Versus Moralism: Conceptions of Nature in Some Literary Texts of Ming-Qing Times,' in *Sediments of Time*, eds. Elvin and Ts'iu-jung, 617–56; Helen Dunstan, 'Official Thinking on Environmental Issues and the State's Environmental Roles in Eighteenth-Century China,' in *Sediments of Time*, eds. Elvin and Ts'iu-jung, 585–614.

[71] *Yuexi wenzai* 粵西文載, by Wang Sen 汪森 (1653–1726), in SKQS, fasc. 1465–1467.

[72] Chao Cuo was a political advisor and official of the Western Han Dynasty (206 BCE–9 CE).

is constantly changing. A proverb says that 'when the four seasons are like summer, once there is rain, it easily converts into fall.' Another proverb says 'to undress and dress again [when the climate is changing], is equal to taking medicine.' When the weather conditions are like this, the vapours of human bodies are circulating between Heaven and Earth. The vapours of Heaven, in the farthest north, are extremely cold; in the farthest south, heat is extreme. [The region] south of the Five-Mountain-Passes, is called the 'hot weather zone' (*yanfang*). The high hills there are a range of peaks, left and right enclosed by water; the vapours are damp and steamy, so that the vapours from mountains are accumulated and become misty (鬱而為嵐). Only in Quanzhou 全州, close to Huxiang 湖湘(= Hunan), in mid-winter, there is lots of snow, and the climate resembles the districts in the central plains (中州). This stops at Guilin 桂林 全州Zimei 子美[Du Fu 杜甫(712–770)?] said: 'In the Five-Mountain-Ranges [the climate] is very hot and only Guilin has pleasant wheather.' He was telling the truth. Left and right there are two rivers (Li and Yijiang River), bordering Zhao and Wu 昭梧Prefectures. Places in the distant wilderness have barely acceptable [weather condition]; but in cities located close to gorges or to low-lying and damp [places] with rough mountain passes, there are (places) where one can see the colour of the sun only close to noon. In the creeks and the uncivilized highlands where the southern barbarians live, with luxuriant vegetation, where large, poisonous snakes come and go, the water of the river contains poison, and the pestilential (malaria) vapours (瘴氣) turn infectious. In the 3rd month (i.e. in spring) this is called 'green-grass miasma' (青草瘴), in the 4th and 5th month (i.e. in the rainy summer season called *mei*) this is called 'yellow-*mei* miasma' (黃梅瘴), in the 6th and 7th month (i.e. in late summer, early fall) this is called 'ripening-crops miasma' (新禾瘴), and in the 8th and 9th month (i.e. fall) it is called 'yellow-reeds miasma' (黃茅瘴); it is also called 'sweet osmanthus miasma' (桂花瘴) or 'chrysanthemums miasma' (菊花瘴)."

This quotation clearly demonstrates the climate consciousness of the author and shows how closely he connected malaria outbreaks to local environment and climate. 'Miasmas' have been discussed in Chinese literature, medicinal, local administrative and statecraft sources for centuries. '*Zhang*,' a pathogenic, atmospheric agent related to 'water caused' diseases in a broader sense, the most important of which is malaria. Although Su Jun did not understand the real causes of malaria, he, like many before him, comprehended that local conditions, a wet and humid climate, that was feared by most Chinese from the north, were very conducive to its spread.[73]

Another author, who lived approximately 100 years later, also discussed the environment and natural phenomena. The Cantonese literati, Qu Dajun 屈大均 (1630–1696), left a collection of notes in his home province Guangdong. In his *Guangdong xinyu* 廣東新語 (New Discourse on Guangdong; around 1680), he wrote not only about miasmas,[74] but also, for example, about the 'typhoon spirit' (*jufeng shen* 颶風神):

> Yue 粵 (= South China Yue region, including Guangdong) is located in the south (離方). As far as typhoons are concerned, the grief of the southern head-winds cannot be escaped there, the fire vapours (火氣) burst out and turn into dangerous calamities. In Yue, typhoons occur every year. Mostly they rise from (the direction of) Qiong[zhou] 瓊 and Lei[zhou] 雷 [i.e. from Hainan and the Leizhou Peninsula], that is, from the utmost point in the south [of China]. This is why Qiong and Lei both have a typhoon temple (颶風祠). Its god is the mother of the typhoons. The local authorities offer sacrifices during the Dragon Boat Festival (端午日). During their procession they offer gifts; they truly fear it. If there is a typhoon, then all the winds [of the four are directions] are being possessive (颶者具也). Once a typhoon rises, the winds from the east, west, south, and north all unite into one single wind. This is why it is called *ju* 颶 (that means, a wind that possessively unites all winds together). As far as the term 'mother (of the typhoons)' is concerned, it is so called, because a typhoon can produce winds of four directions, and it thus becomes the mother of the winds of the four directions; separating the wind of any one single direction, can develop into a storm (大風). This is why it is called the mother. Also *xun* 巽 [i.e. one of the eight diagrams, representing wood and wind] produces wind; the eldest daughter of *qian* 乾 [i.e. the diagrams representing Heaven] rules it. Thunder (雷) is meant to restore the nature (性) of the ten thousand things; it has the way (道) of the father. This is why it is called lord (or father). The wind is meant to restore the fate (命) of the ten thousand things; it has the way of the mother. This is why it is called mother. When a storm (大風) is the mother, then a little breeze (微風) is consequently either a little boy or a little girl. What rises from the marshes (澤) is called "little girls' wind", what rises from the mountains (山) is called "little boys' wind"; and they all have the typhoon as mother. And as ruler of the wind, *Xun* is the root and origin of moon and water.

[73] An excellent overview has recently been provided in: Erhard Rosner, *Miasmen*.

[74] See, for example: Ibid., 29.

Moon and water both are born from wind; this is why it is called mother. There is a saying that the typhoon mother is the goddess of the winds (*mengpo* 孟婆). In spring and summer there are vapours that resemble the halo [or aura] around the moon, [and that is Mengpo]! This saying probably takes a halo as the typhoon mother, but Po [婆; from Mengpo] is in fact Mu 母, namely the mother.

Of the gods of the earth, there are no more powerful (大) than thunder and wind. Thunder and wind give birth to sun and moon. To serve the god of the thunder means, thus, to serve the sun. To serve the god of the winds means, thus, to serve the moon. And the god of thunder resides in Leizhou 雷州 [lit. "Thunder district"], and the god of the winds resides in Qiongzhou. These two districts are, thus, the utmost south. The utmost south, this is the lowest point of the earth (其地最下). Thunder is born at the lowest point of the earth and the winds follow it. This is why the gods of thunder and wind are located there![75]

Qu Dajun sought to understand the origins of and causes of typhoons, which brought so much damage and sorrow to the people, including epidemics. In the passage quoted at the beginning of this chapter, we read further that the historical annals claim that there were annual disasters of 'wind fish' (*fengyu* 風魚) in the Southern Seas. The winds in this context were typhoons, and the fish was possibly some kind of river dolphin, Lipotidae (*ji* 鱀), of which dark and white ones exist.[76] Because they came suddenly with the wind, they are also called 'wind fish.'[77] Interesting in this quotation is also the statement about the frequency of the coming of these fish. This statement attests to shifting cycles of typhoons, and to a direct relation between frequent typhoon disasters and the outbreak of epidemics.

[75] *Guangdong xinyu* 廣東新語, by Qu Dajun 屈大均 (1630–1696) [Lidai shiliao biji congkan 歷代史料筆記叢刊] (Beijing: Zhonghua shuju, 2006), 6.201–202.

[76] The *Hanyu da cidian* 漢語大字典 states for the entry of "*ji*": is the same as 鱀; and the entry explains that the character "*zhu*" 䱵 is identical with the character "*ji*" 鱀, and these are "*baiji*" 白鱀 (white fish). *Hanyu da cidian* also provides Qu Dajun's text on 暨魚 as an example; Qu Dajun namely continues saying that the character is also written as (暨一作) and he states that there are white and dark ones. These white lipotidae (dolphins) actually only lived in the Yangzi River. The observation from *Guangdong xinyu* may thus be interpreted as that, depending on the typhoon cycles and possibly directions, these dolphins (or fish?) actually occurred along the Guangdong coast.

[77] *Guangdong xinyu*, 22.550.

Scholars were also conscious of and sensitive to the impacts of environmental and climatic factors on the health and daily lives of ordinary people. This is, for example, reflected in significant changes in the thinking and theories of contemporary medical theorists and scholars, as Marta H. Hanson has elsewhere shown. Hanson has in detail investigated diseases and geographic imaginations as a core theme in Chinese medicine, and has shown how medicinal doctrines and perceptions changed in the late Ming. She argued, that it was not before 1642 (towards the very end of the Ming dynasty) that 'warm diseases' came to be considered as a separate disease category worthy of analysis: 'Wu Youxing 吳有性 (c. 1582–1652; also Wu Youke 吳又可) argued that a specific pestilential or deviant *qi* (pneuma or vapour) rather than the usual unseasonable *qi* caused 'warm epidemics' (*wenyi* 温疫). His *Treatise on Warm Epidemics* (*Wenyi lun* 温疫論, 1642), gave *wenbing* a contagionist tenor.' As a result of his critique, *wenyi* 瘟疫 (febrile epidemics), defined as the most severe form of *wenbing*, became a new topic of medical analysis. The term itself was already used earlier, and we encounter several records on *wenyi* in the sixteenth century.[78] The fact that the water radical is added to the character '*wen*' may indicate that many epidemics and diseases were in one or the other way related to water, such as from inundations or heavy rainfall. Wu Youxing also discussed the role of poison, pathogenic local *qi*, and person-to-person transmissions. The contagionist view understood epidemics to be caused by human-to-human transmission via some kind of pathogen.[79] This 'contagious turn' in the conception of Chinese, relating the outbreak of diseases rather to pathogenic factors instead of just the environment, only occurred in the late Ming through early Qing dynasties. Hanson also observed that medical essays on the Guangdong region 'reveal a conceptual shift from climate-consciousness to a poison-consciousness.'[80] Chen Sicheng's discussion of 'Cantonese sores,' the *Secret Account of Rotting Sores* (*Meichuang milu* 黴瘡祕錄, 1632) may be taken as a case in point.

[78] Human-to-human transmission certainly occurred, for example, in 1562 Jinjiang 晉江 (Fujian). See: Description, 1257. In 1596, a major smallpox pandemic (痘疹) is mentioned for Shaowu 邵武, Fujian that caused uncountable deaths. See: Description, 1450.

[79] Hanson, *Speaking of Epidemics*, 18.

[80] Ibid., 79.

Conclusion

The period between 1550 and 1640 is significant for a high frequency of La Niña years, according to climatologists. Tentative subsequent research by other climatologists suggests that this may have contributed to the heightened frequency of typhoons making landfall in South China and Southeast Asia. This chapter generally, despite often thin source material, supports this hypothesis: We do indeed encounter many more typhoon incidents for South China: Guangdong province was severely affected, in comparison to China's southeastern coastal provinces, such as Fujian and Zhejiang. But it remains near to impossible to establish any direct correlations between specific El Niño or La Niña years and reduced or increased precipitation and inundations.[81] A tentative reconstruction of ENSO years taken from the KNMI Climate Explorer and from NOAA (National Oceanic and Atmospheric Administration)[82] would actually even suggest the opposite, or at least not confirm any direct correlation. Data recording of these kinds of disasters is simply not even enough for this early period. There are too many holes in our records to be able to provide more reliable dates and estimations concerning correlations between El Niño and La Niña years and typhoon landfalls. We will have to include yet more data from China's northern coastal provinces and Northeast Asia in general, as well as from Southeast Asia, southern Japan, the Ryūkyū Islands, Taiwan, the Philippine Archipelago, as well as Guangxi, Hainan, and northern Vietnam. But these will only be able to show us general tendencies, not the kind of climate reconstruction charts we are used to from NOAA for later periods.

Due to the lack of traceable data, not only on precipitation, we can only select years and locations for which significant rainfalls and inundations are reported in our sources and compare these years with other locations, in order to receive some very general insights into which years and where inundations were particularly frequent. We can then check if or not they

[81] Our first comprehensive data analysis for typhoon landfalls in China's northeastern provinces will only be available after further research. For this reason, I have focused this chapter mainly on where we have already collected data, namely, Fujian, Guangdong, and Zhejiang Provinces.

[82] The World Meteorological Institution (Koninklijk Nederlands Meteorologisch Instituut), see: https://www.knmi.nl/home [Accessed: 7 July 2021].

were related to, for example, typhoons. The result will be a general picture but nothing like what we are used to for the period after ca. 1850.

Generally speaking, as all analyses show, there are by far more reports on inundations than on any other calamity—a fact that has political and ideological reasons, as severe floods and inundations could pose a serious threat to political rule. The concrete causes for the inundations, flooding rivers, tidal floods, or heavy rainfalls, are, however, often not easy to trace back, as sources do not always specify where the water came from and just speak of 'water calamities' (*shuizai*). The fact that Jiangsu province in particular experienced more inundations than, for example, Fujian and Guangdong can certainly be explained by Yangzi River flooding. Both heavy rainfalls and storms have time and again caused catastrophic calamities in the region. The records also demonstrate that nearly all typhoons went along with tidal disasters, which occurred frequently in the coastal regions. Most inundations happened between the late 1530s and the early 1640s, with special peaks during the periods between 1521 and 1540, the 1570s to the early seventeenth century, and the late 1620s to mid-1640s. Jiangsu and Zhejiang province were most affected, followed by Guangdong, where we see a high concentration in and around the Canton River Delta.

As far as epidemics are concerned, Guangdong was, relatively speaking, little affected. Zhejiang in particular was the region experiencing most epidemic outbreaks during the period under investigation. We can observe epidemic peaks across coastal China in the periods 1581–1600 and again 1641–1660. For Zhejiang, the former period was definitely yet more severe than the latter—when, for example, for the northern province of Zhili, the total quantity of reported epidemics was yet higher than in Zhejiang, while we have almost no recorded typhoons.

A relation between some natural catastrophes, such as inundations, and diseases is, however, obvious. As we have seen, in Fujian, for example, almost 80% of the recorded diseases that broke out in Ming Fujian were related to floods and typhoons. Additionally, above we have introduced some snapshots of local typhoon and tide disasters, inundations in general and possible correlations with the outbreak of epidemics and climatic changes. But further research and yet more data are needed to obtain a better picture not only of local micro-histories and crisis management on the ground, but also of possible general tendencies and correlations.

So, how should we proceed in future research? First, as part of our ongoing TRANSPACIFIC and 'Appraising Risk' projects, we will need to expand our research into the larger East Asian maritime space, including Northeast and Southeast Asia, as well as island archipelagos located in the (South-)East Asian area. Second, we will need to cover longer time periods and always thoroughly consider human–environment interactions. Only then will we be able to provide meaningful statements on developments in the spread of diseases in conjunction with changing environments, global travel, and local crisis management. At the same time, due to our uneven and sketchy data for the middle and early modern periods, we need to focus on well-documented cases in varied and multi-lingual sources, apply a historical comparative-analytical approach, examine documents of different contents, provenience, and types (such as diaries, travelogues, local gazetteers, missionary reports, medicinal texts, and administrative and judicial documents), and adopt modern scientific approaches to gain insights into the effectiveness of historically applied 'medications.' Early modern treatments and medications, for example, mostly consisted in a decoction or combination of herbs and other medicinal plants, many of which have undergone a thorough clinical investigation and testing in present times. Camphor, for example, is a case in point, as are *Salvia miltiorrhiza* (*danshen* 丹參) and ginseng Taiyi pills (*taiyi dan* 太乙丹).[83] In an article discussing the relationship between climate, environment, and the spread of diseases in early modern coastal China and (South-)East Asian maritime space, I have selected various case studies from the seventeenth to the nineteenth century to demonstrate which directions our research should take, which sources and methodologies we should use, and what we can learn from

[83] For a detailed discussion of camphor, see my: 'Some Remarks on the Use and Provision of camphor in Early Modern China and in Spanish Asian and American Colonies,' in *From the Steppe to the Sea: A Festschrift for Paul Buell*, eds. Timothy May (forthcoming). Clinical trials, for example, have been conducted to test the efficiency of camphor in treating asthma. See: Rafie Hamidpour, Soheila Hamidpour, Mohsen Hamidpour, and Roxanna Hamidpour, 'The Effect of Camphor Discovery for Treating Asthma,' *Biotechnology Advances*, 1 (2019), 1–4: *Advances in Bioengineering and Biomedical Science Research*, 2019, www.opastonline.com [Accessed: 4 Jan. 2022]. For ginseng Taiyi pills, see: Xijun Yan, ed., *Dan Shen (Salviamiltiorrhiza) in Medicine, Vol. 3, Clinical Research* (Dordrecht: Springer, 2015), 257, table 17.8. These pills could contain a variety of different ingredients, were administered in different preparations, and could have antimicrobial qualities.

them.[84] I hypothesize that there existed a close connection between the increasing global integration of East Asia, environmental problems, and the occurrence of specific diseases, which, in turn significantly influenced risk appraisal and crisis management, including medical treatment.

Bibliography

Angela Ki Che Leung. 'Organized Medicine in Ming-Qing China: State and Private Medical Institutions in the Lower Yangzi Region.' *Late Imperial China*, 8, 1 (1987): 134–66.

Bao Puzi neipian 抱朴子内篇 (ca. 320), by Ge Hong 葛洪 (283–363).

Bo Sen 伯森 et al. *Ming shilu leizuan* 明實錄類纂, *Ziran zaiyi juan*·自然災異卷. Wuhan: Wuhan chubanshe, 1993.

Brook, Timothy. *The Troubled Empire: China in the Yuan and Ming Dynasties*. Cambridge, MA: Belknap Press of Harvard University Press, 2010.

Brook, Timothy. 'Differential Effects of Global and Local Climate Data in Assessing Environmental Drivers of Epidemic Outbreaks.' *PNAS*, 114, 49 (2017): 12845–47.

Brook, Timothy. 'Nine Sloughs: Profiling the Climate History of the Yuan and Ming Dynasties, 1260–1644.' *Journal of Chinese History*, 1 (2017): 27–58.

Caviedes, César. *El Niño in History: Storming Through the Ages*. Gainesville: University Press of Florida, 2001.

Chen Xu 陈旭. *Mingdai yiwen yu Mingdai shehui* 明代瘟疫与明代社会. Chengdu: Xinan caijing daxue chubanshe, 2016.

Chia-Feng Chang. 'Aspects of Smallpox and Its Significance in Chinese History.' Unpublished PhD dissertation, SOAS, 1996.

Dunstan, Helen. 'The Late Ming Epidemics: A Preliminary Survey.' *Ch'ing-shih wen-t'i*, 3, 3 (1975): 1–59.

Dunstan, Helen. 'Official Thinking on Environmental Issues and the State's Environmental Roles in Eighteenth-Century China,' in *Sediments of Time*, eds. Elvin and Ts'ui-jung: 585–614

Elsner, James B., and Kam-biu Liu. 'Examining the ENSO-Typhoon Hypothesis.' *Climate Research*, 25, 1 (2003): 43–54.

Elvin, Mark. *The Retreat of the Elephant: An Environmental History of China*. New Haven, London: Stanford University Press, 2004.

Elvin, Mark, and Liu Ts'ui-jung, eds. *Sediments of Time: Environment and Society in Chinese History*. Cambridge: Cambridge University Press, 1998.

[84] Angela Schottenhammer, 'Climate, Environment, and the Spread of Diseases in Early Modern Coastal China and (South-)East Asian Maritime Space,' *Guojia hanghai* 国家航海, 27 (2021), 161–91.

Gergis, Joëlle L., and Anthony M. Fowler. 'A History of ENSO Events Since A.D. 1525: Implications for Future Climate Change.' *Climatic Change*, 92 (2009): 343–87.

Gong Shengsheng 龚胜生, Wang Xiaowei 王晓伟, and Zhang Chou 张涛. 'Mingdai Jiangnan diqu de yizai dili 明代江南地区的疫灾地理.' *Dili yanjiu* 地理研究 *Geographical Research*, 33, 8 (2014): 1569–78.

Guangdong xinyu 廣東新語, by Qu Dajun 屈大均 (1630–1696) [Lidai shiliao biji congkan 歷代史料筆記叢刊]. Beijing: Zhonghua shuju, 2006.

Haiyanxian tujing 海盐縣圖經 (1624), by Hu Zhenheng 胡震亨 (1549–1645) and Fan Weicheng 樊维城 (?–1643). Reprint of 1624-edition. [Zhongguo difangzhi jicheng 中国地方志集成: Shanben fangzhi ji 善本方志辑; 1, 73]. Nanjing: Fengfeng chubanshe, 2014.

Hamidpour, Rafie, Soheila Hamidpour, Mohsen Hamidpour, and Roxanna Hamidpour. 'The Effect of Camphor Discovery for Treating Asthma.' *Biotechnology Advances*, 1 (2019): 1–4: *Advances in Bioengineering and Biomedical Science Research*. www.opastonline.com [Accessed: 4 Jan 2022].

Hanson, Marta E. *Speaking of Epidemics in Chinese Medicine: Disease and the Geographic Imagination in Late Imperial China*. New York: Routledge, 2013.

von Heidenstam, H. 'The Growth the Yangtze Delta.' *Journal of the Royal Asiatic Society, North China Branch*, LIII (1922): 21–36.

Kam-biu Liu, Caiming Shen, and Kin-sheun Louie. 'A 1,000-Year History of Typhoon Landfalls in Guangdong, Southern China, Reconstructed from Chinese d.' *Annals of the Association of American Geographers*, 91, 3 (2001): 453–64.

Kong Dongyan 孔冬艳, Li Gang 李钢, and Wang Huijuan 王会娟. 'Ming Qing shiqi Zhongguo yanhai diqu haichao zaihai yanjiu 明清时期中国沿海地区海潮灾害研究.' *Journal of Natural Disasters/Ziran zaihai yanjiu* 自然灾害学报, 25, 5 (2016): 89–96.

Lo Jung-pang. *China as a Sea Power, 1127–1368: A Preliminary Survey of the Maritime Expansion and Naval Exploits of the Chinese People During the Southern Song and Yuan Periods*, ed. Bruce A. Elleman. Singapore: National University of Singapore Press, 2011.

Marks, Robert B. '"It Never Used to Snow": Climate Variability and Harvest Yields in Late-Imperial South China, 1650–1850,' in *Sediments of Time*, eds. Mark Elvin and Liu Ts'ui-jung. Cambridge: Cambridge University Press, 1998: 411–46.

Mei Li 梅莉, and Yan Changgui 晏昌贵. 'Guanyu Mingdai chuanranbing chubu kaochao 關於明代传染病的初步考察.' *Hubei daxue xuebao* 湖北大學學報, 5 (1996): 80–88.

Mingshi 明史 [1739], by Zhang Tingyu 張廷玉 [1672–1755] et al. Beijing: Zhonghua shuju, 1974.

Needham, Joseph. *Science and Civilisation in China*, vol. 6, *Biology and Biological Technology*, Part 6, *Medicine*, ed. Nathan Sivin. Cambridge: Cambridge University Press, 2000.

Qiu-Hua Li, Yue-Hai Ma, Ning Wang, Ying Hu, and Zhao-Zhe Liu. 'Overview of the Plague in the Late Ming Dynasty and Its Prevention and Control Measures.' *TMR Journals*, 5, 3 (2020): 136–44.

Quansheng Ge, Jingyun Zheng, Yanyu Tian, Wenxiang Wu, Xiuqi Fang, and Wei-Chyung Wang. 'Coherence of Climatic Reconstruction from Historical Documents in China by Different Studies.' *International Journal of Climatology*, 28 (2008): 1007–24.

Rosner, Erhard. *Miasmen. Studien zur Geschichte der Malaria in Südchina* [Veröffentlichungen des Ostasien-Instituts der Ruhr-Universität Bochum, 69]. Wiesbaden: Harrassowitz Verlag, 2019.

Santangelo, Paolo. 'Ecologism Versus Moralism: Conceptions of Nature in Some Literary Texts of Ming-Qing Times,' in *Sediments of Time*, eds. Elvin and Ts'iu-jung: 617–56.

Schottenhammer, Angela. 'Erdbeben in China: Entzug des "Himmlischen Mandats" oder Verlust des Yin-Yang-Equilibriums,' in *Naturkatastrophen*, eds. Ilja Steffelbauer and Christa Hammerl. Wien: Mandelbaum Verlag, 2014: 90–129.

Schottenhammer, Angela. 'Climate, Environment, and the Spread of Diseases in Early Modern Coastal China and (South-)East Asian Maritime Space.' *Guojia hanghai* 国家航海, 27 (2021): 161–91.

Schottenhammer, Angela. 'Maritime Disasters and Risk Appraisals in the East Asian Waters.' *Études thématiques*, 2022.

Shi Liuqiu lu 使琉球錄, by Chen Kan 陳侃, in ed., Guoli Beiping tushuguan shanben congshu 國立北平圖書館善本. Shanghai: Commercial Press, 1937.

Shi Liuqiu lu 使琉球錄 (1579), by Xiao Chongye 蕭崇業 (*jinshi* 1571) and Xie Jie 謝杰 (*jinshi* 1574), in *Shi Liuqiu lu sanzhong* 使琉球錄三種 [Taiwan wenxian shiliao congkan 台灣文獻史料叢刊, 287]. Taibei: Taiwan datong shuju, 1970: 91 (使疏球錄卷上, 造舟).

Songfeng shuoyi 松峰説疫, by Liu Kui 劉奎, in Xuxiu Siku quanshu 續修四庫全書, zibu 子部, yijia lei 醫家類.

Steffelbauer, Ilja, and Christa Hammerl, eds. *Naturkatastrophen. Dramatische Naturereignisse aus kulturwissenschaftlicher Perspektive*. Wien: Mandelbaum Verlag, 2014.

(Wanli) Tongzhou zhi 通州志 (1577), by Lin Yuncheng 林云程, Shen Mingchen 沈明臣 (fl. mid-sixteenth cent.). [Yuan guoli Beiping tushuguan jiaku shanben congshu 原國立北平圖書館甲庫善本叢書, Shibu 史部, 306]. Beijing: Guojia tushuguan chubanshe, 2013.

Wen Zongdian 閔宗殿. 'MingQing shiqi Dongnan diqi yiqing yanjiu 明清時期東南地區疫情研究.' *Xueshu yanjiu* 學術研究, 10 (2003): 109–15.

Xijun Yan, ed. *Dan Shen (Salviamiltiorrhiza) in Medicine*. Volume 3, *Clinical Research*. Dordrecht: Springer Verlag, 2015.

Xu Zhexin. 'The Environment, Perceptions, and Publication of Medical Texts in Fujian During the Ming Period (1368 to 1644).' Paper presented at the conference: *Seafaring, Trade, and Knowledge Transfer: Maritime Politics and Commerce in Middle Period and Early Modern China*, held at Ghent University, Ghent in 2017, forthcoming in: Wim De Winter, Angela Schottenhammer, Mathieu Torck, eds. *Seafaring, Trade, and Knowledge Transfer: Maritime Politics and Commerce in Early Middle Period to Early Modern China* [CROS Crossroads—History of Interactions Across the Silk Routes]. Leiden: Brill Publishers, 2022/23.

Xu zizhi tongjian changbian 續資治通鑑長編, by Li Dao 李燾 (1115–1184). Taibei: Shijie shuju, 1965.

Yao Shan-yu. 'The Chronological and Seasonal Distribution of Floods and Droughts in Chinese History, 206 B.C.–A.D. 1911.' *Harvard Journal of Asiatic Studies*, 6, 3–4 (1942): 273–312.

Yen-Chu Liu, Huei-Fen Chen, Xingqi Liu, and Yuan-Pin Chang. 'Insight into Tropical Cyclone Behaviour Through Examining Maritime Disasters Over the Past 1000 Years Based on the Dynastic Histories of China—A Dedication to Ocean Researcher V.' *Quaternary International*, 440, A (2017): 72–81.

Zhejiang tongzhi 浙江通志, by Ji Zengjun 嵇曾筠 (1670–1738) and Shen Yiji 沈翼機 (Jinshi 1706), in Siku quanshu-edition, fasc. 519–526: https://www.kanripo.org/text/KR2k0044/062 [Accessed: 7 July 2021].

Zhongyang yanjiuyuan lishi yuyan yanjiusuo 中央研究院歷史語言研究所, ed. *Ming shilu* 明實錄. Taibei: Lishi yuyan yanjiusuo, 1967. 133 vols., esp. *Ming Muzong shilu* 明穆宗實錄; *Ming Shenzong shilu* 明神宗實錄, *Ming Xianzong shilu* 明憲宗實錄.

Zhou Zhiyuan 周致元. 'Mingdai Dongnan diqu de haichao zaihai 明代东南地区的海潮灾害.' *Shikue jikan* 史学集刊/ *Collected Papers of History Studies*, 2 (2005): 83–93.

CHAPTER 4

The El Nino of 1685–1687 in Golconda and Northern Coromandel, South Asia: Drought, Famine, and Mughal Wars

Archisman Chaudhuri

Methodologically, this chapter interrogates the archives of the Dutch East India Company (VOC) and draws upon Richard Grove's theses on coeval climatic anomalies that struck South Asia and Southeast Asia in the seventeenth century, as well as Joëlle L. Gergis and Anthony M. Fowler's chronology of El Niño Southern Oscillation (ENSO) events since 1525, to explore the impact of a severe climatic anomaly, the El Nino of 1685–1687, in the region of Golconda and northern Coromandel, located in the northern part of the South Asian littoral of Coromandel, which was a major maritime and industrial hub of the early modern Indian Ocean. The first section introduces readers to the concepts of Little Ice Age (LIA) and El Nino and their implications, especially with regard to South Asia,

A. Chaudhuri (✉)
Indian Ocean World Centre, McGill University, Montreal, QC, Canada
e-mail: archisman.chaudhuri0190@gmail.com

P. Gooding (ed.), *Droughts, Floods, and Global Climatic Anomalies in the Indian Ocean World*, Palgrave Series in Indian Ocean World Studies, https://doi.org/10.1007/978-3-030-98198-3_4

and it sets out the chapter's thematic and methodological frameworks; the second section familiarizes readers with the maritime region in question and its politico-economic scenario in the seventeenth century; and the third section discusses the impact of the 1685–1687 El Nino in Golconda and northern Coromandel to finally conclude that this climatic anomaly in question—along with coeval warfare and political instability—marked a turning point in northern Coromandel's economy.

The VOC had large commercial stakes in South Asia throughout the seventeenth and eighteenth centuries. Coromandel, a major component of their intra-Asian trade, provided textiles that were used to procure spices in Southeast Asia. The Dutch managed their extensive business operations in Coromandel from the coast to the interior through layers of officials, middlemen, textile merchants, and weavers.[1] The mentality reflected in the VOC archives (shared for the most part with fellow Europeans) was characterized by Eurocentrism, Christ-centricism (for the VOC, the Dutch Reformed Protestantism), climatic determinism, and proto-Orientalism, where the Dutch Republic was the reference point of all comparisons, including physical features of a land, and Christian western Europe was the yardstick for civilization.[2] As James Tracy argues, a crucial feature of the VOC archives was the tendency of VOC officials to pin the blame of their failure to achieve good trade results on Asian despotism in the person of Mughal officials or the emperor himself.[3] Nonetheless, due to the extensive business operations of the VOC and their diligent record-keeping of all major politico-economic, military, and environmental developments that affected their trade, the VOC archives are a major source to study early modern South Asia and the wider Indian Ocean World (IOW). Finally, the most important consideration

[1] For more details, see: Om Prakash, *European Commercial Enterprise in Pre-Colonial India, The New Cambridge History of India*, II.5 (Cambridge: Cambridge University Press, 2008).

[2] For a summary of Dutch perceptions of India, see: Markus Vink, *Mission to Madurai: Dutch Embassies to the Nayaka Court in the Seventeenth Century. Dutch Sources on South Asia, 1600–1825*, vol. 4 (New Delhi: Manohar, 2012), 35–37, 86–87, 89, 92. See also: Om Prakash, 'Dutch Source Material on Indian Maritime History in the Early Modern Period: An Evaluation,' *Indian Historical Review*, 8, 1–2 (1981–1982), 35–43.

[3] James Tracy, 'Asian Despotism? Mughal Government as Seen from the Dutch East India Company Factory in Surat,' *Journal of Early Modern History*, 3, 3 (1999), 256–80.

that shaped the VOC discourse on South Asia was what they stood to gain or lose from any politico-economic, military, and environmental developments.

The Little Ice Age and El Nino Events

In 1939, Francois E. Matthes, a glaciologist, used the term LIA to describe the development of glaciers in Sierra Nevada and other mountain ranges in the western USA. Based on his research into moraines from lakes, especially the Owen Lake, fed by glaciers in the western USA, Matthes argued that the glaciers in Sierra Nevada and other mountain ranges in the western USA developed during a period of renewed but moderate glaciation—one that had lasted for around 4000 years. Proffering evidence for coeval glaciation elsewhere in the world, Matthes referred to the glaciers around Chamonix in the French Alps, which had recorded their greatest extension for about 250 years, before beginning to recede around the mid-nineteenth century.[4] The term LIA, however, has increasingly been understood to centre on the period of three hundred odd years that Matthes had tangentially alluded to in the context of the French Alps. Jean Grove, in her study of glaciation entitled *The Little Ice Age*, draws upon geographers, geologists, glaciologists, and climatologists to understand the LIA as a period of glacial advance from about 1550 to 1800, when glaciers in many parts of the world expanded and fluctuated about more advanced positions than those that they occupied before or after this cooler interval in earth's climate.[5] More recent research, including by paleoclimatologists, has pushed the LIA's onset back by another three hundred years. A decline in solar activity, orbital factors, and volcanic eruptions combined to cool average annual temperatures on each continent during c.1300–1840. Spatially and temporally, the LIA was considerably variant; but, a globally synchronous trend of cooling in both the northern and southern hemispheres from the late sixteenth century to the late seventeenth century marked a more severe

[4] Francois E. Matthes, 'Report of Committee on Glaciers,' *Transactions, American Geophysical Union*, 20, 4 (1939), 518–20.

[5] Jean M. Grove, *The Little Ice Age* (London: Methuen, 1988), 3–4.

part of the LIA.[6] In the tropics, the LIA led to a reduced migration of the Intertropical Convergence Zone (ITCZ) and a sharp increase in ENSO anomalies, producing droughts and frequent failure of South and East Asian monsoons.[7] For instance, Joëlle L. Gergis and Anthony M. Fowler suggest that 213 ENSO anomalies (El Niño and La Nina combined) of varying intensity occurred from 1525 to 1799.[8]

ENSO events often begin between March and May and can extend for twelve months or more, running into subsequent calendar years. Usually, the sea-surface air pressure on the eastern side of the Pacific Ocean is higher than that on its western side along the equator. This difference in air pressure causes east to west surface winds (easterlies) to blow across the Pacific Ocean from the South American coast to Australia and Indonesia. During an El Niño year, this wind pattern is reversed: air pressure in the eastern Pacific drops and air pressure in the western Pacific rises. Due to the absence of air pressure, the easterly winds that blow across the Pacific Ocean from the South American coast weaken and retreat to the east. Consequently, tropical rainfall patterns move away from Australia and monsoon Asia towards the central and eastern Pacific. The disappearance of easterlies produces dry monsoon seasons and droughts in Australia, Indonesia, and other Southeast Asian countries, the Indian subcontinent, the highlands of Ethiopia, and southern Africa.[9] Indeed, some of the most severe droughts that struck South Asia during the LIA occurred during El Nino years. El Nino is one extreme of the ENSO phenomenon; its other extreme is the La Nina, which often follows El Nino years. The La Nina years see a reversal of the aforementioned El Nino impacts, although

[6] For a summary and its effects on world history, see: Geoffrey Parker, *Global Crisis: War, Climate Change and Catastrophe in the Seventeenth Century* (New Haven: Yale University Press, 2013).

[7] Christian Pfister, Rudolf Brázdil, Jürg Luterbacher, Astrid E.J. Ogilvie, and Sam White, 'Early Modern Europe,' in *The Palgrave Handbook of Climate History*, eds. Sam White, Christian Pfister, and Franz Mauelshagen (London: Palgrave Macmillan, 2018), 268.

[8] Joëlle L. Gergis, and Anthony M. Fowler, 'A History of ENSO Events Since A.D. 1525: Implications for Future Climate Change,' *Climatic Change*, 92, 3–4 (2009), 369–72.

[9] John F. Richards, *The Unending Frontier: An Environmental History of the Early Modern World* (Berkeley: California University Press, 2003), 82–83.

Table 4.1 Select ENSO events, 1600–1710[12]

El Nino	*La Nina*
1614	
1620–1621	1622–1624
1630	1629–1632
1659–1661	1663
1684, 1687	1685–1686
1694, 1695	1696
1707, 1709	1709, 1710

the inverse relationship—as Richard Grove and George Adamson have noted—is not entirely uniform.[10]

In the context of early modern South and Southeast Asia, Richard Grove argues severe climatic anomalies in the seventeenth century were coeval: Strong El Nino-induced droughts struck South Asia, Burma, and the Indonesian archipelago in 1614–1616, 1623–1624, 1629–1632, 1660–1662, and 1685–1688. Globally, this pattern of droughts had parallels with Mexico, where the 1624 drought was severe, and South Asian monsoon failure (at least) used to be preceded by forerunning cycles of cold winters in Europe, such as in 1684–1685.[11] Following Gergis and Fowler, Table 4.1 lists these coeval ENSO events with their corresponding La Nina years from the seventeenth century to the early eighteenth century.

The use of VOC archives for this history opens two core methodological opportunities. Firstly, coeval climatic anomalies forged connections between South Asia and Southeast Asia, two primary blocs in the VOC's intra-Asian trade. Secondly, the VOC was careful to record major developments—political, economic, military, and environmental—that affected their commercial operations in the IOW. In the context of Coromandel, this is borne out by Table 4.2, which lists the occurrences of poor monsoons in the late seventeenth century and the early eighteenth

[10] Richard Grove and George Adamson, *El Niño in World History* (London: Palgrave Macmillan, 2018), 4–7.

[11] Richard Grove, 'El Nino Chronology and the History of Socio-economic and Agrarian Crisis in South and Southeast Asia 1250–1990,' in *Land Use-Historical Perspectives: Focus on Indo-Gangetic Plains*, eds. Yash P. Abrol, Satpal Sangwan, and Mithilesh K. Tiwari (New Delhi: Allied Publishers, 2002), 141, 148, 154.

[12] Adapted from: Gergis and Fowler, 'A History of ENSO Events,' 370.

Table 4.2 Climatic anomalies in Coromandel, 1680–1710[14]

Year	*Climatic anomaly*
1682	Poor rainfall
1685–1686	Poor rainfall and famine (in northern Coromandel)
1687	Famine and flood (in northern Coromandel)
1693	Poor rainfall
1695	Poor rainfall
1698	Poor rainfall
1705	Poor rainfall
1707	Flood (southern Coromandel)
1708	Poor rainfall

century, according to VOC sources. The ENSO event of 1685–1688 in Coromandel, South Asia, moreover coincided with droughts and heavy rainfall in island Southeast Asia, including parts of Sumatra, Java, Borneo, and the Moluccas.[13] Table 4.2 also highlights that poor monsoons in late seventeenth-century Coromandel, reflected in the VOC's correspondence, corroborate many of the ENSO events listed in Table 4.1.

[13] Peter Boomgaard, 'Crisis Mortality in Seventeenth-Century Indonesia,' in *Asian Population History*, eds. Liu Ts'ui-jung, James Lee, David Sven Reher, Osamu Saito, and Wang Feng (New York: Oxford University Press, 2004), 205, 210.

[14] The rainfall data presented in Table 4.2 is mainly culled from the author's unpublished PhD thesis, 'From Camp to Port: Mughal Warfare and the Economy of Coromandel, 1682–1710' (Leiden University, 2019). The data is based on unpublished VOC archival records kept at the National Archive (henceforth NA), The Hague, The Netherlands, and a translated Mughal chronicle written in Persian, whose details are as follows: NA VOC 8808, Willem Hartsinck (director in Masulipatnam) to Cornelis Speelman (governor-general in Batavia), 17 Mar. 1682, f.61v.–f.62r.; NA VOC 1411, Hartsinck to Gentlemen Seventeen (in Amsterdam), 8 Oct. 1685, f.61v.–f.62r.; NA VOC 1411, J.J. Pits (governor in Pulicat) to Gentlemen Seventeen, 12 Nov. 1685, f.5r.; NA VOC 1423, Pits to Joannes Camphuis (governor-general in Batavia), 27 June 1686, f.121v., f.138r.; NA VOC 1423, Pits to Camphuijs, 26 Aug. 1686, f.168v.; NA VOC 1423, Pits to Camphuijs, 14 Sept. 1686, f.176r., f.177v.; NA VOC 1438, Pits to Gentlemen Seventeen, 14 Nov. 1686, f.1066v., f.1076v.; NA VOC 1438, Laurens Pit (governor in Pulicat) to Camphuijs, 6 Aug. 1687, f.1176r., f.1182r., f.1185v.; NA VOC 1438, Joannes Huijsman (director in Masulipatnam) to Camphuijs, 19 Aug. 1687, f.1243r.; NA VOC 1537, Bruijnig Wildelant (director in Masulipatnam) to Willem van Outhoorn (governor-general in Batavia), 19 Sept. 1693, f.643v.; NA VOC 1570, Wildelant to van Outhoorn, 8 Oct. 1695 f.343.; NA VOC 1610, Laurens Pit (governor in Nagapatnam) van Outhoorn, 18 May 1698, f.14–f.16, f.36–f.37; NA VOC 8824, Joannes van Steelanţ (governor in Nagapatnam) to

Richard Grove argues that there are close connections between ENSO events and major epidemics, especially plague, malaria, cholera, and influenza. ENSO events can also strongly influence diseases like smallpox, yellow fever, Rift Valley Fever, Japanese encephalitis, Ross River Fever, Murray Valley Fever, typhus, dengue fever, hantavirus, erythermalgia, as well as diseases that affect animals, such as rinderpest, African Horse Sickness, and anthrax. The changed hydrological conditions during ENSO years are often advantageous for mosquitoes, a major vector of several of the aforementioned diseases. Stagnant water remnants of normally perennial streams that turn dry during ENSO years are conducive for a sharp increase in the number of vectors, while arid regions that are suddenly flooded during La Nina years allow water to expand and help vectors breed. Rodents—much like insects—respond to temperature and hydrological changes, acting as the vector for plague and other rodent-borne diseases. In such cases too, the vector population increases as ENSO events continue.[15] Peter Boomgaard opines that in seventeenth-century island Southeast Asia, as a rule, smallpox originated in Batavia (in Java) and then spread to other islands, including Ternate/Tidore, Ambon, Banda, and the Lesser Sunda Islands of Solor, Flores, and Timor. While droughts and subsequent crop failures did not cause smallpox, climatic anomalies could have prolonged the disease, vulnerability to it, and its lethality. All these could weaken a labour force and partly disrupt the next harvest too.[16]

Eventually, the impact of ENSO events in South Asia and Southeast Asia created an intriguing component of this connected history of coeval severe climatic anomalies in early modern South and Southeast Asia. In seventeenth-century Coromandel, famines caused by droughts and/or wars depopulated lands, while survivors often migrated to regions that had not experienced crop failures or entered into bondage to escape starvation. Trafficking of enslaved labour was a corollary of coeval

Joan van Hoorn (governor-general in Batavia), 25 May 1706, f.387; NA VOC 8686, van Steelant to van Hoorn, 09 Sept. 1707, f.466–f.468; NA VOC 8373, van Steelant to van Hoorn, 7 May 1709, f.104–f.105; Jadunath Sarkar (trans.), *Maasir-i-Alamgiri: A History of the Emperor Aurangzib-Alamgir (reign 1658–1707 A.D.)* (Calcutta: The Asiatic Society, Reprint Edition 1990), 178.

[15] Grove and Adamson, *El Niño in World History*, 159–60. See also: Chapters by Campbell, Gooding, Schottenhammer, Warren, and Williamson, this volume.

[16] Boomgaard, 'Crisis Mortality,' 200.

climatic anomalies and the expansion of the Dutch colonial empire during the LIA. For example (and as will be seen), in the aftermath of the synchronous 1660–1662 and 1685–1687 ENSO events, the VOC shipped enslaved survivors from the Coromandel Coast in South Asia to Ceylon, and more particularly to repopulate their colonies in island Southeast Asia: Batavia and their spice gardens in the Moluccas suffered from depopulation at regular intervals due to the outbreak of epidemics. In other words, climatic anomalies contributed to the creation of what might loosely be described in the present as 'climate refugees' in the seventeenth-century IOW.

Seventeenth-Century Golconda and Northern Coromandel: Politico-economic Landscape

The sultanate of Golconda, located in the eastern part of the Deccan plateau in South India, was founded in the sixteenth century with its capital at the fort city of Golconda. Ruled by the Qutb Shahi dynasty, the sultanate attracted a steady influx of migrants from other parts of South India, Central Asia, and Arabia, throughout the sixteenth century, and a new city, Hyderabad, was completed in the early seventeenth century to accommodate the new arrivals.[17] State-formation in early modern South Asia, following Jos Gommans' 'high roads controlling the empire' hypothesis, hinged upon successful connections that states established between different geographic zones by controlling the highways. States consciously chose sites for capitals that were close to rivers, thus enabling them to control the highways of commerce—the key to the success of the Mughal Empire in northern India during the sixteenth century.[18] However, this argument could be extended to Golconda and northern Coromandel, too. For instance, Hyderabad, the Qutb Shahi capital by the river Musi, linked the cotton-growing tracts of Maharashtra in western Deccan, which lay to its west, and with the textile weaving villages of eastern Deccan to its east, where the maritime outlet to the Indian Ocean was Masulipatnam—the principal Qutb Shahi port-city in northern Coromandel.

[17] H.K. Sherwani, *History of the Qutb Shahi Dynasty* (New Delhi: Munshiram Manoharlal Publishers, 1974), vii–xiii, 2–3, 6–9, 14–16, 199–202.

[18] See: Jos Gommans, *Mughal Warfare: Indian Frontiers and High Roads to Empire, 1500–1700* (London: Routledge, 2002), 7–37.

The Golconda sultanate expanded eastwards in the 1560s towards the littoral of northern Coromandel along the Bay of Bengal and brought under its control Masulipatnam, which was until then a minor port featuring in coastal trade along the Coromandel Coast and an outlet for locally produced textiles. Contemporaneous Portuguese records point to the rapid rise of Masulipatnam as an oceanic port after its annexation by the Golconda sultanate. In the 1570s and 1580s, ships from Masulipatnam departed to Pegu and Arakan in mainland Southeast Asia, Aceh in Sumatra, and ports on the Malay Peninsula. By the early seventeenth century, Masulipatnam had developed trading connections with the Red Sea, and textiles from northern Coromandel had found a market in the Middle East.[19] We may fairly safely assume that the stability and prosperity in the Golconda sultanate provided Masulipatnam with a powerful springboard from which to expand, and architectural expressions in the sultanate celebrated this connection with Masulipatnam. The main attraction of Hyderabad was the monument of Char Minar, out of which four roads jutted; the eastern road led to the ports of the eastern coast, including Masulipatnam (Fig. 4.1).

As in other parts of India, rainfall in Coromandel depends on the annual cycle of monsoons. The western littoral of India primarily receives rainfall from the south-west monsoon which lasts from June to August. For example, the Malabar Coast in south-western India (modern-day Kerala) receives most of its rains during the south-west monsoon season and some more during the retreating north-east monsoon season, from October to December. However, the eastern littoral—the Coromandel Coast—receives less rainfall in the south-west monsoon season and more rainfall in the retreating north-east monsoon season. During the south-west monsoon, the monsoon winds enter India through the Malabar Coast, causing heavy rainfall there and further north along the Kanara and Konkan coasts. But the winds begin to lose moisture as they cross over the Western Ghats (a mountain range that runs parallel to the western littoral of India from modern-day Maharashtra to Kerala and parts of Tamil Nadu) and blow across the Deccan plateau to the east, causing less rainfall in Coromandel. During the retreating north-east monsoon season, the winds blow from north to south and gain moisture as they cross the Bay of Bengal, and they cause heavy rainfall in Coromandel.

[19] Sanjay Subrahmanyam, *The Political Economy of Commerce Southern India 1500–1650* (Cambridge: Cambridge University Press, 1990), 148–51, 154, 157–58.

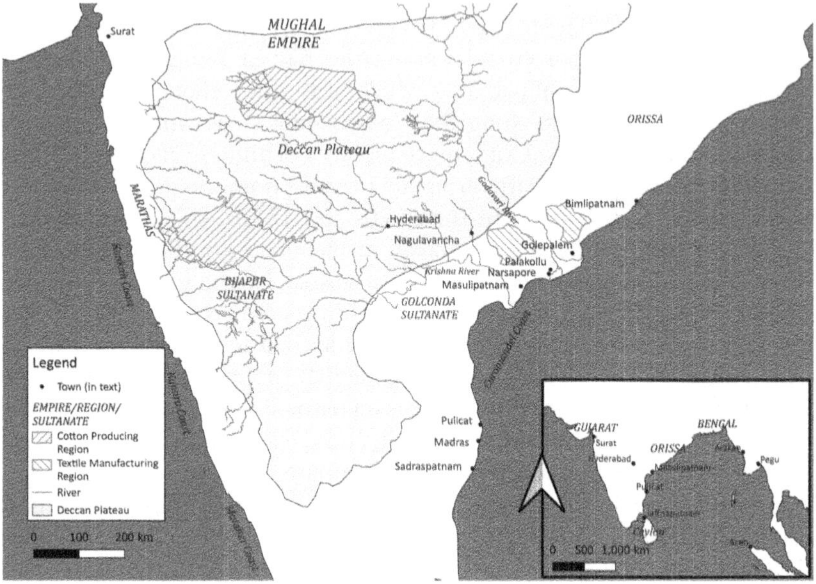

Fig. 4.1 Map of South Asia with notable places mentioned in-text marked. Owing to the changeable political situation during the seventeenth century, borders between empires and sultanates are unmarked. Drawn by Philip Gooding

Two major rivers in the Deccan plateau flow west to the east across Coromandel into the Bay of Bengal, the Godavari, and the Krishna; the port-city of Masulipatnam was situated near the mouth of the Krishna. The principal subsistence crop in Coromandel was rice, produced along the coastal plains and in the river valleys. Away from these zones, agriculture was supported by tank irrigation.[20] The Golconda sultanate operated a system called revenue farming, through which revenue officials leased out their lands to local entrepreneurs who bid for the office, and the position went to the highest bidder. They were then entrusted with the organization of revenue collection in villages and cities, before crediting

[20] For illustrative maps with indexes on the economy of the Deccan and South India see: Irfan Habib, *An Atlas of the Mughal Empire: Political and Economic Maps with Detailed Notes, Bibliography and Index* (New Delhi: Oxford University Press, 1982), 14B, 15B, 16B.

an annual figure to the treasury in Golconda. The ruler's claim to a share of agricultural production was the greatest single source of revenue for the Golconda sultanate, and this appropriated a large share of every harvest—although we must remember there would have been differences between what was assessed and what was collected (or expected to be collected). Tax-farming involved an elaborate system of lessees and sub-lessees. This system operated at port towns and their hinterlands, with export centres along the coast. Undoubtedly harsh, it did not slow agricultural production. Contemporary observers simultaneously deplored the severe tax system and marvelled at the prosperity in the Golconda countryside, which was partly caused by fertile agriculture and partly by some checks that the state itself had introduced in the taxation system.[21]

By the late sixteenth century, the Mughals had secured two gateways to the Indian Ocean—in Gujarat to the west and Bengal to the east, with Surat in Gujarat as their premier port. From this time, they then began to make inroads into the sultanates of the Deccan plateau through a combination of diplomatic and military manoeuvres. During the reign of Shah Jahan (1627–1657), the sultanates of Bijapur and Golconda became tributaries of the Mughals in the 1630s. Aurangzeb, then a prince and the eventual successor to Shah Jahan as the Mughal emperor, served two tenures as the governor of the Deccan, 1636–1644 and 1653–1657. In the second of these two tenures, Aurangzeb—confronted by budgetary deficits—unsuccessfully sought permission to annex Golconda.[22]

Aurangzeb's insistence on annexing Golconda was pragmatic in the context of acquiring control over its political economy. Golconda was a lucrative asset with its rich agriculture, diamond mines, and thriving maritime trade with the Indian Ocean. Annexing Golconda would have allowed the Mughals to control the Deccan and its commerce from the west to the east: from the port-city of Surat in Gujarat by the Arabian Sea in the west, through the cotton-growing tracts of modern-day Maharashtra to Hyderabad. In other words, it would have connected two of the major ports in the seventeenth-century IOW under Mughal authority.

[21] John F. Richards, *Mughal Administration in Golconda*, 21–23, 25–26.

[22] For an overview of Mughal expansion in the Deccan before Aurangzeb became the Mughal emperor, see: John F. Richards, *The Mughal Empire*, 52–54, 112–13, 120–21, 137–38, 154–58; Muzaffar Alam and Sanjay Subrahmanyam, 'The Deccan Frontier and Mughal Expansion, ca. 1600: Contemporary Perspectives,' *Journal of the Economic and Social History of the Orient*, 47, 3 (2004), 357–89.

Aurangzeb, who became the Mughal emperor in 1658, resumed his incomplete project of annexing the Deccan sultanates of Bijapur and Golconda during the 1680s. Although Bijapur (1686) and Golconda (1687) fell to the Mughals in succession, the war with the Marathas—another major adversary who were primarily based in the west of the Deccan but had also expanded their control to parts of the southern Coromandel Coast by the late seventeenth century—kept Aurangzeb occupied in the Deccan and elsewhere in South India until his death in 1707. This background of Mughal expansion and consequential warfare heightened vulnerability to the effects of climatic extremes, which is evident below in the context of the effects of the 1685–1687 ENSO event in Golconda and northern Coromandel.

The El Nino of 1685–1687 in Golconda and Northern Coromandel

South and Southeast Asia experienced five major coeval climatic anomalies in the seventeenth century. Strong El Nino-related droughts struck South Asia, Burma, and the Indonesian archipelago in 1614–1616, 1623–1624, 1629–1632, 1660–1662, and 1685–1688. As noted in Table 4.2, 1682 was also a year of poor monsoon in northern Coromandel. A July 1682 letter from the director of the Dutch factory in Masulipatnam to Batavia noted that a poor monsoon had increased the prices of grains; but the company's trade had not been much affected by it. We may fairly safely assume the poor monsoon conditions to have existed over the next few months too. Another letter from Masulipatnam to Batavia, written in October 1682, pointed out that a consignment of iron that was being brought from inland to Masulipatnam for export took an inordinately long time *en route* because extreme heat had made it difficult to find water and grass, which were provisions for draught animals.[23] The Mughal emperor Aurangzeb's southern military campaigns also began in 1682. While the VOC correspondence from northern Coromandel during 1683 and 1684 did not speak of climatic anomalies, they did express apprehensions about the imminent threat of a Mughal invasion of Golconda. The VOC director in Masulipatnam wrote to Batavia that

[23] NA VOC 8808, Hartsinck to Speelman, 17 Mar. 1682, f.152r., f.158v; NA VOC 1378, Hartsinck to Speelman, 12 Sept. 1682, f.1749v.–f.1750r.

the ruler of Golconda had been trying to ward off a Mughal invasion by regularly paying money to Aurangzeb.[24] By 1684, conditions for trade deteriorated quickly, as merchants stopped coming to Masulipatnam due to unsafe highways. Thus, the VOC director in Masulipatnam could not assure Batavia that the orders for textiles for 1686 would be fulfilled.[25] Such was the politico-economic scenario in Golconda and Masulipatnam before the ENSO event of 1685–1687 struck the region.

As early as October 1685, the effects of a monsoon failure had begun to appear in northern Coromandel. The director of the VOC enclave in Masulipatnam wrote to the Gentlemen Seventeen in Amsterdam that people were afraid of scarcity (alluding to expensive grains and famine), as it had rained too little. Moreover, a fever epidemic had broken out, killing many people (it is unclear if this was malaria or typhus). As a result of the failure of rains, raging fevers, and Mughal wars, business around Masulipatnam and the subordinate Dutch factories in the region came to a standstill.[26] A November 1685 letter from the Dutch governor in Pulicat to the VOC's Gentlemen Seventeen in Amsterdam added that food grains and crops planted in the fields had wilted because of a severe drought, and people feared that scarcity (implying a famine) was imminent.[27] These observations by the Dutch factors in Coromandel tell us that both the south-west and north-east monsoons had been poor in 1685. We may fairly safely assume this would have affected agriculture also in areas away from coastal plains and river valleys, where tank irrigation utilized the two monsoons for harvests.

The drought continued through to 1686, as is evidenced by VOC correspondence from June of that year, which stated that rice had become expensive around Pulicat.[28] Correspondence from later months, moreover—August and September—between the VOC governor in Pulicat and the governor-general in Batavia kept on lamenting about 'expensive' times (meaning high prices of commodities), hunger, and the Mughal

[24] NA VOC 8809, Hartsinck to Camphuijs, 13 Aug. 1683, f.101r.

[25] NA VOC 8811, Hartsinck to Camphuijs, 7 Oct. 1684, f.174r.

[26] NA VOC 1411, Hartsink to Gentlemen Seventeen, 08 Oct. 1685, f.61v.–f.62r.

[27] NA VOC 1411, Pits to Gentlemen Seventeen, 12 Nov. 1685, f.5r.

[28] NA VOC 1423, Pits to Camphuijs, 27 June 1686, f.121v., 138r.

wars that had been affecting the VOC's commerce.[29] This suggests that a famine might have become ubiquitous in northern and central Coromandel by September. Also, an October 1686 letter from Masulipatnam to Amsterdam tells us about the intensity of the famine in and around Masulipatnam. Except Orissa to the north, the littoral of northern Coromandel and the kingdom of Golconda had been struck by a massive shortage of grains and food. People died from starvation and collapsed on streets, while surviving parents sold or gave away their children to save the latter from dying. Initially, the VOC was thrifty in the sale of rice to the inhabitants of Masulipatnam. However, later the VOC was relieved with supplies of rice traded by ships from the north (presumably Orissa and Bengal) and they could sell rice at seven different places in Masulipatnam, although the famine conditions in the region hardly improved. The VOC wrote about a regular influx of people who thronged near Masulipatnam in search of food, and who even pushed each other aside and overran each other. Expecting the misery to continue, the VOC lamented that people were desperate to buy rice with all that they possessed, and would be barely left with anything more.[30] VOC correspondence from November 1686 noted that miseries caused by the famine in northern Coromandel continued to increase, as thousands of people died from starvation and crops could not be planted due to the failure of rains, which further increased the price of grains. The primary cause behind the famine, the VOC concluded, was almost a year-long drought which, by November 1686, had engulfed the whole of Coromandel. Nonetheless, the Dutch also noted a slight silver lining, as, by mid-November 1686, they reported a darkening of skies and thunder,[31] implying the imminent but much delayed arrival of the north-east monsoon (October to December) in Coromandel. However, the conditions of the famine around Masulipatnam still remained critical. By mid-December 1686, the VOC reported that the price of rice in Masulipatnam was at 120 pagodas per last (each last equalled about 1250 kg), because no vessels from Orissa carrying rice had plied to Masulipatnam for the past month.[32]

[29] NA VOC 1423, Pits to Camphuijs, 26 Aug. 1686, f.168v.; NA VOC 1423, Pits to Camphuijs, 14 Sept. 1686, f.176r., f.177v.

[30] NA VOC 1424, Laurens Pit (governor-elect of Coromandel in Masulipatnam) to Gentlemen Seventeen 13 Oct. 1686, f.847v.–f.848v.

[31] NA VOC 1438, from Pits to Gentlemen Seventeen, 14 Nov. 1686, f.1066v. f.1076v.

[32] NA VOC 1438, Pit to Gentlemen Seventeen 12 Dec. 1686, f.1079v.

It was in this environmental context that the Mughals, under Aurangzeb, besieged Golconda, exacerbating the effects of drought on the region's population. Already by 1685, unsafe highways resulting from Mughal warfare had begun to adversely affect the VOC's trade from Masulipatnam. In 1686, the Mughals invaded Hyderabad and conquered Bijapur (the latter after a long siege), and in early 1687, they besieged Golconda. The VOC in Pulicat wrote to Batavia in March that they were closely observing the conditions in Golconda, where the Mughals had besieged the Qutb Shahi king and his nobility in the Golconda fort and had bombarded the fort with little effect.[33] Mughal histories of Aurangzeb's reign, such as the *Maasir-i-Alamgiri*, also allude to such unsuccessful attempts to capture the fort.[34] In other words, the siege was a long-drawn affair. Thus, in June 1687, the VOC representatives in Masulipatnam wrote to Batavia that rice, fish, meat, and vegetables were scarce, and whatever was available cost ten times higher than earlier times. As a result of this, 15 or 20 people died from starvation and were found lying dead on streets almost every day.[35] In late 1686, as we noted, the VOC reported that the price of rice in Masulipatnam was 120 pagodas per last—this implies that normal prices of rice were often under 20 pagodas per last. While data from the early 1680s is inconclusive, in a normal year, for example, 1680, when there was neither drought, warfare, nor any serious political instability, the cost of rice shipped from Bimlipatnam—a minor port in northern Coromandel where the VOC had a factory—was around 9 ½ pagodas per last.[36] A comparable high inflation in rice prices around Masulipatnam is reported by Daniel Havart, a Dutch employee posted in Masulipatnam, in his account of Coromandel too: During 1686–1687, the cost of rice in Masulipatnam was around 2000 Dutch guilders per last, compared to an average of 100 Dutch guilders in normal years.[37] Apart from inflation and starvation-induced deaths, the company's merchants (those who undertook to supply cloth to the

[33] NA VOC 1438, Pits and Pit to Camphuijs, 27 Mar. 1687, f.1049r.

[34] Sarkar, *Maasir-i-Alamgiri*, 175–77.

[35] NA VOC 1438, Huijsman to Camphuijs, 12 June 1687, f.1169r.–f.1169v.

[36] NA VOC 1360, Willem Carel Hartsinck (President in Pulicat) to Rijcklof van Goens (governor-general in Batavia), 23 Feb. 1680, f.1452r.

[37] Daniel Havart, *Op-en ondergangh van Cormandel, Eerste Deel* (Amsterdam: Jan ten Hoorn, 1693), 214–15.

company) entered bankruptcy and incurred serious losses. In Masulipatnam, the number of personnel employed at the Dutch lodge became depleted.[38] The severe famine, deaths, and migration stripped this part of the littoral of people, as it were, and only a few remained alive. As early as May 1687, the Dutch wrote of mass depopulation in northern Coromandel.

Following these depredations, new ones developed in the second half of 1687, partly stemming from a La Nina anomaly in that year, which contributed to floods. That the region had had good rainfall in the first half of 1687 in Coromandel was alluded to by the Dutch in their correspondence with Batavia. Around Pulicat in central Coromandel, the Dutch wrote, agriculture suffered despite enough rainfall because of the oppressive practices of revenue farmers who had driven peasants off the land, resulting in depopulation of villages; while famine conditions and deaths continued to plague northern Coromandel.[39] A labour force weakened by famine would have found it extremely difficult to undertake agricultural work—this, besides the impact of Mughal wars, explains why the Dutch repeatedly lamented famine and depopulation without harbouring much hopes for recovery, especially in the context of northern Coromandel.[40] In Golconda, where the siege had turned into a protracted military encounter by mid-1687, the Mughals faced acute logistical challenges and encountered pestilential disease(s), as the region became flooded.

Saqi Mustaid Khan, the author of *Maasir-i-Alamgiri*, described the conditions around Golconda, especially of the Mughal camp:

> At this time owing to excess of rain the river Manjera raged in flood. No provision could come from the neighbourhood. Famine prevailed; wheat, pulse and rice disappeared. Cries of grief at the disappearance of grain rose from the famished on all sides of the camp. Of the men of Haidarabad, not a soul remained alive; houses, river and plain became filled with the dead. The same was the condition of the camp. At night piles of the dead were formed round the Emperor's quarters. Daily sweepers dragged them

[38] NA VOC 1438, Pit to Camphuijs, 31 May 1687, f.1113r., f.1117v.

[39] NA VOC 1438, Pit to Camphuijs, 6 Oct. 1687, f.1176r., f.1182r., f.1185v.

[40] NA VOC 1438, Huijsman to Camphuijs, 19 Oct. 1687, f.1242 v.

> and flung them on the bank of the river from sunrise to sunset. The same thing happened every day and night. The survivors did not hesitate to eat the carrion of men and animals. Kos after kos the eye fell only on mounds of corpses. The incessant rain melted away the flesh and the skin; otherwise the putrid air would have finished the business of the survivors. After some months when the rains ceased, the white ridges of bones looked from a distance like hillocks of snow. Through the grace of God to the survivors, the rains abated, the violence of the river ceased, and provisions came from the surrounding country.[41]

The heavy rains in northern Coromandel during the south-west monsoon of 1687 affected the Dutch enclaves too. In Palakollu, a textile weaving village near Masulipatnam (and already affected by the impact of the ENSO), the continual daily rainfall made it difficult to prepare consignments of textiles for export. A large part of the Dutch lodge in Nagulavancha collapsed from heavy rainfall, and many fell ill.[42]

The combined effects of ENSO-related drought and floods contributed to the spread of epidemics. The port-city of Masulipatnam was surrounded with swamps,[43] which, in general, provided ideal conditions for water-borne diseases, such as cholera, and ideal breeding grounds for vectors, such as mosquitoes. Daniel Havart, a Dutch employee posted in Masulipatnam, wrote of a pestilential fever epidemic that claimed the lives of several Dutchmen from late 1686 to late 1687. High prices (and the famine), Havart lamented, had made it difficult to procure animal protein, such as chicken and eggs for the sick.[44] In 1687, the VOC continued to report on the pestilential air (swamps) of Masulipatnam and raging fevers (malaria) that had caused deaths among Dutchmen, including Joannes Huijsman—the director of the Dutch enclave in Masulipatnam—and his family.[45] Presumably in this case, heavy rainfall during the south-west monsoon of 1687 had flooded the swamps around Masulipatnam, triggering a malaria epidemic. If it was the unhealthy environs of Masulipatnam that had caused fevers during the drought-famine cycle of 1685–1687, in and around Golconda, as the

[41] Sarkar, *Maasir-i-Alamgiri*, 178.

[42] NA VOC 1438, Huijsman to Camphuijs, 19 Oct. 1687, f.1243r.

[43] Havart, *Op-en ondergangh, Eerste Deel*, 142.

[44] Ibid., 215–18.

[45] NA VOC 1438, Pit to Camphuijs, 09 Oct. 1687, f.1247v.

Maasir-i-Alamgiri tells us, the Mughals strove hard to contain contagions from corpses, which hints at airborne transmission. While the nature of the diseases that had struck during the siege of Golconda is not clear, Dutch reports from 1687 on drought-famine induced bondage do tell us about the diseases like diarrhoea and smallpox, which afflicted enslaved people who were about to be transported to Batavia and Ceylon.

The combination of climatic, environmental, and military factors additionally contributed to a growth in slave trading. As has been observed elsewhere, famines, often related to climatic anomalies, and other hardships, such as military conquests and debts, have driven people into bondage as a means of survival.[46] Similar to the impact of the 1659–1661 ENSO event when the VOC shipped enslaved survivors from Coromandel to Ceylon and Batavia as a result of drought-and-famine induced bondage,[47] the climatic anomaly during 1685–1688 (and military factors) produced opportunities for slave trading. In 1686, for example, the VOC reported that hundreds of people had sold themselves off into slavery in Coromandel to escape death from starvation. Factories in Coromandel were thus instructed to buy a good number of them.[48] Adult male slaves aged from 18 to 26 years could be bought at 4 pagodas per slave, female slaves for 2 pagodas, and children in proportion to their age. In January 1687, the VOC shipped 88 slaves to Ceylon, and Pulicat still had at its disposal 360 slaves comprising 143 males, 143 females, and children including 32 girls and 42 boys. However, a significant number of the enslaved who remained in Pulicat, the VOC noted, died from diarrhoea and smallpox. The sheer magnitude of the drought-famine-induced bondage during the 1685–1688 ENSO event was alluded to by the VOC in two of its observations. Firstly, they noted that there were more slaves who had been bought by private slave traders around Madras

[46] For an overview of how climatic, environmental, economic, and military factors shaped patterns of bondage in the IOW, see: Gwyn Campbell (ed.), *Bondage and the Environment in the Indian Ocean World* (Cham, CH: Palgrave Macmillan, 2018).

[47] W.Ph. Coolhaas (ed.), *Generale Missiven van Gouverneurs-generaal en Raden aan Heren XVII der Verenigde Oostindische Compagnie, Deel III* (The Hague: Martinus Nijhoff, 1968), 338, 355, 357–58. J.A. van der Chijs (ed.), *Dagh-Register gehouden int Casteel Batavia vant passerende daer ter plaetse en het geheel Nederlandts-India Anno 1661* (The Hague: Martinus Nijhoff and Batavia: Landsdrukkerij, 1889), 325.

[48] W.Ph. Coolhaas (ed.), *Generale Missiven van Gouverneurs-generaal en Raden aan Heren XVII der Verenigde Oostindische Compagnie, Deel V* (The Hague: Martinus Nijhoff, 1975), 57.

and other places. The Dutch had ordered the private traders to bring those they enslaved to Nagapatnam (a port-city in southern Coromandel that the VOC had conquered in 1658) for sale. Should the Dutch and these private traders fail to strike a good bargain, it was decided that the VOC would allow the private traders to sell the slaves to a second buyer, provided he did not rent them again without orders from the Dutch. Secondly, the VOC lamented that while thousands of people had entered into slavery in Coromandel to escape from starvation, Dutch participation in the slave trade had been late primarily because of their low provisions of rice in Pulicat, and issues around their trade in Golconda. The Dutch had planned to ship 360 slaves to Batavia, hoping that smallpox would stop spreading by the time of their departure. In case the disease did not abate, the Dutch planned to send at least 100 of those slaves to Batavia. Following upon the orders of Hendrik Adriaan van Reede, the commissioner-general of the VOC on an inspection of the VOC establishments in South Asia, the Dutch decided that all purchases of people for enslavement would be divided equally between Ceylon and Batavia.[49]

The VOC continued to buy enslaved people during 1687, which hints at how serious drought-famine-induced bondage was in Coromandel at that juncture. In May 1687, the Dutch wrote to Batavia that 162 slaves in Pulicat had died from fevers and diarrhoea. The Dutch bought another 71 slaves and received a shipment of 32 slaves from Sadraspatnam (a port in central Coromandel). By mid-1687, the Dutch had 298 slaves in Pulicat, comprising 118 adult males, 102 adult females, 39 boys, and 32 girls, whom they hoped to transport to Batavia or Ceylon.[50] In June 1687, the Dutch factor in Masulipatnam noted that 100 slaves (67 males and 33 females) had been shipped to Malacca. Of another 100 slaves who remained in Masulipatnam, 87 died, including males and females.[51] In August 1687, the VOC in Pulicat reported the shipment of 150 slaves to Ceylon, comprising 65 males, 50 females, 20 boys, and 15 girls. The shipment in question was to take on board another 100 slaves from Sadraspatnam and then sail to Jaffnapatnam in Ceylon. The VOC governor in Pulicat had earmarked another 250 slaves, who were to be

[49] NA VOC 1438, Pits to Camphuijs, 27 May 1687, f.1059r.–f.1059v., f.1063v.–f.1064r.

[50] NA VOC 1438, Pit to Camphuijs, 31 May 1687, f.1131v.

[51] NA VOC 1438, Huijsman to Camphuijs, 12 June 1687, f.1169v.

transported to Batavia.[52] The latter acknowledged the arrival of 100 slaves from Coromandel in Malacca. While slaves had been cheap in the aftermath of the 1685–1687 drought-famine cycle and Mughal wars, the VOC noted that southern Coromandel provided less of them than northern Coromandel.[53] By late 1687, neither the Mughal wars nor the famine had affected southern Coromandel as severely as it had northern Coromandel, which perhaps explains the low number of people for enslavement that the VOC could procure from the former region.

Of course, the regions to which the VOC took slaves were also affected by the 1685–1687 ENSO event. Southeast Asia is key in this context, containing, as it did, the VOC's main export centres of spices and Batavia, its Indian Ocean capital. Drought struck various Southeast Asian regions from 1686 to 1688. The eastern part of Java and Mataram in south-central Java suffered from a drought in 1686, and the region could not ship rice. The VOC's stocks of rice in Batavia fell because of the poor harvest in Mataram, where the drought had continued through to 1687, while poor harvests were reported from Makassar (Sulawesi) and Bima (Sumbawa). Maros in Sulawesi had a poor rice harvest too because of the drought. In the Moluccas, further to the east, the Banda islands primarily reported heavy rains in 1685 and 1687, which damaged crops of nutmeg and mace; while drought conditions were reported in Banda in September and November 1687. Kisar in the Moluccas experienced a ten-month-long drought in 1687, followed by a shortage of food and deaths of cattle due to lack of pastures. In 1688, the pepper harvest in Palembang (Sumatra) was low, owing to drought.[54] However, the VOC did not report any deaths in these regions because of famine(s) induced by the drought (Fig. 4.2).

Instead, deaths were attributed to epidemics, such as fevers, measles, and smallpox. As discussed above, anomalous levels of rainfall may have contributed to the spread of such diseases. In 1687 in Wayer (Greater Banda), the VOC reported deaths among its personnel due to fevers. In 1688, in Lampung of Palembang (Sumatra), the VOC reported deaths

[52] NA VOC 1438, Pit to Camphuijs, 6 Aug. 1687, f.1186v.–f.1187r.

[53] Coolhaas (ed.), *Generale Missiven, Deel V*, 133, 137.

[54] Coolhaas (ed.), *Generale Missiven, Deel V*, 30–31, 64–65, 104, 115–16, 147, 251. For heavy rainfall in Banda in 1685, see: W.Ph. Coolhaas (ed.), *Generale Missiven van Gouverneurs-Generaal en Raden aan Heren XVII der Verenigde Oostindische Compagnie, Deel IV* (The Hague: Martinus Nijhoff, 1968), 792.

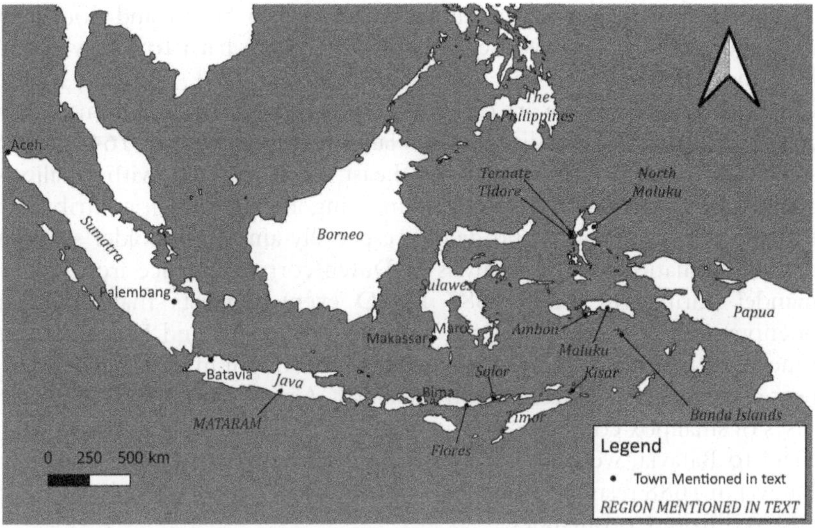

Fig. 4.2 Map of Southeast Asia, with places mentioned in-text marked. Drawn by Philip Gooding

from fevers, where the climate was especially unhealthy during the change of monsoons, during which time pepper was harvested. In Lampung, among 100 personnel in 1688, 40 Europeans and 25 mardijkers (freed slaves) died of fevers in nine months. As we have noted, 1687 was a year of heavy rainfall in the Banda islands, interspersed by droughts in September and November, and in Lampung, it was dominated by drought. We can only wonder if these were cases of malarial fevers. In 1688, the VOC in Batavia noted the deaths of many children and adults due to an outbreak of measles and smallpox during August and September.[55] Contagion from smallpox continued in Batavia in 1689 too. Havart, in his history of Coromandel, refers to a Dutch employee in Masulipatnam who, while losing most of his family members to the fevers (1685–1687) in Masulipatnam, had himself survived the outbreak of diseases and had migrated to Batavia, where his youngest daughter died of smallpox in September 1689.[56]

[55] Coolhaas (ed.), *Generale Missiven, Deel V*, 250, 255.

[56] Havart, *Op-en ondergangh, Eerste Deel*, 216–18.

The links between coeval rainfall anomalies in South and Southeast Asia and outbreaks of disease in both regions are hard to firmly establish, though possibilities are gleanable from the VOC sources. In the context of Batavia, there are earlier examples of how infected humans had acted as carriers of the disease over large areas, such as in 1644, when slaves from Arakan (mainland Southeast Asia) infected with smallpox arrived in Batavia (with many of them dying aboard ships), contributing to widespread infections and deaths, especially among the older sections of the population.[57] Our analysis of Dutch correspondence from Coromandel during the 1685–1687 ENSO event suggests that a similar phenomenon may have occurred linking Coromandel and Batavia at this time as well. Many of the slaves that the VOC bought in Pulicat and transported to Batavia suffered from smallpox. Given such coeval recorded cases of smallpox contagion in Coromandel and movement of slaves from there to Batavia, we can thus also suggest that the smallpox outbreak in Batavia in 1688 resulted from the VOC shipping infected people they had enslaved from Coromandel.

Returning to South Asia, the effects of the 1685–1688 ENSO event on northern Coromandel were long-lasting. About 500,000 people, Daniel Havart estimated, died around Masulipatnam.[58] Usually, in a chain reaction, monsoon failures in India affected the agrarian sector first, followed by allied sectors, such as textile production. Failure to grow crops resulted in loss of employment and depletion of stocks of rice from previous harvests, leading to inflation in prices of food grains and fodder for livestock. Textile production was hit next as diminishing food security and imminent famine-like conditions prompted weaving populations to migrate to places that offered better food security. The VOC correspondence reveals this familiar pattern in northern Coromandel in 1685–1686. Crop failures meant a major section of the peasant population either migrated to areas with better food security or died from starvation or

[57] W.Ph. Coolhaas (ed.), *Generale Missiven van Gouverneurs-Generaal en Raden aan Heren XVII der Verenigde Oostindische Compagnie, Deel II* (The Hague: Martinus Nijhoff, 1968), 223–24. H.T. Colenbrander (ed.), *Dagh-Register gehouden int Casteel Batavia vant passerende daer ter plaetse als over geheel Nederlandts-Indie Anno 1643–1644* (The Hague: Martinus Nijhoff, 1902), 33, 41.

[58] Havart, *Op-en ondergangh, Eerste Deel*, 214.

entered into bondage. The Mughal invasion of Golconda too interrupted planting of crops in Hyderabad.[59] The weavers met a similar fate, if not worse. By 1685, when the drought began, conditions for trade around Masulipatnam had deteriorated in the aftermath of the onset of Aurangzeb's southern campaigns, and the weaving villages around Masulipatnam were depopulated during 1685–1686 El Nino years. The VOC, who relied on shipments of grains from Bengal and Orissa to adapt to the famine, complained of poor sales of their imports and low levels of textile production due to the death and migration of weavers (along with military factors); the textile merchants who contracted to supply cloth to the company entered into bankruptcy and incurred serious losses.

Combatants during the siege of Golconda (1687) also experienced vulnerabilities to climatic extremes. These applied primarily in terms of logistical challenges related to poor supplies and a devastating epidemic. Failure of crops, high inflation of food grains and fodder, deaths from famine, diseases, and migration of peasants implied food security was low by early 1687 when the Mughals besieged Golconda. Conditions worsened as the siege of Golconda continued and the 1687 south-west monsoon flooded the region. This, as we noted, obstructed supply of provisions to the Mughal camp and an epidemic, intensified by airborne contamination from corpses, killed many. But while the surviving non-combatant population adapted to such conditions through migration or entering into bondage, combatants remained in the region, and though weakened, resumed the siege once supplies were available again after the flood(s). So, when the Mughals captured the fort of Golconda after a protracted siege in October 1687[60] and annexed the former Qutb Shahi sultanate into the Mughal Empire, they encountered a heavily depleted region where their priority was to rebuild it. The scale of this project for the Mughal rulers was in some ways unprecedented in northern Coromandel's seventeenth-century history. Although the region had experienced similarly adverse climatic and environmental conditions related to ENSO anomalies in, for example, 1630–1632 and 1659–1661, the coalescence of climatic and military factors in 1685–1687 increased levels of vulnerability to scarcity in distinct ways. The pressure on food resources affected non-combatants and combatants alike, contributing to

[59] See also: Richards, *Mughal Administration in Golconda*, 69.

[60] Sarkar, *Maasir-i-Alamgiri*, 182.

widespread deaths from starvation and epidemic diseases, such as cholera and/or diarrhoea, fevers, and smallpox.

Following the Mughal conquest of Golconda, the VOC sent an envoy to the Mughal emperor Aurangzeb's court to have their trading privileges in the region reconfirmed. The terms of the firman (royal order) that Johannes Bacherus, the Dutch envoy to Aurangzeb, secured from the Mughal emperor point out that the Mughals were keen to resettle the depopulated region around Masulipatnam, especially for its textile industry and its maritime trade. For instance, Mughal officials were instructed to not impose levies on textile washers employed by the VOC in villages like Golepalem, Gondewarom, and Draksharama; while carpenters and/or labourers working for the VOC at the shipyard of Narsapore near Masulipatnam were exempted from any kind of charges.[61] Although the Mughals appeared keen to rebuild the famine-stricken, war-battered region of northern Coromandel, recovery was slow. The Dutch tone of pessimism with regard to the textile trade continued during the early 1690s. Representatives wrote to Batavia in 1690, for example, that there was hardly any trade in textiles in Masulipatnam or at its other factories in northern Coromandel, owing to weavers finding it difficult to produce textiles because of famines, pestilence, and general devastation caused by wars.[62] Conditions slightly improved by 1691 as the VOC in Masulipatnam was able to procure textiles for export to Ceylon, Southeast Asia, Japan, and the Netherlands.[63] By 1692, a vague semblance of normalcy had returned, as the VOC governor wrote that production of textiles for Batavia and the Netherlands had resumed in northern Coromandel; while the VOC also sounded upbeat as they had encountered fewer difficulties than in previous year(s) to meet demand for cloth from Asia and Europe.[64]

Nevertheless, the devastating intensity of the 1685–1687 ENSO event was vivid in and around Masulipatnam even after a decade, as the account

[61] NA VOC 1510, Mughal Emperor Aurangzeb to Johannes Bacherus (decree), 24 Oct. 1689, f.377r.–f.377v.

[62] NA VOC 1473, Pit to Camphuijs, 23 July 1690, f.299r.–f.299v.

[63] NA VOC 1499, Barent Wildelant (director in Masulipatnam) to Camphuijs, 08 Oct. 1691, f.272v.–f.273r.

[64] NA VOC 1508, Pit to Gentlemen Seventeen, 26 Jan. 1692, f.127r.–f.127v.; NA VOC 1508, Pit to van Oudhoorn, 10 Oct. 1692, f.148r.

of William Norris, the English ambassador to the Mughal emperor Aurangzeb, demonstrates. Writing around 1699–1700, Norris observed:

> In the year 1686 a serious famine had occurred at Masulipatnam and in the surrounding country ... Thousands of people died of starvation and many families sold themselves to the Dutch for bread. The latter took advantage of the catastrophe and transported a large number of famine-stricken people to Batavia and the Spice Islands, where they remained in a state of slavery. The famine was followed the next year by an outbreak of plague, which carried away most of the survivors, enfeebled by their privations. In consequence of those disasters ... the town had never recovered its former importance, both its population and its trade being much diminished. The famine had caused a great increase in the prices of all provisions. Most of the factories had been removed or closed, except that belonging to the Dutch. But the most serious loss to the town was the disappearance of its artificers and other workmen, as well as the art of chintz, a famous product of Masulipatnam.

However, Norris also noted that when he visited Masulipatnam, the art of chintz had begun to revive itself.[65] A coalescence of climatic and military factors represented a major 'turning point' in northern Coromandel's economy and its connections to the wider IOW.[66]

Conclusion

Richard Grove argues that the ferocity of the climatic anomalies in Southeast India during 1685–1687 could be compared only with the El Nino droughts of 1790–1794, which had global implications.[67] Part of a global story of climatic anomalies, including coeval rainfall anomalies (both droughts and heavy rains) in island Southeast Asia, the ENSO event

[65] Harihar Das, *The Norris Embassy to Aurangzib (1699–1702)* (Calcutta: Firma K. L. Mukhopadhyay, 1959), 125–26.

[66] Greg Bankoff and Joseph Christensen (eds.), *Natural Hazards and Peoples in the Indian Ocean World: Bordering on Danger* (New York: Palgrave Macmillan, 2016); Sam White, *The Climate of Rebellion in the Early Modern Ottoman Empire* (New York: Cambridge University Press, 2011).

[67] Grove and Adamson, *El Niño in World History*, 67–68.

of 1685–1687 had a terrible impact in Golconda and northern Coromandel—a region that had already began to endure economic distress caused by Aurangzeb's southern campaigns. The preliminary effects of these campaigns, which created an unsafe environment for trade, worsened as crops failed in the region during the prolonged drought of 1685–1686. Floods further exacerbated the situation in 1687. Food security drastically fell as a devastating famine struck the region, contributing to deaths from starvation and epidemic diseases. The port-city of Masulipatnam turned into a shadow of its former self.

In this context, famine-induced bondage became prevalent, as a desperate attempt at survival by people who could neither migrate to other regions without crop failures nor find sustenance around Masulipatnam. Although Dutch reports from late 1686 stated that the whole of Coromandel had experienced a drought for a year, the impact of the drought-famine chain reaction seems to have been less devastating in southern Coromandel. This is borne out by two arguments that the VOC made: one, peasants fleeing oppressive tax farmers around Pulicat in 1687 (central Coromandel) migrated southwards; and two, the VOC in Batavia acknowledged that it could buy fewer people for enslavement in southern Coromandel than in the north, which alludes to the idea that the intensity of drought- and war-induced famine was relatively low in southern Coromandel.

On the whole, resilience against famines, either induced by droughts and/or wars, was low on the Coromandel Coast—especially when climatic extremes and warfare occurred together. The repeated cases of deaths due to starvation and diseases, and the high incidence of the slave trade point to low levels of food security in the region during extreme droughts. Malnourished people low on immunity succumbed to epidemics that followed droughts and famines, and as a result mortality rates were quite high in South Asia during ENSO event(s). In Southeast Asia, food security might have been comparatively high due to alternative sources of food; as noted, the VOC reports of droughts in island Southeast Asia during ENSO events (in this case, 1685–1688) hardly refer to deaths caused by starvation. But they do speak of high mortalities from disease epidemics in island Southeast Asia, which may have links with broader climatic teleconnections, as well as their effects on parts of South Asia. Finally, the author hopes to direct more future research on this connected and coeval pattern of major ENSO episodes across South Asia and island Southeast Asia during the LIA, to bring the effects of environmental anomalies in the IOW in sharp relief against each other.

Acknowledgements I thank the editor, reviewer(s), and Dr. Manjusha Kuruppath for their helpful comments on previous drafts of this chapter; and Dr. Monisha Sanyal, McGill University, and Anasuya Moitra, University of Tübingen, who educated me on the workings of epidemic diseases.

Bibliography

Archival Sources

Inventory Number 1.04.02 of the Archive of the VOC (the Dutch East India Company), 1602–1795. National Archive, The Hague, the Netherlands.

Published Sources

Abrol, Yash P., Satpal Sangwan, and Mithilesh K. Tiwari, eds. *Land Use-Historical Perspectives: Focus on Indo-Gangetic Plains*. New Delhi: Allied Publishers, 2002.

Alam, Muzaffar, and Sanjay Subrahmanyam. 'The Deccan Frontier and Mughal Expansion, ca. 1600.' *Journal of the Economic and Social History of the Orient*, 47, 3 (2004): 357–89.

Bankoff, Greg and Joseph Christensen, eds. *Natural Hazards and Peoples in the Indian Ocean World: Bordering on danger*. New York: Palgrave Macmillan, 2016.

Boomgaard, Peter. 'Crisis Mortality in Seventeenth-Century Indonesia,' in *Asian Population History*, eds. Ts'ui-jung, et al.: 191–220.

Campbell, Gwyn, ed. *Bondage and the Environment in the Indian Ocean World*. Cham, CH: Palgrave Macmillan, 2018.

Chijs, J.A. van der ed., *Dagh-Register gehouden int Casteel Batavia vant passerende daer ter plaetse en het geheel Nederlandts-India Anno 1661*. The Hague, Martinus Nijhoff and Batavia, Landsdrukkerij, 1889.

Colenbrander, H.T., ed. *Dagh-Register gehouden int Casteel Batavia vant passerende daer ter plaetse als over geheel Nederlandts-Indie Anno 1643–1644*. The Hague: Martinus Nijhoff, 1902.

Coolhaas, W.Ph., ed. Generale Missiven van Gouverneurs-Generaal en Raden aan Heren XVII der Verenigde Oostindische Compagnie, Deel II. The Hague: Martinus Nijhoff, 1964.

Coolhaas, W.Ph., ed. Generale Missiven van Gouverneurs-Generaal en Raden aan Heren XVII der Verenigde Oostindische Compagnie, Deel III. The Hague, Martinus Nijhoff, 1968.

Coolhaas, W.Ph., ed. Generale Missiven van Gouverneurs-Generaal en Raden aan Heren XVII der Verenigde Oostindische Compagnie, Deel IV. The Hague: Martinus Nijhoff, 1971.

Coolhaas, W.Ph., ed. *Generale Missiven van Gouverneurs-Generaal en Raden aan Heren XVII der Verenigde Oostindische Compagnie, Deel V*. The Hague: Martinus Nijhoff, 1975.

Das, Harihar. *The Norris Embassy to Aurangzib*. Calcutta: Firma K. L. Mukhopadhyay, 1959.

Gergis, Joëlle L., and Anthony M. Fowler. 'A History of ENSO Events since A.D. 1525: Implications for Future Climate Change.' *Climatic Change*, 92, 3–4 (2009): 343–87.

Gommans, Jos. *Mughal Warfare: Indian Frontiers and High Roads to Empire, 1500–1700*. London: Routledge, 2002.

Grove, Jean M. *The Little Ice Age*. London: Methuen, 1988.

Grove, Richard. 'El Nino Chronology and the History of Socio-economic and Agrarian Crisis in South and Southeast Asia 1250–1990,' in *Land Use-Historical Perspectives*, eds. Abrol and Tiwari: 133–72.

Grove, Richard and George Adamson. *El Niño in World History*. London: Palgrave Macmillan, 2018.

Habib, Irfan. *An Atlas of the Mughal Empire: Political and Economic Maps with Detailed Notes, Bibliography and Index*. New Delhi: Oxford University Press, 1982.

Havart, Daniel. *Op-en ondergangh van Cormandel*. 3 vols. Amsterdam: Jan Ten Hoorn, 1693.

Ts'ui-jung, Lui, James Lee, David Sven Reher, Osamu Saito, and Wang Feng, eds. *Asian Population History*. New York: Oxford University Press, Reprint Edition, 2004.

Matthes, Francois E. 'Report of Committee on Glaciers.' *Transactions, American Geophysical Union*, 20, 4 (1939): 518–23.

Parker, Geoffrey. *Global Crisis: War, Climate Change and Catastrophe in the Seventeenth Century*. New Haven and London: Yale University Press, 2013.

Pfister, Christian, Rudolf Brázdil, Jürg Luterbacher, Astrid E.J. Ogilvie, and Sam White. 'Early modern Europe,' in *The Palgrave Handbook of Climate History*, eds. White, Pfister, and Mauelshagen: 265–95.

Prakash, Om. 'Dutch Source Material on Indian Maritime History in the Early Modern Period—An Evaluation.' *Indian Historical Review*, 8, 1–2 (1981–1982): 35–43.

Prakash, Om. *The New Cambridge History of India, II.5: European Commercial Enterprise in Pre-Colonial India*. Cambridge: Cambridge University Press, 2008.

Richards, John F. *Mughal Administration in Golconda*. Oxford: Clarendon Press, 1975.

Richards, John F. 'The Seventeenth-Century Crisis in South Asia.' *Modern Asian Studies*, 24, 4 (1990): 625–38.

Richards, John F. *The New Cambridge History of India, I.5: The Mughal Empire*. Cambridge: Cambridge University Press, Reprint Edition 1995.

Richards, John F. *The Unending Frontier: An Environmental History of the Early Modern World*. Berkeley and Los Angeles: California University Press, 2003.

Tracy, James D. 'Asian Despotism? Mughal Government as Seen from the Dutch East India Company Factory in Surat.' *Journal of Early Modern History*, 3, 3 (1999): 256–80.

Sarkar, Jadunath, translated. *Tarikh–i–Dilkasha (Memoirs of Bhimsen Relating to Aurangzeb's Campaigns): Sir Jadunath Sarkar Birth Centenary Commemoration Volume with an Introduction by V.G. Khobrekar*. Bombay: Department of Archives, Maharashtra, 1972.

Sarkar, Jadunath, translated. *Maasir-i-Alamgiri: A History of the Emperor Aurangzib-Alamgir (reign 1658–1707 A.D.)*. Calcutta, The Asiatic Society, Reprint Edition 1990.

Sherwani, H.K. *History of the Qutb Shahi Dynasty*. New Delhi: Munshiram Manoharlal Publishers, 1974.

Subrahmanyam, Sanjay. *The Political Economy of Commerce Southern India 1500–1650*. Cambridge: Cambridge University Press, 1990.

Vink, Markus. *Dutch Sources on South Asia c. 1600–1825, Vol. 4. Mission to Madurai: Dutch Embassies to the Nayaka Court of Madurai in the Seventeenth Century*. New Delhi: Manohar, 2012.

VOC–Glossarium: Verklaringen van termen, verzameld uit de rijks geschiedkundige publicatiën die betrekking hebben op de Verenigde Oost-Indische Compagnie. The Hague, Instituut voor Nederlandse Geschiedenis, 2000.

White, Sam. *The Climate of Rebellion in the Early Modern Ottoman Empire*. New York: Cambridge University Press, 2011.

White, Sam, Christian Pfister, and Franz Mauelshagen, eds. *The Palgrave Handbook of Climate History*. London: Palgrave Macmillan, 2018.

CHAPTER 5

Rainfall and Floods in the Upper Zambezi Basin, 1680s to 1910s

William Gervase Clarence-Smith

Despite lying far inland in Central Africa, the Upper Zambezi Basin is climatically aligned with the Indian Ocean, receiving only weak influences from the South Atlantic. In terms of its rainfall regime, the region therefore needs to be considered as part of the Indian Ocean World.

Evidence for precipitation and inundations is sparse for the 1680s to the 1830s. To the fore are Portuguese references to conditions in the South Central highlands of Angola. These matter greatly, as rivers from this area probably supply most of the waters that accumulate in the floodplains of the Upper Zambezi. Tree-rings can further be pressed into service for rainfall from the mid-1790s, despite the eccentric location of the study, and bearing in mind the small size of the sample for the first few

W. G. Clarence-Smith (✉)
Department of History, Religions, and Philosophies,
SOAS University of London, London, UK
e-mail: wc2@soas.ac.uk

P. Gooding (ed.), *Droughts, Floods, and Global Climatic Anomalies in the Indian Ocean World*, Palgrave Series in Indian Ocean World Studies, https://doi.org/10.1007/978-3-030-98198-3_5

decades. Oral traditions and personal testimonies provide further minor contributions.

The quality of information improves markedly from the 1840s. Written accounts multiply, and the tree-ring sample expands from the 1870s. Following the implantation of mission stations in the early 1880s, it even becomes possible to chart the evolution of rainfall and floods year by year, with some degree of precision. The gathering of statistics, continuous from 1907, marks the final stage in the emergence of the modern climate history of the region.

Rainfall in the Upper Zambezi Basin

Sharon Nicholson, a doyenne of climate historians of Africa, considers that the average amount of rainfall has been 'remarkably stable' over the nineteenth and twentieth centuries, notwithstanding sharp differences from year to year, and considerable variation between decades and sets of decades.[1] This overall steadiness probably ranges back much further in time, as evidence for the extent of forest in the Congo Basin testifies to broadly constant rainfall over the past two millennia.[2] This contrasts with arguments, current since the middle of the nineteenth century, that a process of gradual desiccation has long been under way in Central and Southern Africa.[3]

[1] Sharon E. Nicholson, 'Spatial Teleconnections in African Rainfall: A Comparison of Nineteenth and Twentieth Century Patterns,' *Holocene*, 24, 12 (2014), 1846. See also: Sharon E. Nicholson, Chris Funk, and Andreas H. Fink, 'Rainfall over the African Continent from the Nineteenth Through the Twenty-First Century,' *Global and Planetary Change*, 165, (2018), 114, 119; Mark R. Jury, 'The Coherent Variability of African River Flows: Composite Climate Structure and the Atlantic Circulation,' *Water SA*, 29, 1 (2003), 4.

[2] Jean Maley, 'Synthèse sur l'histoire de la végétation et du climat en Afrique centrale au cours du quaternaire recent,' in *Peuplements anciens et actuels des forêts tropicales*, eds. Alain Froment and Jean Guffroy (Paris: IRD, 2003), 53–75.

[3] Frederick C. Selous, *A Hunter's Wanderings in Africa, Being a Narrative of Nine Years spent amongst the Game of the Far Interior of South Africa* (London: Richard Bentley & Son. 1881), 392; Kabunda Kayongo, 'The Social Evolution of the Barotse Society in the Nineteenth Century' (Unpublished MA diss., University of Lund, 1983), 47–49; Paul Shaw, 'The Desiccation of Lake Ngami: An Historical Perspective,' *Geographical Journal*, 151, 3 (1985), 318–26; Lawrence S. Flint, 'Historical Constructions of Postcolonial Citizenship and Subjectivity: The Case of the Lozi Peoples of Southern Central Africa' (Unpublished PhD diss., University of Birmingham, 2004), 29.

The desiccation hypothesis has been reinforced by a tendency to investigate the incidence of rainfall through the lens of drought. Both Joseph Miller and Jill Dias portray a fragile ecosystem, constantly threatened by the failure of the rains, even though they acknowledge in principle that excessive precipitation could be just as disruptive.[4] In a similar vein, a University of London research project of the early 1970s had 'drought' in its title.[5] And it is undeniable that Africans in the region keenly sought out rain-makers.[6] However, the historical record suggests that 'fat' and 'lean' years were relatively balanced over the *longue durée*, as in the famous portrayal of Ancient Egypt in chapter 41 of the Book of Genesis.

Precipitation in the Upper Zambezi Basin is chiefly determined by the annual migration of the Intertropical Convergence Zone (ITCZ), broadly correlating with the position of the sun. The ITCZ moves down from equatorial latitudes in the hot season, bringing rain, and retreats again as the weather turns colder.[7] A northwest to southeast configuration of the ICTZ during the rainy season, more marked than in South America or Australia, may reflect topographical features. Another theory is that the Botswana High, hovering over central Namibia and western Botswana, may hinder precipitation from penetrating into the Kalahari Basin.[8]

[4] Joseph C. Miller, 'The Significance of Drought, Disease and Famine in the Agriculturally Marginal Zone of West-Central Africa,' *Journal of African History*, 23, 1 (1982), 17–61; Joseph C. Miller, *Way of Death: Merchant Capitalism and the Angolan Slave Trade, 1730–1830* (London: James Currey, 1988); Jill R. Dias, 'Famine and Disease in the History of Angola c. 1830–1930,' *Journal of African History*, 22, 3, (1981), 349–78.

[5] No general publication resulted. For the purposes of this chapter, see: A.D. Roberts, 'A Note on Drought, Flood, Famine, and Pestilence in and Around Zambia,' *African History Seminar* (SOAS, 15 May 1974); A. Livneh, 'Some Notes on Drought, Famine, and Pestilence in Rhodesia,' *African History Seminar* (SOAS, 12 June 1974); William G. Clarence-Smith, 'Drought in Southern Angola and Northern Namibia, 1837–1945,' *African History Seminar* (SOAS, 12 June 1974); William G. Clarence-Smith, 'Climatic Variations and Natural Disasters in Barotseland, 1847–1907,' *History Staff Seminars* (University of Zambia, 1 June 1977).

[6] For example, see: Gwyn Prins, *The Hidden Hippopotamus: Reappraisal in African History; the Early Colonial Experience in Western Zambia* (Cambridge: Cambridge University Press, 1980), 128; Meredith McKittrick, 'Making Rain, Making Maps: Competing Geographies of Water and Power in Southwestern Africa,' *Journal of African History*, 58, 2 (2017), 187–212.

[7] Peter Hutchinson, *The Climate of Zambia* (Lusaka: Zambia Geographical Association, 1974), 4, 9, 18.

[8] Nkosinathi G. Xulu, Hector Chikoree, Mary-Jane M. Bopape, and Nthaduleni S. Nethengwe, 'Climatology of the Mascarene High and Its Influence on Weather and

Annual variations in this pattern of rainfall are mostly affected by conditions in the Indian Ocean. This results from the interplay between the Mascarene (Indian Ocean Subtropical) High, and the Angolan Low, which forms around the Angolan Highlands during the hot season. When a deep Angolan Low coincides with an enhanced Mascarene High, trade winds blow forcefully from the Indian Ocean, and precipitation increases markedly.[9] Indeed, the remains of tropical cyclones may sweep far inland from December to February.[10]

Changing sea surface temperatures affect the location and intensity of the Mascarene High, bringing into play the vagaries of the Indian Ocean Dipole (IOD), which correlate to some degree with those of the better studied El Niño Southern Oscillation (ENSO). As in areas to the southeast of the Upper Zambezi Basin, and contrary to Equatorial Eastern Africa, strong El Niños seem broadly to coincide with dry years in the Upper Zambezi Basin. Conversely, marked La Niñas are associated with wet years, even though causal connections remain to be fully understood.[11]

By way of contrast, rainfall in the Upper Zambezi Basin is hardly at all affected by conditions to the west, because the 'South Atlantic High essentially exports moisture away from the landmass.'[12] The broad and powerful Benguela Current, running north from South Africa to the equator, has a cold-wind-induced upwelling along its eastern edge, so that desert and semi-desert conditions prevail in the coastal zone of Namibia and Angola. Precipitation very rarely reaches the continent from the west, and, when it does, the high escarpment usually causes much of it to fall in the narrow coastal strip.[13] That said, the climate of the Upper Zambezi

Climate over Southern Africa,' *Climate*, 8, 86 (2020), 6; Chris J.C. Reason and Sandi Smart, 'Tropical South East Atlantic Warm Events and Associated Rainfall Anomalies over Southern Africa,' *Frontiers in Environmental Science*, 3 (2015), 10.

[9] Xulu et al., 'Climatology of the Mascarene High,' 1–11.

[10] Hutchinson, *The Climate of Zambia*, 9, 18–19.

[11] Joëlle L. Gergis and Anthony M. Fowler, 'A History of ENSO Events Since A.D. 1525: Implications for Future Climate Change,' *Climatic Change*, 92, 3 (2009), 343–87, esp. Fig. 5; Matthew D. Therrell, David W. Stahle, Lydia P. Ries, and Herman H. Shugart, 'Tree-Ring Reconstructed Rainfall Variability in Zimbabwe,' *Climate Dynamics*, 26, (2006), 677–85; Philip Gooding, email, 11 Mar. 2021.

[12] Reason and Smart, 'Tropical South East Atlantic Warm Events,' 2.

[13] Ibid., *passim*; Brian J. Huntley, 'Angola in Outline: Physiography, Climate, and Patterns of Biodiversity,' in *Biodiversity of Angola: Science and Conservation, a Modern*

Basin is not entirely exempt from influences emanating from the South Atlantic, as the interaction between the South Atlantic High and the Angolan Low plays some role in drawing down the Congo Air Boundary, part of the ITCZ.[14]

The Upper Zambezi Basin: Floods and Rainfall

Extending over more than 3.5 million square kilometres, the Upper Zambezi Basin is largely smothered in deep layers of infertile wind-blown Kalahari sands, which allow for little water to run off. Within this sparsely populated zone, limited areas have attracted the lion's share of scholarly attention. The main Bulozi, or Barotseland, floodplain, a former lake that contains much of the scarce population, lies at the heart of today's Western Province of Zambia. Narrowly conceived, the floodplain extends over some 5000 square kilometres, but that roughly doubles if inundated zones along the main tributaries are included. Flooded zones penetrate particularly deeply into the west because the margins of the plain are generally lower on that side.[15] To the north are the seasonally flooded areas of Luvale country and Angola's Cazombo Salient, although this area is not considered here. Downriver, the smaller Caprivi floodplain straddles the Western and Southern Provinces of Zambia, Namibia's eastern Caprivi Strip, and a small part of northern Botswana.[16] Rulers of Bulozi directly administered the main and Caprivi floodplains before

Synthesis, eds. Brian J. Huntley; Vladimir Russo; Fernanda Lages; and Nuno Ferrand (Cham: Springer Open, 2019), 26–28; Angola, *O Clima de Angola* (Luanda: Serviços Meteorológicos, 1955).

[14] Hutchinson, *The Climate of Zambia*, 4, 9, 18.

[15] See, among many works: Max Gluckman, *The Economy of the Central Barotse Plain* (Livingstone: Rhodes-Livingstone Institute, 1941); Eugene Hermitte, 'An Economic History of Barotseland, 1800–1940' (Unpublished PhD diss., Northwestern University, 1974), Prins, *The Hidden Hippopotamus*; Jack Hogan, 'The Ends of Slavery in Barotseland, Western Zambia, c.1800–1925' (Unpublished PhD diss., University of Kent, 2014); John Mendelsohn and B. Weber, *Atlas e Perfil do Moxico, Angola—An Atlas and Profile of Moxico, Angola* (Windhoek: Raison, 2015).

[16] Sakwiba Muyunda, 'Agricultural Change in Sesheke District of Western Zambia, 1899–1964' (Unpublished MA diss., University of Zambia, 2019).

colonial times, while raiding and collecting tribute in the thinly inhabited surrounds.[17] Further to the southeast, spectacularly demarcating the border between the basins of the Upper and Middle Zambezi, lie the Victoria Falls, or Mosi-oa-Tunya, 'the smoke that thunders'[18] (Fig. 5.1).

There is a certain natural balancing between rainfall and floods in Bulozi. Abundant local precipitation stimulates agricultural output, but outsized inundations are destructive. Historically, people were most wary of early, sudden, high, and late-falling floods. A high flood drowned crops and destroyed buildings and trees, especially on the mounds that dotted the main floodplain, which were partly natural and partly made by humans. A lengthy flood reduced the growing season. Moreover, the condition of cattle deteriorated due to extended grazing on poor pastures in the wooded savanna, where they were also vulnerable to tsetse-borne trypanosomes and the predation of lions. A sudden and early flood swept away immature crops in fertile depressions and left people stranded on mounds, with insufficient canoes to reach the margins of the plain. That said, such floods also trapped wild animals, especially antelopes, on high ground, yielding a bountiful harvest for hunters.

Conversely, late, low, gradual, and short floods came with their own distinct consequences. They insufficiently renewed gardens, pastures, and fishing sites with water and silt. They also reduced the employment of canoes, which were a rapid and effective means of transport. And they made for poor hunting, as fewer wild animals were trapped by the floodwaters on high ground, although severe droughts might lead to concentrations of wild animals at water points.

[17] Some historical overviews include: Gerald L. Caplan, *The Elites of Barotseland, 1878–1969: A Political History of Zambia's Western Province* (London: Christopher Hurst & Co., 1970); Mutumba Mainga, *Bulozi Under the Luyana Kings: Political Evolution and State Formation in Pre-colonial Zambia* (London: Longman. 1973); Hermitte, 'An Economic History of Barotseland'; William G. Clarence-Smith, 'Slaves, Commoners and Landlords in Bulozi, c. 1875 to 1906,' *Journal of African History*, 20, 2 (1979), 219–34; Prins, *The Hidden Hippopotamus*; Kayongo 'The Social Evolution of the Barotse Society;' Flint, 'Historical Constructions of Postcolonial Citizenship;' Hogan 'The Ends of Slavery in Barotseland.'

[18] Brett Hilton-Barber and Lee R. Berger, 'Victoria Falls Seasonal Weather Calendar' (2010): http://www.siyabona.com/explore-victoria-falls-seasonal-weather-calendar.html [Accessed: 20 Aug. 2020].

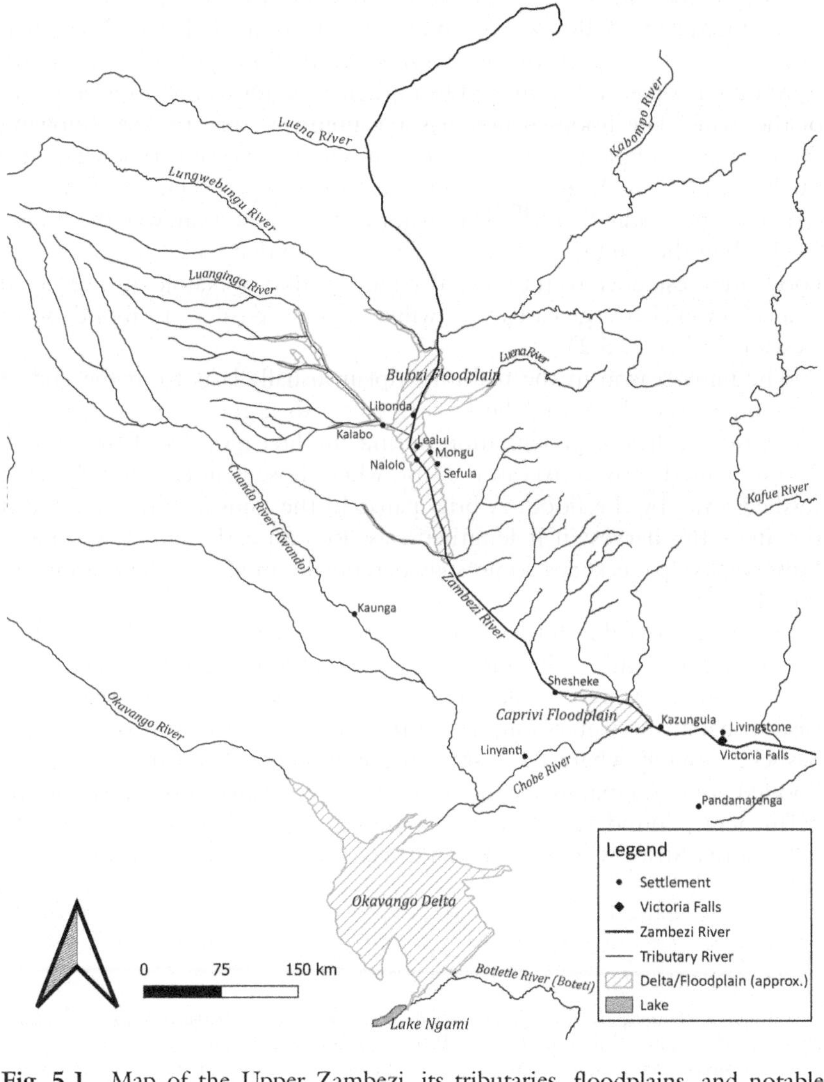

Fig. 5.1 Map of the Upper Zambezi, its tributaries, floodplains, and notable settlements. Drawn by Philip Gooding

As described in the first half of the twentieth century, waters in the main floodplain of Bulozi rose quite fast from around late December, causing cattle to depart for the margins. As the flood peaked in February to March, people in turn moved from their mounds to the forested edges of the plain. The flood turned at some point in April to May, allowing the king and people to return to their mounds, followed somewhat later by their cattle. By August, the plain was largely dry, though shallow lakes remained here and there.[19] The major event of the year was the *kuomboka*, when the king's *nalikwanda*, a large decorated barge, set off for his flood-time residence to the east. The female ruler of Nalolo also engaged in a ceremonial voyage eastwards, while the female ruler of Libonda went westwards[20] (Fig. 5.2).

The rainy season in the main floodplain usually lasts from the end of September to early April. The Lozi word for drought, *linanga*, generally refers to insufficient precipitation during the first months of the season. Crops planted early shrivel in the sun, while those planted late risk being washed away by the flood. A brief pause in the rains is common, and it threatens the harvest if it lengthens, as does an early end to the rains. However, seepage keeps some soils perennially moist on the margins of the plain.[21]

In the Caprivi floodplain, where there are neither mounds nor ritual journeys, the Zambezi begins to rise in January or February. The flood spreads out in March, peaks in June, and ebbs in July. On the left or eastern bank, the waters inundate a thin strip downstream from modern Sesheke, notably where three small tributaries enter from the north. The flooded zone lies mainly on the right or western bank, today forming the permanently inundated Linyanti Swamps and the eastern tip of Namibia's Caprivi Strip, where the Kwando contributes some additional water.

[19] Eugène Béguin, *Les Ma-Rotse: Étude géographique et ethnographique du Haut-Zambèze* (Lausanne: Librairie Benda, 1903), 37–38; Gluckman, *The Economy of the Central Barotse Plain*, 53–66; Hermitte, 'An Economic History of Barotseland,' 88–93; Flint, 'Historical Constructions of Postcolonial Citizenship,' 29.

[20] Lawrence S. Flint, 'Contradictions and Challenges in Representing the Past: the Kuomboka Festival of Western Zambia,' *Journal of Southern African Studies*, 32, 4 (2006), 701–17; Prins, *The Hidden Hippopotamus*, 115–23.

[21] Hermitte 'An Economic History of Barotseland,' 73–88.

Fig. 5.2 Arrival of the ***nalikwanda***. François Coillard, Arrival of the Lewanika's Nalikwanda (n.d.). Reproduced with the permission of: Défap-service protestant de mission, Paris. The original image is viewable at: https://catalogue.defap-bibliotheque.fr/stock/Arrivee-de-la-Nalikwanda-barque-royale-du-Litunga-roi-des-Lozi-Lewanika;id=8516.jpg

In addition, a little overspill from the Okavango (Cubango) occasionally flows along the Selinda or Magwekana Spillway.[22]

Rainfall is less abundant in the south than in the main floodplain, and it is subject to lengthier interruptions. The wet season usually begins at the end of November and often ends in March. It can become quite cold in the dry season. Even manioc (cassava), a reserve crop against hunger, is threatened by a combination of severe drought and harsh frost.[23]

[22] Muyunda, 'Agricultural Change in Sesheke District,' 32–34, 76.

[23] Ibid.; Hermitte 'An Economic History of Barotseland,' 74–76, 81–85, 150 (fn. 9), 342.

The volume of water hurtling over the Victoria Falls may provide further indications of rainfall in the catchment area, although only a small proportion of water reaches the falls. Most of it vanishes long before, as the two floodplains retain moisture like giant sponges, accentuating percolation and evaporation.[24] The flow rises at the Victoria Falls from December to January, peaks in April to May, and descends thereafter to a minimum in October to November.[25]

The correlation between floods and catchment precipitation is high, albeit not absolute.[26] Abrupt yearly alternations show that rainfall to the west and north mainly governs the flood. In 1893–1896, for example, an extremely low flood was sandwiched between two very high ones.[27] Discrepancies arise, however, due to the height of the water table, which is affected by previous rainfall and floods, and by unexplained changes in subterranean temperatures. Other natural factors include vegetative cycles and fluctuations in numbers of hippopotamuses, which graze vegetation and form water channels. The main human contribution is clearing forest in the catchment area or encouraging forest re-growth there. In addition, people build and maintain, or neglect, fishing bunds, canals, ditches, roads, and causeways in and around the plain.[28]

[24] Lawrence S. Flint, 'Socio-ecological Vulnerability and Resilience in an Arena of Rapid Environmental Change: Community Adaptation to Climate Variability in the Upper Zambezi Valley Floodplain,' *Working Paper of the Research Institute for Humanities and Nature, Kyoto* (2008): http://www.chikyu.ac.jp/resilience/files/WorkingPaper/WP2008-004.Flint.pdf [Accessed: 10 Aug. 2020].

[25] Hilton-Barber and Berger 'Victoria Falls Seasonal Weather Calendar.'

[26] Henry Zimba, Banda Kawawa, Anthony Chabala, Wilson Phiri, Peter Selsam, Markus Meinhardt, and Imasiku Nyambe, 'Assessment of Trends in Inundation Extent in the Barotse Floodplain, Upper Zambezi River Basin: A Remote Sensing-Based Approach,' *Journal of Hydrology: Regional Studies*, 15 (2018), 149–70; J.H. Chaplin, 'On Some Aspects of Rainfall in Northern Rhodesia,' *Northern Rhodesia Journal*, 2, 6 (1954–1955), 17–18.

[27] *Journal des Missions Évangéliques*, 69 (1894), 521–23; 70 (1895), 389–91; 71 (1896), 422–23.

[28] Murray Armor, 'Notes on the Abnormal Flood Conditions Experienced in Barotseland During Recent Years' (Unpublished paper by District Commissioner of Kalabo, n.d.). See also: Hogan, 'The Ends of Slavery in Barotseland,' 50.

Tentative Evidence on Floods from Angola, 1680s–1790s

Portuguese observers mentioned climatic factors in northern Angola from the 1550s, but only from the late seventeenth century did they begin to refer to conditions in and around the South Central (Bihe, or Bié) Highlands. Rising up to the culminating point in Southwestern Africa, these highlands attract copious rainfall, mainly from the northeast, and have a high run off rate.[29] Providing clues for the amounts of rainfall are small rivers that flood the coastal oases and the Ovambo Plain, as well as the great Cuanza (Kwanza) River running northwest. Modern scholars thus dub this high ground the 'water tower' of Southwestern Africa, whereas, more poetically, local Ovimbundu people call it 'the mother of all the waters'[30] (Fig. 5.3).

Rivers originating in this part of Angola probably supply the majority of the water that ends up in the floodplains of the Upper Zambezi.[31] Indeed, the Lungwebungu River, the largest tributary in the main floodplain, should perhaps be recognised as the Zambezi's mainstream.[32] J.H. Chaplin, who studied water flows in the early 1950s, was acutely aware of the significance of rivers coming from Angola.[33] Later writers affirm that the northern headwaters provide the majority of the floodwaters, but without presenting statistical evidence.[34] In addition, within these

[29] Angola, *O Clima de Angola*.

[30] Huntley, 'Angola in Outline,' 22–28; Judith Listowel, *The Other Livingstone* (Lewes: Julian Friedmann, 1974), 116, reporting Lázló Magyar's observations from the 1850s.

[31] For partial figures, see: Kawawa Banda, emails, 8 and 11 Mar. 2021; Innocent C. Chomba, Victoria Ngwnya, and Mulema Mataa, *First Field Data Campaign on the Barotse Floodplain: WASP/WeMAST Project Report* (Lusaka: School of Mines, University of Zambia, 2019); Zimba, 'Assessment of Trends in Inundation Extent;' Elias A. Mohammed, 'Hydrodynamics Flood Modelling in Barotse Floodplain, Zambia: Effectiveness of Digital Elevation Models' (Unpublished MSc diss., Abba Minch University, Ethiopia, 2015), 25, 27; Eric Deneut, Charles K. Chileya, and Christophe Nativel, *Environmental and Social Impact Assessment for the Improved Use of Priority Traditional Canals in the Barotse Sub-Basin of the Zambezi* (Sunningdale: NIRAS, 2014), 53. For rivers of Southeast Angola, see: John Mendelsohn and Antonio Martins, 'River Catchments and Development Prospects in South-Eastern Angola' (2018), 32–34.

[32] John Mendelsohn, email, 30 Sept. 2020.

[33] Chaplin, 'On Some Aspects of Rainfall,' 2, 6 (1954–1955), 22, fn. 2.

[34] Hermitte, 'An Economic History of Barotseland,' 88–89; Prins, *The Hidden Hippopotamus*, 20; Hogan, 'The Ends of Slavery in Barotseland,' 50.

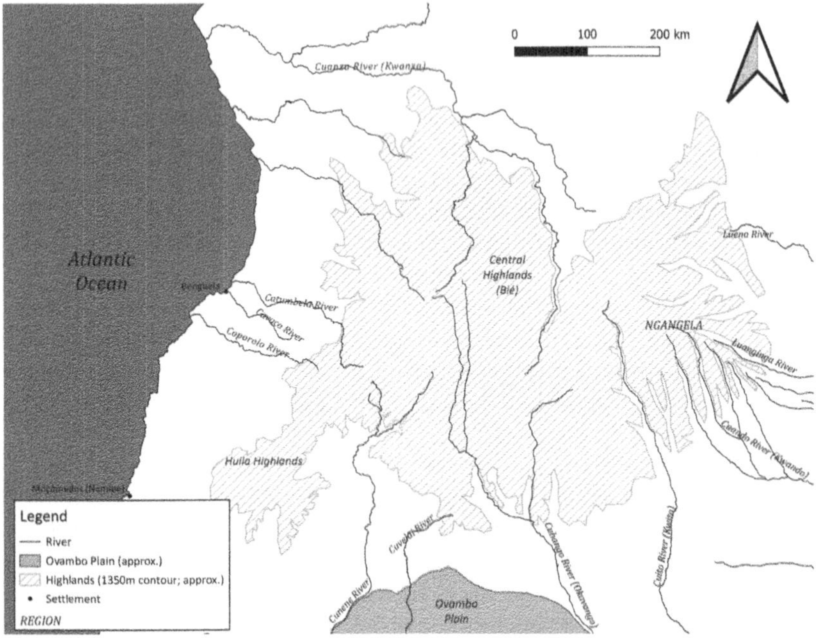

Fig. 5.3 Map of Angolan Highlands and Rivers. Drawn by Philip Gooding

northern headwaters, the understudied northwestern or Angolan Luena may supply more water than the Zambian Kabompo entering from the northeast.[35]

From the 1680s to the 1730s, the Portuguese, based in the port of Benguela, reported low rainfall in South Central Angola. In the early 1680s, a severe drought desolated the Ngangela lands, to the east of the highlands. Another dry episode, lasting from 1713 to 1728 in Luanda, was felt in the south. During a patchy set of droughts from 1731 to 1741, the governor sought food in Brazil in 1735 for the 'whole kingdom.'[36]

[35] José Pedro Gamito de Saldanha Matos, 'Hydraulic-Hydrologic Model for the Zambezi River, Using Satellite Data and Artificial Intelligence Techniques' (Unpublished PhD diss., École Polytechique Fédérale de Lausanne 2014), Appendices, A-215–A-217.

[36] Miller, 'The Significance of Drought,' 21, 46–51; Miller, *Way of Death*, 150 (Ngangela).

There followed some four decades of average to high rainfall, from the mid-1740s through to the early 1780s, and Joseph Miller specifically mentions the Benguela area when referring to the 'rainy 1770s.'[37] Furthermore, governors of Benguela encouraged the cultivation of food crops in the oasis of the Catumbela River from the late 1760s, suggesting that annual floods were propitious.[38]

Although Lozi oral traditions mostly treat political and cosmological matters and are chronologically hard to interpret, some may refer back to these wet decades of the eighteenth century. Queen Mbuywamwamba, the founding ancestress, reportedly experienced disastrously high floods during her reign, convincing the elite to replace her with a male ruler. Other traditions recall that rainfall and floods were high in these foundational times.[39] Queen Mbuywamwamba perhaps reigned around the mid-eighteenth century, in view of the short reigns of some of her successors, although these accounts may refer to the seventeenth century.[40] While such traditions may simply seek to legitimise men's political domination, they may also preserve memories of climatic conditions.

The decade from 1785 witnessed a return to dry conditions in Angola, culminating in the worst famine of the century from 1790 to 1793, which affected Benguela.[41] Similarly, there was a severe drought in Namibia in 1792.[42] There are references to food shortages in Kazembe and the

[37] Miller, 'The Significance of Drought,' 21, 46–51; Miller, *Way of Death*, 701 (Benguela). See also: Elias Alexandre da Silva Corrêa, *História de Angola* (Lisbon: Agência Geral das Colônias.) I, 111, and II, 13–14. This latter text was likely composed in the 1790s.

[38] José C. Curto, 'Women Along the Catumbela River, 1797: Land Ownership, Agricultural Production, Labour and Trade,' *Canadian Journal of African Studies*, 54, 3 (2020), 373–93. The author kindly let me see a pre-publication file.

[39] Flint, 'Historical Constructions of Postcolonial Citizenship,' 29.

[40] Lawrence Flint, emails 17–18 December 2020. For a list of rulers, see: Ibid., 278; Mainga, *Bulozi under the Luyana Kings*, 215.

[41] Miller, 'The Significance of Drought,' 21, 51–54. See also: Corrêa, *História de Angola*, I, 50–51 (fn. 3), II, 111, 118.

[42] J. van Reenen, *Report of the Drought Investigation Commission of South West Africa, 1924* (Pretoria: Weather Bureau, 1949), 23.

Lower Zambezi at this time, though they do not specifically mention the lack of rain.[43] In a wider context, volcanic eruptions and El Niños contributed to global climatic volatility in the 1780s and 1790s.[44]

Out of Kilter with the Wider Region, Mid-1790s to Mid-1840s

The Upper Zambezi region may have formed an exception to a very dry period in southern and eastern Africa. Some scholars claim that drought was a significant factor in the emergence and spread of the *mfecane*, the wars that radiated out from the Zulu heartlands over Southern Africa from the late 1810s to around the 1840s.[45] According to Nicholson, this dry period culminated in a very unusual decade-long drought across almost the whole of Africa, from 1825 to 1834, although she also noted there was a limited exception in the northwestern coastal zone of Angola, where rainfall was higher than usual.[46]

The Angolan exception may in fact have been more extensive, although Miller's evidence is equivocal.[47] For South Central Angola, the most significant pointer was that plantations in coastal oases flourished from the 1790s to the early 1840s.[48] A heavy flood damaged the Catumbela saltpans in October 1813, and work began in 1824 to alter the course

[43] Roberts, 'A Note on Drought,' 1–2. For Kazembe, see: Richard F. Burton (ed.), *The Lands of Cazembe: Lacerda's journey to Cazembe in 1798; also journey of the pombeiros, P.J. Baptista and Amaro José, across Africa from Angola to Tette on the Zambeze; and a résumé of the journey of MM. Monteiro and Gamitto* (London: John Murray, 1873), 78, 92–93.

[44] See: Richard H. Grove, 'The Great El Niño of 1789–93 and Its Global Consequences: Reconstructing an Extreme Climate Event in World Environmental History,' *The Medieval History Journal*, 10, 1–2 (2006), 75–98; Vinita Damodaran, Rob Allan, Astrid E.J. Ogilvie, Gaston R. Demarée, Joëlle Gergis, Takehiko Mikami, Alan Mikahil, Sharon E. Nicholson, Stefan Norrgård, and James Hamilton, 'The 1780s: Global Climate Anomalies, Floods, Droughts, and Famines,' in *The Palgrave Handbook of Climate History*, eds. Sam White, Christian Pfister, and Franz Mauelshagen (London: Palgrave Macmillan, 2018), 517–50.

[45] Michael Garstang, Anthony D. Coleman, and Matthew Therrell, 'Climate and the Mfecane,' *South African Journal of Science*, 110, 5–6 (2014), 1–6.

[46] Nicholson, 'Spatial Teleconnections in African Rainfall,' 1841, 1846.

[47] Miller, 'The Significance of Drought,' 21.

[48] Curto, 'Women along the Catumbela River,' 1–5. See also: Antonio de Carvalho e Menezes, *Memoria Geografica e Politica das Possessões Portuguezas n'Affrica Occidental, que*

of the smaller Cavaco River because it was causing frequent and costly inundations in the town of Benguela.[49] A high flood also changed the course of the Coporolo River, to the south of Benguela, in 1832–1833.[50] There was even a proposal, in 1836, to move the capital of the south to the flourishing Catumbela oasis.[51] Extraordinary torrential rains fell in the arid town of Benguela, causing damage in 1796–1797, 1821–1822, 1829–1930, and 1832–1833, although this probably reflected unusual conditions in the South Atlantic.[52]

There were certainly dry episodes in Southern Angola in these years, but they were brief, for example in 1807–1808, 1816–1817, and 1828–1829.[53] There was a further report of a four-year drought in the Huila Highlands from 1837 to 1841. Confusingly, however, the author simultaneously states that there were abundant floods in the neighbouring Moçâmedes (Namibe) coastal plain during the same period.[54]

The Southern Angolan exception probably extended further inland, over the southern stretches of the Upper Zambezi Basin, according to a tree-ring study. The authors of this study collected specimens of

diz respeito aos Reinos de Angola, Benguela, e suas dependencias (Lisbon: Typografia Carvalhense, 1834), 40–41; Tito Omboni, *Viaggi nell'Africa Occidentale* (Milan: Stabilimento Civelli e Comp., 1845), 80; G. Tams, *Visit to the Portuguese Possessions in South-Western Africa* (London: T. C. Newby, 1845), I, 80, 94, 101, 193–94, 199–200, 205–7, 214, and II, 78; Joaquim António de Carvalho e Menezes, *Demonstração Geográphica e Política do Territorio Portuguez na Guiné Inferior, que abrange o Reino de Angola, Benguella, e suas dependencias; causas da sua decadencia e atrasamento, suas conhecidas producções, e os meios que se podem applicar para o seu melhoramento e utilidade geral da nação* (Rio de Janeiro: Typ. Clássica de F. A. de Almeida, 1848), 80–81.

49 Ralph Delgado, *A Famosa e Histórica Benguela: Catálogo dos governadores, 1779–1940* (Lisbon: Edições Cosmos, 1940), 56, 91.

50 Francisco Xavier Lopes, 'O Dombe Grande da Quisamba,' *Annaes do Conselho Utramarino, Parte Não-Oficial, janeiro de 1859 a dezembro de 1861*, 2 (1867), 180.

51 José Joaquim Lopes de Lima and Francisco Maria Bordalo, *Ensaios Sobre Statística das Possessões Portuguesas na África Occidental e Oriental, na Ásia Occidental, na China, e na Oceania* (Lisbon: Imprensa Nacional, 1844–1862), III, Part 2, 40.

52 Delgado, *A Famosa e Histórica Benguela*, 27–28.

53 Ralph Delgado, *Ao Sul do Cuanza: Ocupação e aproveitamento do antigo reino de Benguela, 1483–1942* (Lisbon: author's edition, 1944), 285, 588–89; Miller, 'The significance of drought,' 57.

54 João Francisco Garcia, 'Explorações no sertão de Benguella: Derrota que fez o Tenente de Artilheria João Francisco Garcia, commandante do novo estabelecimento da Bahia de Mossâmedes,' *Annaes Marítimos e Coloniaes*, 4, 6 (1844), 240, 243–45.

mukwa (*Pterocarpus angolensis*) from three places in the western point of Zimbabwe, plus an adjacent site in northern Botswana, below the Caprivi floodplain. Correlation with material from twentieth-century rain-gauges is far from exact, droughts are better revealed than wet phases, and the small sample for early decades can only be indicative. Bearing these caveats in mind, the dendrochronology suggests that rainfall was somewhat above average from 1796 to 1810, followed by a slightly dry period from 1811 to 1824. Then, from 1825 to 1839, there was a phase of almost consistently high rainfall, including some exceptionally wet years. This was in turn followed by drier conditions in the early 1840s.[55]

Personal testimonies from the southern marches of Bulozi may reinforce this hypothesis, though the chronology is problematic. Locals told European visitors in the middle of the century that they remembered Lake Ngami's waters as being much higher, and the flow of the Botletle (Boteti) River out of the Okavango Delta as being much larger and more regular. Sharon Nicholson places such memories before 1820, and yet she states that one informant, a chief, was in office from about 1830 to 1840.[56] James Chapman, one of her main sources, wrote in 1853 that Lake Ngami had grown smaller over the past twenty years or so, which would mean since the early 1830s.[57] However, in 1862, Chapman declared that the main 'desiccation' of the area had occurred over the past ten years.[58] In 1879, Frederick Selous collected African testimonies recalling high floodwaters in the times of Sibitwane, who ruled Bulozi from the late 1830s to 1851.[59] It is thus possible that accounts of falling lake and river levels date from the 1840s, which would concur with the dendrochronology. That said, the changing density of reed growth may have affected water levels.[60]

[55] Therrell, Stahle, Ries, and Shugart, 'Tree-Ring Reconstructed Rainfall Variability,' Fig. 5a. The raw figures are at: https://www.ncdc.noaa.gov/paleo-search/study/6297 [Accessed: 1 Apr. 2021].

[56] Sharon E. Nicholson, 'The Nature of Rainfall Variability over Africa on Time Scales of Decades to Millennia,' *Global and Planetary Change*, 26 (2000), 152.

[57] James Chapman, *Travels in the Interior of South Africa, 1849–1863* (London: Bell and Daldy, 1868), I, 203.

[58] Ibid., II, 61–64.

[59] Selous, *A Hunter's Wanderings*, 392. The Start Date for Sibitwane's Reign is Uncertain.

[60] Shaw, 'The Desiccation of Lake Ngami,' 318–26.

If the Upper Zambezi Basin remained free from the crippling drought that gripped most of Southern and Eastern Africa from the mid-1820s to the mid-1830s, it would help to explain why the Kololo, a group of Sotho-speakers from far to the south, were attracted to the region. Mutumba Mainga states that the newcomers, from a malaria-free area, encountered health problems, which could have reflected wet conditions.[61] In contrast, scholars have generally suggested that the Kololo were taking refuge from powerful Ndebele enemies to the southeast, who pursued and harried them. Unfortunately, the chronology remains stubbornly obscure. Kabunda Kayongo notes that writers have tentatively dated the arrival of the Kololo, and their conquest of Bulozi, between 1824 and 1840, with the late 1830s favoured.[62]

Differing Experiences in North and South, Mid-1840s to Mid-1860s

Direct written references to rainfall in Bulozi appear from the mid-1840s, although reports are sporadic. Antonio Ferreira da Silva Porto, based in the South Central Highlands of Angola, began to tap the ivory resources of the Upper Zambezi at this time, with connections to the Portuguese in Mozambique. His copious writings, derived partly from reports received from his African employees, were later partially published, and complete publication is expected.[63] The Hungarian traveller, Lázló

[61] Mainga, *Bulozi Under the Luyana Kings*, 69. Derived from: David Livingstone, *Livingstone's Private Journals, 1851–1853*, I. Schapera (London: Chatto and Windus), 163–64.

[62] Kayongo, 'The Social Evolution of the Barotse Society,' 55.

[63] António F. Ferreira da Silva Porto, *Silva Porto e Livingstone: Manuscripto de Silva Porto encontrado no seu espólio* (Lisbon: Sociedade de Geographia de Lisboa, 1891); António F. Ferreira da Silva Porto, *Silva Porto e a Travessia do Continente Africano*, ed. G. de Sousa Dias (Lisbon: Agência Geral das Colonias, 1938); António F. Ferreira da Silva Porto, *Viagens e Apontamentos de um Portuense em África: Excerptos do 'diário' de António Francisco Ferreira da Silva Porto*, ed. G. de Sousa Dias (Lisbon: Agência Geral das Colonias, 1942); António F. Ferreira da Silva Porto, *Viagens e Apontamentos de um Portuense em África: diário de António Francisco Ferreira da Silva Porto*, ed. Maria-Emília Madeira Santos (Coimbra: Biblioteca Geral da Universidade de Coimbra, 1986); António F. Ferreira da Silva Porto, *The Lands of the Lui: The Upper Zambezi journals of António Francisco Ferreira da Silva Porto, 1847–1884*, ed. Jack Hogan and Ana Rita Amaral (Oxford: Oxford University Press, forthcoming). My thanks to Rita Amaral and Jack Hogan for extracts of their draft of the Portuguese text.

Magyar, explored the country around his base in the South Central Highlands for many years from 1849, although most of his writings sadly disappeared.[64] From 1849, missionaries, hunters, traders, and explorers arrived in Bulozi from the south and southwest. David Livingstone, the trailblazer for this latter group, was an indefatigable chronicler, and the first Westerner to visit the Victoria Falls.[65]

Natural phenomena influenced the timing and nature of the arrival of Europeans in Bulozi. A receding elephant frontier, caused by rising Western demand for ivory combined with increasingly lethal firearms, pulled traders ever further inland.[66] Visitors from the south, reliant on oxen and horses, had to learn how to circumvent dense belts of tsetse flies, which transmitted deadly trypanosomes to their animals, whereas caravans from Angola consisted essentially of head-porters.[67]

Reports by foreign visitors suggest that rainfall was at or above-average levels in the main floodplain in these years. Silva Porto wrote about floods occurring regularly from the mid-1840s to the mid-1860s.[68] During the flood of 1854–1855, Livingstone stated that 'the water approached nearer to an entire submergence of the whole valley, than has been known in the

[64] Ladislaus Magyar, *Reisen in Süd-Afrika in den Jahren 1849 bis 1857* (Pest: Lauffer & Stolp, 1859); Listowel, *The Other Livingstone.*

[65] David Livingstone, *Missionary Travels and Researches in South Africa* (London: John Murray, 1857); David Livingstone and Charles Livingstone, *Narrative of an Expedition to the Zambezi and Its Tributaries, and the Discovery of the Lakes Shirwa and Nyassa, 1858–1864* (London: John Murray, 1865); Francis Galton, *The Narrative of an Explorer in Tropical South Africa* (London: John Murray, 1853); Charles John Andersson, *Lake Ngami, or, Explorations and Discoveries during Four Years' Wanderings in the Wilds of South Western Africa* (London: Hurst and Blackett, 1856); Charles John Andersson, *The Okavango River: A Narrative of Travel, Exploration, and Adventure* (New York: Harper and Brothers, 1861); Thomas Baines, *Explorations in South-West Africa, being an Account of a Journey in the Years 1861 and 1862 from Walvisch Bay, on the Western Coast, to Lake Ngami and the Victoria Falls* (London: Longmans, Roberts and Green, 1864); Chapman, *Travels in the Interior*; W.E. Oswell, *William Cotton Oswell, Hunter and Explorer: The Story of His Life* (London: William Heinemann, 1900).

[66] Hogan, 'The Ends of Slavery in Barotseland,' 91–113.

[67] William G. Clarence-Smith, 'A Note on Tsetse Fly and Rinderpest in Barotseland, 1850s–1900s,' *History Staff Seminars* (University of Zambia, 1 June 1977).

[68] Silva Porto, *Silva Porto e Livingstone*, 43.

memory of man.'[69] And the flood of 1863–1864 was both high and very late to subside.[70]

This impression gains support from South Central Angola. Augusto Bastos, a long-term resident, held that the floods of the Catumbela River reached their zenith in the period from 1856 to 1864. They assured the agricultural prosperity of the oasis and even changed the course of the river.[71]

The Caprivi floodplain experienced divergent conditions towards the end of the period, with low floods and droughts prevailing in 1859–1862.[72] Furthermore, the tree-ring study implies that there was quite a severe drought in the region in the early 1860s.[73] And missionary correspondence reflected a dry period from 1858 to 1863 in Southern Africa, up to what is today northern Botswana.[74]

This divergence may have influenced the outcome of the 1864 revolution, which ended the Kololo regime. Bovine pleuropneumonia (lung sickness) raged among cattle from 1862 to 1864 in both floodplains.[75] However, the epizootic may have been more deadly in the south, the Kololo heartlands, due to drier conditions. Drought, led to the concentration of cattle at water points, which increased rates of infection.[76] That

[69] Livingstone, *Missionary Travels*, 495.

[70] Adolphe Jalla, 'History, Traditions and Legends of the Barotse Nation,' Livingstone Museum, Livingstone, Zambia (1916: Typescript of 1909 published edition, with later additions), 33; Silva Porto, *The Lands of the Lui*, IV, 8 May 1864.

[71] Augusto S. Bastos, *Monographia de Catumbella* (Lisbon: Tipografia Universal, 1912), 13–14.

[72] Livingstone and Livingstone, *Narrative of an Expedition*, 272; Chapman, *Travels in the Interior* I, 445, and II, 5, 61–64, 70, 304; Baines, *Explorations in South-West Africa*, 411–12.

[73] Therrell, Stahle, Ries, and Shugart, 'Tree-Ring Reconstructed Rainfall Variability,' Fig. 5a.

[74] Georgina H. Endfield and David J. Nash, 'Missionaries and Morals: Climatic Discourse in Nineteenth-Century Central Southern Africa,' *Annals of the Association of American Geographers*, 92, 4 (2002), 730.

[75] Silva Porto, *The Lands of the Lui*, IV, 14 May 1864; John Ford, *The Role of the Trypanosomiases in African Ecology* (Oxford: Clarendon Press, 1971), 337.

[76] For a South African example, see: J.B. Peires, *The Dead will Arise: Nongqawuse and the Great Xhosa Cattle-Killing Movement of 1856–7* (Johannesburg: Ravan Press, 1989).

said, Mainga stresses political factors, notably worsening relations between the Kololo and their subjects after the death of Sibitwane, as the main cause for upheaval at this time.[77]

An Atypically Calm Period, Mid-1860s to Late 1870s

What evidence there is for the 1870s suggests that the Upper Zambezi Basin escaped the momentous, and divergent, climatic events affecting its neighbours. Rainfall in equatorial eastern Africa was extremely heavy in the late 1870s, as indicated by both written reports and lake levels.[78] In contrast, abnormal drought wracked Northern Angola in the early 1870s, stretching into the latter part of the decade.[79] Southern Angola also experienced a very dry decade, and a three-year drought in Namibia culminated in a serious famine in 1879–1880.[80] Drought was also widespread in other parts of the IOW, especially in 1875–1878.[81]

By contrast, the weather was relatively uneventful in the main floodplain of the Upper Zambezi. Oral testimonies, written down by missionaries a little later, attest that the flood occurred every year, except in 1867–1868, when it was so low that king and people did not leave their mounds.[82] This exceptionally low inundation was probably the one that King Lewanika (b. 1842) remembered from his youth.[83] As for a late-falling flood in 1869–1870, it may have been the one noted in

[77] Mainga, *Bulozi Under the Luyana Kings*, 90.

[78] Sharon E. Nicholson, 'Historical and Modern Fluctuations of Lakes Tanganyika and Rukwa, and Their Relationship to Rainfall Variability,' *Climatic Change*, 41, 1 (1999), 53–71; Chapter by Gooding, this volume.

[79] Dias, 'Famine and Disease in the History of Angola,' 354.

[80] Raúl J. Candeias da Silva, 'Subsídios para a história da colonização do Distrito de Moçâmedes,' *Studia* (Lisbon), 32, 371–78; 33, 341–72; 34, 481–534; 35, 421–39; 36, 293–390 (1971–1973), *passim*; Harri Siiskonen, *Trade and Socioeconomic Change in Ovamboland* (Helsinki: SHS, 1990), 123.

[81] Deepti Singh, Richard Seager, Benjamin I. Cook, Mark Cane, Mingfang Ting, Edward Cook, and Mike Davis, 'Climate and the Global Famine of 1876–1878,' *Journal of Climate*, 31, 23 (2018), 9445–67; Chapters by Gooding and Williamson, this volume.

[82] Jalla, 'History, Traditions and Legends,' 35–42. See also: Jalla and Jalla, *Pionniers Parmi les Ma-Rotse*, 328–29. This may have been associated with El Niño. See: Chapter by Warren, this volume.

[83] Gervas Clay, *Your Friend Lewanika: The Life and Times of Lubosi Lewanika, Litunga of Barotseland, 1842–1916* (London: Chatto & Windus, 1968), 155.

traditions concerning the reign of Sipopa (r. 1864–1876).[84] Silva Porto confirms both these events, but his diaries then stop from 1871 to 1882.[85] However, Alexandre da Serpa Pinto's travel account suggests that the flood of 1877–1878 was high and slow to fall.[86]

Problems with rainfall caused tensions between the Lozi and outsiders in the main floodplain on at least one occasion. In mid-December 1869, the local rains were very late, despite a normal rise of the Zambezi floodwaters. Sipopa then ordered Silva Porto to tell his men not to employ sorcery to stop the rains. Fortunately, it rained abundantly soon after.[87]

As for the Caprivi floodplain, the tree-ring study indicates that the drought of the early 1860s receded in the latter part of the decade. The sample of wood is larger for the 1870s, when rainfall rose a little above average, albeit with a slight dip in the middle of the decade.[88] And Western visitors described nothing out of the ordinary in the mid-1870s.[89]

From Dry to Wet in the Early Colonial Period, 1880 to 1904

Coinciding with the ending of the Lozi civil wars in the early 1880s, the establishment of the Paris Evangelical Mission, a Protestant society, greatly expanded the quantity and quality of available information.[90] The missionaries even produced irregular rainfall statistics, although cattle

84 *Journal des Missions Évangéliques*, 77 (1902), II, 202.

85 Silva Porto, *The Lands of the Lui*, VIII; Silva Porto, *Silva Porto e Livingstone*, 42.

86 Alexandre de Serpa Pinto, *Como eu atravessei África do Atlántico ao Mar Indico: Viagem de Benguella à contra-costa, atravès regiões desconhecidas* (London: Sampson, Low, Marston, Searle, and Rivington, 1881), I, 282, 296, 328.

87 Silva Porto, *The Lands of the Lui*, VIII, 11 Oct. 1869, 18 Dec. 1869, 23 Dec. 1869.

88 Therrell, Stahle, Ries, and Shugart, 'Tree-Ring Reconstructed Rainfall Variability,' Fig. 5a.

89 Frederick C. Selous, 'Journeys into the Interior of South Central Africa,' *Proceedings of the Royal Geographical Society*, 2nd series, 3, 3 (1881), 172; Emil Holub, *Seven Years in South Africa: Travels, Researches, and Hunting Adventures, between the Diamond Fields and the Zambesi, 1872–1879* (London: Sampson, Low, Marston, Searle, and Rivington, 1881), 273–77.

90 *Journal des Missions Évangéliques*; *Nouvelles du Zambèze* (from 1898); Livingstone Museum, Manuscript Collection. The Paris archives of the society were closed in the late 1970s.

sometimes knocked over their gauges.[91] Competing missionaries supplied further information, notably the Spiritans and the Mission Philafricaine to the west, the Plymouth Bretheren to the north, and the Jesuits to the southeast.[92] In 1909, Father Edmund Goetz, a Jesuit, published a major study on rainfall in Central Africa, which included scattered statistics from Bulozi.[93] Explorers, traders, elephant hunters, and, from 1890, colonial officials, added to this growing presence of literate observers.[94]

After quite a wet start to the 1880s, a pronounced dry period set in from 1885 to 1893, marked by poor harvests and hunger. The flood almost failed altogether in 1886–1887 and again for two years in succession in 1888–1890. During these low floods, the king and people did not leave their mounds. South Central Angola also suffered from low floods from 1886 to 1893, causing a decline of oasis agriculture on the coast. In addition, there were long pauses in the rainy season in Bulozi's main floodplain, notably in January to February 1886–1887, December to January 1891–1892, and December to January 1892–1893. And overall rainfall at Sefula in 1888–1889 was nearly 10 inches below the twentieth-century yearly average of 38 inches. This engendered bitter popular memories of drought. In contrast, the Caprivi floodplain experienced somewhat higher rainfall than usual, especially in the late 1880s.[95]

[91] *Journal des Missions Évangéliques*, 61 (1886), 175.

[92] For the Spiritans, see: William G. Clarence-Smith, *Slaves, Peasants and Capitalists in Southern Angola, 1840–1926* (Cambridge: Cambridge University Press, 1979); For the Mission Philafricaine, see: Alida Chatelain, *Héli Chatelain, l'Ami de l'Angola, 1859–1908: Fondateur de la Mission Philafricaine, d'après sa correspondence* (Lausanne: Mission Philafricaine, 1918); For the Plymouth Bretheren, see: Frederick S. Arnot, *Missionary Travels in Central Africa* (London: Holness, 1914); For the Jesuits, see: Gelfand Michael (ed.), *Gubulawayo and Beyond: Letters and Journals of the Early Jesuit Missionaries to Zambesia, 1879–1887* (London: G. Chapman, 1968).

[93] E. Goetz, 'The Rainfall of Rhodesia,' *Proceedings of the Rhodesian Scientific Association*, 8, 3 (1909), 1–129.

[94] Caplan, *The Elites of Barotseland*.

[95] The main source is the *Journal des Missions Évangéliques*. For other Paris Mission materials, see: François Coillard, *Sur le Haut Zambèze: Voyages et travaux de mission* (Paris: Berger-Levrault 1899); Jalla, *Pionniers parmi les Ma-Rotse*; Jalla, 'History, Traditions and Legends;' E. Favre, *François Coillard, Missionaire au Zambèze* (Paris: Société des Missions Évangeliques, 1913), III; William Waddell's Journal in Cambridge University Library. For other missions, see: Frederick S. Arnot, *From Natal to the Upper Zambesi: First Year among the Barotsi*, 3rd ed. (Glasgow: The Publishing Office, 1884); Frederick S. Arnot, *Garenganze, or Seven Years' Pioneer Mission Work in Central Africa* (London:

Some blamed newly established missionaries for the failure of the rains. In early 1887, it was alleged that the missionaries had caused an eclipse of the sun, leading to a long pause in the rains.[96] Two years later, in south-eastern Angola, local Africans attacked the Cubango Spiritan missionaries for allegedly stopping the rains.[97] In April 1891, after two consecutive years of inundations too low to float canoes, Lozi royal counsellors chided the king for having moved his flood-time capital to be close to the mission station at Sefula.[98] And François Coillard, leader of the Paris Mission, was personally accused of sorcery in early December 1892, not only for stopping the rains but also for propagating smallpox.[99]

Hawkins, 1889); Arnot, *Missionary Travels*; Henri Depelchin and Charles Croonenberghs, *Trois Ans dans l'Afrique Australe, 1879–1881* (Brussels: P. Imbreghts, *1882–1883*); Joseph Spillmann (comp.), *Vom Cap zum Sambesi: Die Anfänge der Sambesimission, aus den Tagebüchern des P. Terörde, S.J., und aus den Berichten der andern Missionäre* (Freiburg im Breisgau: Herder'sche Verlagshandlung, 1882); Gelfand, *Gubulawayo and Beyond*. For others, see: Hermenegildo Capello and Roberto Ivens, *De Angola á Contra-Costa: Descripção de uma Viagem Atravez do Continente Africano* (Lisbon: Imprensa Nacional, 1886); Hermenegildo Capello and Roberto Ivens, *Diários da Viagem de Angola à Contra-Costa*, ed. Francisco de Assis de Oliveira Martins (Lisbon: Agência Geral das Colónias, 1951); [Henrique M. de] Paiva Couceiro, *Relatorio de Viagem entre Bailundo e as Terras do Mucusso* (Lisbon: Imprensa Nacional, 1892); Edward C. Tabler, *Trade and Travel in Early Barotseland: The Diaries of George Westbeech, 1885–1888, and Captain Norman MacLeod, 1875–1876; Illustrated with the Sketches of Lieutenant William Fairlie* (London: Chatto & Windus, 1963)—for Westbeech; Emil Holub, *Von der Capstadt ins Land der Maschukulumbe: Reisen im Südlichen Afrika in den Jahren 1883–1887* (Vienna: Alfred Hölder, 1890); Aurel Schulz and August Hammar, *The New Africa: A Journey Up the Chobe and Down the Okavango Rivers; A Record of Exploration and Sport* (London: William Heinemann, 1897); James Johnston, *Reality Versus Romance in South Central Africa* (London: Hodder and Stoughton, 1893); Lionel Dècle, *Three Years in Savage Africa* (London: Methuen & Co., 1898). This section also draws on: Interview with N. Simalumba, 27 Apr. 1977; Therrell, Stahle, Ries, and Shugart, 'Tree-Ring Reconstructed Rainfall Variability,' 677–85; Bastos, *Monographia de Catumbella*, 23, 50–51. The latter citation is for southern Angola.

96 Coillard, *Sur le Haut Zambèze*, 294–96.

97 Archives Générales de la Congrégation du Saint-Esprit, Chevilly-la-Rue, *Bulletin Général de la Congrégation du Saint-Esprit*, 15, 42 (1890), 651.

98 *Journal des Missions Évangéliques*, 66 (1891), 375.

99 *Journal des Missions Évangéliques*, 68 (1893), 254. The links between smallpox and drought are understudied for precolonial Africa, although a wider, more global view may suggest correlation between incidences. Although other factors are almost always at play, drought may encourage migration, for work and food, spreading outbreaks across larger areas and among more people. See: Chapters by Gooding and Williamson, this volume.

The weather then swung back to the opposite extreme from 1893 to 1904. Although the flood of 1894–1895 was feeble, four inundations were among the highest in living memory in the main plain, those of 1893–1894, 1895–1896, 1899–1900, and 1901–1902. They caused extensive damage and popular anguish. Soaked ground may have encouraged locusts to hatch, as swarms devastated the country from 1893 to 1900. There were also deadly spikes of malaria in the late 1890s and early 1900s. And yet, local rainfall was generally below average. The floods reflected rains in the catchment area, particularly in South Central Angola, which recorded several exceptionally wet years between 1894 and 1903.[100]

The blame for extreme climatic events now shifted away from missionaries, who were not held responsible for high floods, possibly reflecting a deepening process of conversion to Christianity. Africans even asked missionaries in the Caprivi floodplain to hold special services for rain, when pauses threatened the harvest in December 1901 and January 1905.[101]

Harsh conditions may have undermined African resistance to colonial annexation. In the words of Mutumba Mainga, the Lozi were 'frightened, sick, and famished,' and in no position to resist the demands of the British South Africa Company, with which they signed protectorate treaties in 1890 and 1898.[102] Human and animal diseases, as well as locusts, heightened Lozi distress in the 1890s, and their incidence was probably aggravated by climatic stresses.[103] However, Gwyn Prins argues that King Lewanika skilfully strengthened his political position by effectively overcoming natural misfortunes and thus turning them to his advantage.[104]

[100] For an overview, see: Bastos, *Monographia de Catumbella*, 51. See also: Alfredo [A. Freire] de Andrade, *Relatório da Viagem de Exploração Geographica no Districto de Benguela e Novo Redondo, 1898–1899* (Lisbon: Imprensa Nacional, 1902); Alexandre Malheiro, *Chronicas do Bihé* (Lisbon: Livraria Ferreira, 1903); Alfredo [A. Freire] de Andrade, *A Bacia Hydrographica do Rio Cuanza, desde a nascente á confluencia do Rio Gango* (Lisbon: Imprensa Nacional, 1905); Chatelain, *Héli Chatelain*; *Le Philafricain*.

[101] Livingstone Museum, MSS collection, 'Rapport, Sesheke 1901'; *Nouvelles du Zambèze*, 8, 2 (1905), 53.

[102] Mainga, *Bulozi under the Luyana Kings*, 194–95.

[103] Clarence-Smith, 'Climatic Variations and Natural Disasters.'

[104] Prins, *The Hidden Hippopotamus*, 60–70.

Enhanced Volatility in the Early Statistical Age, 1904 to 1919

Colonial administrators began to collate rainfall statistics regularly from 1907 in Northern Rhodesia, and a little earlier further south.[105] They also measured flood levels for the Zambezi slightly upstream of the Victoria Falls and for the Kafue River near Lusaka. Chaplin's graph of flood levels for these two rivers measures the rise of the flood from trough to peak in each season, rather than the maximum height in any given year.[106]

The period threw up puzzlingly sharp variations between precipitation in different geographical areas. The rains from 1907 to 1919 in Mongu, in the main plain of Bulozi, were slightly below average, and without major deviations in either direction. In contrast, the rains in Sesheke, in the Caprivi floodplain, and in neighbouring Livingstone, were well above average in many seasons, albeit also below in others.[107] This southern reality is reflected in the tree-ring study, which exhibits more volatility than for any other period in two centuries. Statistics from Southern Rhodesia (Zimbabwe) tell much the same story.[108]

The southern half of Angola experienced yet another pattern. After a few years of rainfall that dipped somewhat below average from 1904, extremely wet weather dominated in 1908–1910. However, severe drought and famine then gripped all of southern Angola in 1910–1911, which was dubbed the 'Year of the Great Hunger' in the Ngangela lands to the east of the highlands. Although 1911–1912 witnessed very high rainfall, from 1912 drought returned. This culminated, in 1914–1915, in the worst famine ever recorded in Southern Angola and Northern Namibia, accentuated by Portuguese military campaigns against the Germans and the Ovambo. Rainfall bounced back up to excessive and

[105] Zambia, *Totals of Monthly and Annual Rainfall for Selected Stations in Zambia* (Lusaka: Department of Meteorology, 1972).

[106] Chaplin, 'On Some Aspects of Rainfall,' 2, 6 (1954–55), 16–23.

[107] Figures from: Zambia, *Totals of Monthly and Annual Rainfall.*

[108] Therrell, Stahle, Ries, and Shugart, 'Tree-Ring Reconstructed Rainfall Variability,' 677–85.

destructive levels in 1915–1917, before slipping once more into drought conditions by the end of the decade.[109]

This chaotic pattern of Angolan rainfall correlated relatively well with flood levels in the Upper Zambezi, although peaks were less marked. Measured at Livingstone from 1907 to 1919, the average annual rise of the Zambezi was about 15 inches below the longer-term average, only rising above this level in two years. In comparison, the Kafue River's annual rise, roughly reflecting conditions in the Zambezi's northeastern catchment, was about 4 inches below average.[110] An alternative proxy for the Zambezi's northeastern catchment is rainfall in Solwezi, close to the Kabompo River. The rains here were fairly constant from 1907 to 1919, registering 90% of their average longer-term values.[111]

The year 1914–1915 even witnessed the unusual spectacle of the complete absence of the annual flood in the main plain.[112] Local rainfall that year was also about 10 inches below average. In contrast, the Kafue's flow was normal. And rain in and around the Caprivi floodplain was above average, in places far above.[113] This implies that it was the catastrophic drought in Angola that caused the failure of the flood in the main plain of Bulozi.

[109] *Le Philafricain* provides an almost yearly record. See also: Bastos, *Monographia de Catumbella*; Gladwyn Murray Childs, *Umbundu Kinship and Character: Being a Description of Social Structure and Individual Development of the Ovimbundu of Angola, with Observations Concerning the Bearing on the Enterprise of Christian Missions of Certain Phases of the Life and Culture Described* (London: International African Institute, 1949); F. Rudolph Lehmann, 'Die Politische und Soziale Stellung der Häuptlinge im Ovamboland während der Deutschen Schutzherrschaft in Südwest-Afrika,' *Tribus*, NS, 4–5 (1954–1955), 265–328; Jan-Bart Gewald, 'Near Death in the Streets of Karibib: Famine, Migrant Labour, and the Coming of the Ovambo to Central Namibia,' *Journal of African History*, 44, 2 (2003), 211–39; Maria da Conceição Neto, 'In Town and Out of Town: A Social History of Huambo, Angola, 1902–1961' (Unpublished PhD diss., SOAS University of London, 2012); and documents from the Spiritan Archives, many of which are reprinted in the *Bulletin Général de la Congégation du Saint-Esprit*. Again, these patterns may be linked to El Niño. See: Chapter by Ventura, this volume.

[110] Chaplin, 'On some aspects of rainfall,' 2, 6 (1954–55), 17.

[111] Zambia, *Totals of Monthly and Annual Rainfall.*

[112] Clay, *Your Friend Lewanika*, 155.

[113] Zambia, *Totals of Monthly and Annual Rainfall*; Chaplin, 'On Some Aspects of Rainfall,' 2, 6 (1954–55), 17.

Abnormal climatic conditions in 1915 were credited with pushing up the number of recruits for the Rhodesia Native Labour Board to three times earlier levels.[114] A severe outbreak of bovine pleuropneumonia, probably worsened by drought, further contributed to this outflow.[115] However, there were also social and political factors at play, notably the colonial abolition of slavery, together with the imposition of new taxes.[116]

Conclusion

The Upper Zambezi Basin remains an understudied region in the context of the climate history of the wider Indian Ocean World, as it is geographically so remote from the sea. One argument made here is that the influences of the Indian Ocean spread surprisingly far inland in Central Africa and that this would warrant the inclusion of the region within the broader story.

While this chapter makes a start in attempting to recover historical variations in rainfall and floods over nearly two and a half centuries, it remains for other researchers to ascertain how these fluctuations correlated with climatic drivers. Sea surface temperature anomalies in the Indian Ocean may well have impacted the changing climatic fortunes of the Upper Zambezi Basin, but these need to be more exhaustively charted, and the causal dynamics of these processes remain to be determined. The suggested correlations between global climatic anomalies, such as El Niño events, and droughts and floods over the *longue durée*, are only tentative at this stage.

In addition, a better understanding of weather patterns in the Indian Ocean might help to explain and fill out the climatic history of the Upper Zambezi Basin, and this could shed new light on other historical phenomena. It has been suggested here that early political structures in the Lozi kingdom, as well as the rise and fall of the Kololo state,

[114] Laurel Van Horn, 'The Agricultural History of Barotseland, 1840–1964,' in *The Roots of Rural Poverty in Central and Southern Africa*, eds. Robin Palmer and Neil Parsons (London: Heinemann, 1977), 144–69.

[115] Hogan, 'The Ends of Slavery in Barotseland,' 210–11.

[116] Ibid., Ch. 9; Clarence-Smith, 'Slaves, Commoners and Landlords in Bulozi,' 219–34.

may have owed something to climatic phenomena. Similarly, Lozi relations with early Western traders, missionaries, colonial administrators, and labour recruiters were affected by patterns of rainfall and floods.

A fuller record of past rainfall and floods might also assist in making better predictions for future use. Short runs of figures for rainfall and floods have hindered effective economic and social planning. For example, an otherwise thorough study of the water resources of Western Province, compiled in 1968, underestimated the probability of extremely low floods.[117] Conversely, in 2004, construction of a new bridge, on the Mongu-Kalabo road, ground to a halt, when the enormous volume of two impressive floods astonished the Kuwaiti builders.[118] Finally, although the dependence on precipitation falling in South Central Angola for the degree of flooding in the Upper Zambezi plains remains to be statistically determined by modern methods of measuring river flows, events in the historical period covered in this chapter suggest a close relationship. This raises delicate problems as to the potential impact on neighbouring countries of building dams in Angola or pumping and diverting water for irrigation purposes. A better comprehension of the causes and patterns of inundations, over a long historical horizon, would be vital for any future political negotiations over such issues.

Acknowledgements My thanks are due to the University of Zambia, which provided funding for research trips to Livingstone and Mongu, and the staff of the Livingstone Museum. The late Mrs. 'Paddy' Radunski, and Robin and Marguerite Derricourt, graciously put me up in Livingstone. During the pandemic, the following kindly supplied me with materials: Rita Amaral, Kawawa Banda, Innocent Chomba, José Curto, Lawrence Flint, Aida Freudenthal, Philip Gooding, Jack Hogan, Brian Huntley, John Mendelsohn, Elias Awol Mohammed, Richard Moorsom, Sharon Nicholson, Neil Parsons, Matthew Therrell, Estevam Thompson, Adrian Wood, and Henry Zimba.

[117] Parsons Corporation, 'Final Report: The Water Resources of Barotse Province' (Los Angeles, Unpublished).

[118] Flint, 'Historical Constructions of Postcolonial Citizenship,' 221.

Bibliography

Archival Collections

Archives Générales de la Congrégation du Saint-Esprit, Chevilly-la-Rue, Angola series.

Cambridge University Library, England, Royal Commonwealth Society Papers, 'The Journal of William Waddell.'

Livingstone Museum, Livingstone, Zambia, Manuscript Collection, Paris Evangelical Mission Papers.

National Archives of Zambia, Lusaka, British South Africa Company Papers.

SOAS University of London, Council for World Mission, London Missionary Society Papers.

Oral Interview

N. Simalumba (Acting District Secretary), 27 Apr. 1977, Mongu.

Published Materials

Andersson, Charles John. *Lake Ngami, or, Explorations and Discoveries During Four Years' Wanderings in the Wilds of South Western Africa*. London: Hurst and Blackett, 1856.

Andersson, Charles John. *The Okavango River: A Narrative of Travel, Exploration, and Adventure*. New York: Harper and Brothers, 1861.

Andrade, Alfredo [A. Freire] de. *Relatório da Viagem de Exploração Geographica no Districto de Benguela e Novo Redondo, 1898–1899*. Lisbon: Imprensa Nacional, 1902.

Andrade, Alfredo [A. Freire] de. *A Bacia Hydrographica do Rio Cuanza, desde a Nascente á Confluencia do Rio Gango*. Lisbon: Imprensa Nacional, 1905.

Angola. *O clima de Angola*. Luanda: Serviços Meteorológicos, 1955.

Armor, Murray. 'Notes on the Abnormal Flood Conditions Experienced in Barotseland During Recent Years.' Unpublished paper by District Commissioner of Kalabo, n.d. (Supplied by courtesy of Gwyn Prins).

Arnot, Frederick S. *From Natal to the Upper Zambesi: First Year among the Barotsi*, 3rd ed. Glasgow: The Publishing Office, 1884.

Arnot, Frederick S. *Garenganze, or Seven Years' Pioneer Mission Work in Central Africa*. London: Hawkins, 1889.

Arnot, Frederick S. *Missionary Travels in Central Africa*. London: Holness, 1914.

Baines, Thomas. *Explorations in South-West Africa, Being an Account of a Journey in the Years 1861 and 1862 from Walvisch Bay, on the Western Coast, to Lake Ngami and the Victoria Falls*. London: Longmans, Roberts and Green, 1864.

Bastos, Augusto S. *Monographia de Catumbella*. Lisbon: Tipografia Universal, 1912.

Béguin, Eugène. *Les Ma-Rotse: Étude géographique et ethnographique du Haut-Zambèze*. Lausanne: Librairie Benda, 1903.

Bertrand, Alfred. *Au Pays des Ba-Rotsi, Haut-Zambèze: Voyage d'exploration en Afrique et retour par les Chutes Victoria, le Matébéléland, le Transvaal, Natal, Le Cap*. Paris: Hachette et Cie, 1898.

Burton, Richard F., ed. *The Lands of Cazembe: Lacerda's journey to Cazembe in 1798; also journey of the pombeiros, P.J. Baptista and Amaro José, across Africa from Angola to Tette on the Zambeze; and a résumé of the journey of MM. Monteiro and Gamitto*. London: John Murray, 1873.

Capello, Hermenegildo, and Roberto Ivens. *De Angola á Contra-Costa: Descripção de uma viagem atravez do continente africano*. Lisbon: Imprensa Nacional, 1886.

Capello, Hermenegildo, and Roberto Ivens. *Diários da Viagem de Angola à Contra-Costa*, ed. Francisco de Assis de Oliveira Martins. Lisbon: Agência Geral das Colónias, 1951.

Caplan, Gerald L. *The Elites of Barotseland, 1878–1969: A Political History of Zambia's Western Province*. London: Christopher Hurst & Co., 1970.

Chaplin, J.H. 'On Some Aspects of Rainfall in Northern Rhodesia.' *Northern Rhodesia Journal*, 2, 4 (1954–1955), 32–38; and 2, 6 (1954–1955), 16–23.

Chapman, James. *Travels in the Interior of South Africa, 1849–1863*. London: Bell and Daldy, 1868.

Chatelain, Alida. *Héli Chatelain, l'Ami de l'Angola, 1859–1908: Fondateur de la Mission Philafricaine, d'après sa correspondance*. Lausanne: Mission Philafricaine, 1918.

Childs, Gladwyn Murray. *Umbundu Kinship and Character: Being a Description of Social Structure and Individual Development of the Ovimbundu of Angola, with Observations Concerning the Bearing on the Enterprise of Christian Missions of Certain Phases of the Life and Culture Described*. London: International African Institute, 1949.

Chomba, Innocent C., Victoria Ngwnya, and Mulema Mataa. *First Field Data Campaign on the Barotse Floodplain: WASP/WeMAST Project Report*. Lusaka: School of Mines, University of Zambia, 2019.

Clarence-Smith, William G. 'Drought in Southern Angola and Northern Namibia, 1837–1945.' *African History Seminar*. SOAS, 12 June 1974.

Clarence-Smith, William G. 'Climatic Variations and Natural Disasters in Barotseland, 1847–1907.' *History Staff Seminars*. University of Zambia, 1 June 1977.

Clarence-Smith, William G. 'A Note on Tsetse Fly and Rinderpest in Barotseland, 1850s–1900s.' *History Staff Seminars*, University of Zambia, 1 June 1977.

Clarence-Smith, William G. *Slaves, Peasants and Capitalists in Southern Angola, 1840–1926*. Cambridge: Cambridge University Press, 1979.

Clarence-Smith, William G. 'Slaves, Commoners and Landlords in Bulozi, c. 1875 to 1906.' *Journal of African History*, 20, 2 (1979): 219–34.

Clark, Percy M. *The Autobiography of an Old Drifter: The Life-Story of Percy M. Clark of Victoria Falls*. London: George G. Harrap, 1936.

Clay, Gervas. *Your Friend Lewanika: The Life and Times of Lubosi Lewanika, Litunga of Barotseland, 1842–1916*. London: Chatto & Windus, 1968.

Coillard, François. *Sur le Haut Zambèze: Voyages et travaux de mission*. Paris: Berger-Levrault, 1899.

Corrêa, Elias Alexandre da Silva. *História de Angola*. Lisbon: Agência Geral das Colônias, 1937 [probably composed in the 1790s].

Couceiro, [Henrique M. de] Paiva. *Relatorio de Viagem entre Bailundo e as Terras do Mucusso*. Lisbon: Imprensa Nacional, 1892.

Curto, José C. 'Women Along the Catumbela River, 1797: Land Ownership, Agricultural Production, Labour and Trade.' *Canadian Journal of African Studies*, 54, 3 (2020): 373–93.

Damodaran, Vinita, Rob Allan, Astrid E.J. Ogilvie, Gaston R. Demarée, Joëlle Gergis, Takehiko Mikami, Alan Mikahil, Sharon E. Nicholson, Stefan Norrgård, and James Hamilton. 'The 1780s: Global Climate Anomalies, Floods, Droughts, and Famines,' in *The Palgrave Handbook*, eds. White, Pfister, and Mauelshagen: 517–50.

Dècle, Lionel. *Three Years in Savage Africa*. London: Methuen & Co., 1898.

Delgado, Ralph. *A Famosa e Histórica Benguela: Catálogo dos governadores, 1779–1940*. Lisbon: Edições Cosmos, 1940.

Delgado, Ralph. *Ao Sul do Cuanza: Ocupação e aproveitamento do antigo reino de Benguela, 1483–1942*. Lisbon: author's edition, 1944.

Deneut, Eric, Charles K. Chileya, and Christophe Nativel. *Environmental and Social Impact Assessment for the Improved Use of Priority Traditional Canals in the Barotse Sub-Basin of the Zambezi*. Sunningdale: NIRAS, 2014.

Depelchin, Henri, and Charles Croonenberghs. *Trois Ans dans l'Afrique Australe, 1879–1881*. Brussels: P. Imbreghts, 1882–1883.

Dias, Jill R. 'Famine and Disease in the History of Angola c. 1830–1930.' *Journal of African History*, 22, 3 (1981): 349–78.

Endfield, Georgina H., and David J. Nash. 'Missionaries and Morals: Climatic Discourse in Nineteenth-Century Central Southern Africa.' *Annals of the Association of American Geographers*, 92, 4 (2002): 727–42.

Favre, E. *François Coillard, Missionaire au Zambèze*. Paris: Société des Missions Évangeliques, 1913.

Flint, Lawrence S. 'Historical Constructions of Postcolonial Citizenship and Subjectivity: The Case of the Lozi Peoples of Southern Central Africa.' Unpublished PhD Dissertation: University of Birmingham, 2004.

Flint, Lawrence S. 'Contradictions and Challenges in Representing the Past: The Kuomboka Festival of Western Zambia.' *Journal of Southern African Studies*, 32, 4 (2006): 701–17.

Flint, Lawrence S. 'Socio-Ecological Vulnerability and Resilience in an Arena of Rapid Environmental Change: Community Adaptation to Climate Variability in the Upper Zambezi Valley Floodplain.' *Working Paper of the Research Institute for Humanities and Nature*. Kyoto, 2008. http://www.chikyu.ac.jp/resilience/files/WorkingPaper/WP2008-004.Flint.pdf [Accessed: 10 Aug. 2020].

Ford, John. *The Role of the Trypanosomiases in African Ecology*. Oxford: Clarendon Press, 1971.

Froment, Alain, and Jean Guffroy, eds. *Peuplements Anciens et Actuels des Forêts Tropicales*. Paris: IRD, 2003.

Galton, Francis. *The Narrative of an Explorer in Tropical South Africa*. London: John Murray, 1853.

Garcia, João Francisco. 'Explorações no sertão de Benguella: Derrota que fez o Tenente de Artilheria João Francisco Garcia, commandante do novo estabelecimento da Bahia de Mossâmedes.' *Annaes Marítimos e Coloniaes*, 4, 6 (1844): 240–54.

Garstang, Michael, Anthony D. Coleman, and Matthew Therrell. 'Climate and the Mfecane.' *South African Journal of Science*, 110, 5–6 (2014): 1–6.

Gelfand, Michael, ed. *Gubulawayo and Beyond: Letters and Journals of the Early Jesuit Missionaries to Zambesia, 1879–1887*. London: G. Chapman, 1968.

Gergis, Joëlle L., and Anthony M. Fowler, Anthony M. 'A History of ENSO Events Since A.D. 1525: Implications for Future Climate Change.' *Climatic Change*, 92, 3 (2009): 343–87.

Gewald, Jan-Bart. 'Near Death in the Streets of Karibib: Famine, Migrant Labour, and the Coming of the Ovambo to Central Namibia.' *Journal of African History*, 44, 2 (2003): 211–39.

Gibbons, Alfred St. H. *Exploration and Hunting in Central Africa, 1895–1896*. London: Methuen & Co., 1898.

Gibbons, Alfred St. H. *Africa from South to North Through Marotseland*. London: J. Lane, 1904.

Gluckman, Max. *The Economy of the Central Barotse Plain*. Livingstone: Rhodes-Livingstone Institute, 1941.

Goetz, E. 'The Rainfall of Rhodesia.' *Proceedings of the Rhodesian Scientific Association*, 8, 3 (1909): 1–129.

Grove, Richard H. 'The Great El Niño of 1789–93 and Its Global Consequences: Reconstructing an Extreme Climate Event in World Environmental History.' *The Medieval History Journal*, 10, 1–2 (2006): 75–98.

Harding, Colin. *In Remotest Barotseland: Being an Account of a Journey of over 8,000 Miles Through the Wildest and Remotest Parts of Lewanika's Empire*. London: Hurst and Blackett, 1905.

Harding, Colin. *Far Bugles*. London: Simpkin Marshall, 1933.

Hermitte, Eugene. 'An Economic History of Barotseland, 1800–1940.' Unpublished PhD dissertation: Northwestern University, 1974.

Hilton-Barber, Brett, and Lee R. Berger. 'Victoria Falls Seasonal Weather Calendar.' 2010. http://www.siyabona.com/explore-victoria-falls-seasonal-weather-calendar.html [Accessed: 20 Aug. 2020]

Hogan, Jack. 'The Ends of Slavery in Barotseland, Western Zambia, c.1800–1925.' Unpublished PhD Dissertation: University of Kent, 2014.

Holub, Emil. *Seven Years in South Africa: Travels, Researches, and Hunting Adventures, Between the Diamond Fields and the Zambesi, 1872–1879*. London: Sampson, Low, Marston, Searle, and Rivington, 1881.

Holub, Emil. *Von der Capstadt ins Land der Maschukulumbe: Reisen im Südlichen Afrika in den Jahren 1883–1887*. Vienna: Alfred Hölder, 1890.

Huntley, Brian J. 'Angola in Outline: Physiography, Climate, and Patterns of Biodiversity,' in *Biodiversity of Angola*, eds. Huntley, Russo, Lages, Ferrand: 15–42.

Huntley Brian J., Vladimir Russo; Fernanda Lages; and Nuno Ferrand, eds. *Biodiversity of Angola: Science and Conservation, a Modern Synthesis*. Cham: Springer Open, 2019.

Hutchinson, Peter. *The Climate of Zambia*. Lusaka: Zambia Geographical Association, 1974.

Jalla, Adolphe. 'History, Traditions and Legends of the Barotse Nation.' Livingstone, Zambia: Livingstone Museum, Livingstone, Zambia, 1916 (Typescript of 1909 published edition, with later additions).

Jalla, Adolphe and Emma Jalla. *Pionniers Parmi les Ma-Rotse*. Florence: Imprimerie Claudienne, 1903.

Johnston, James. *Reality Versus Romance in South Central Africa*. London: Hodder and Stoughton, 1893.

Jury, Mark R. 'The Coherent Variability of African River Flows: Composite Climate Structure and the Atlantic circulation.' *Water SA*, 29, 1 (2003): 1–10.

Kayongo, Kabunda. 'The Social Evolution of the Barotse Society in the Nineteenth Century.' Unpublished MA Dissertation: University of Lund, 1983.

Lehmann, F. Rudolph. 'Die politische und soziale Stellung der Häuptlinge im Ovamboland während der deutschen Schutzherrschaft in Südwest-Afrika.' *Tribus*, NS, 4–5 (1954–1955): 265–328.

Liénard, Jacques. *Lettres et Fragments*. Cahors: A. Coueslant, 1902.

Lima, José Joaquim Lopes de, and Francisco Maria Bordalo. *Ensaios Sobre Statística das Possessões Portuguesas na África Occidental e Oriental, na Ásia Occidental, na China, e na Oceania*. Lisbon: Imprensa Nacional, 1844–1862.

Listowel, Judith. *The Other Livingstone*. Lewes: Julian Friedmann, 1974.

Livingstone, David. *Missionary Travels and Researches in South Africa*. London: John Murray, 1857.

Livingstone, David, and Charles Livingstone. *Narrative of an Expedition to the Zambezi and Its Tributaries, and the Discovery of the Lakes Shirwa and Nyassa, 1858–1864*. London: John Murray, 1865.

Livingstone, David. *Livingstone's Private Journals, 1851–1853*, ed. I. Schapera. London: Chatto and Windus, 1960.

Livneh, A. 'Some Notes on Drought, Famine, and Pestilence in Rhodesia.' *African History Seminar*. SOAS, 12 June 1974.

Lopes, Francisco Xavier. 'O Dombe Grande da Quisamba.' *Annaes do Conselho Utramarino, Parte Não-Oficial, janeiro de 1859 a dezembro de 1861*, 2 (1867): 179–86.

Luck, Reginald A. *A Visit to Lewanika, King of the Barotse*. London: Simpkin, Marshall, Hamilton, Kent, & Co. Ltd, 1902.

Magyar, Ladislaus. *Reisen in Süd-Afrika in den Jahren 1849 bis 1857*. Pest: Lauffer & Stolp, 1859.

Mainga, Mutumba. *Bulozi under the Luyana Kings: Political Evolution and State Formation in Pre-colonial Zambia*. London: Longman, 1973.

Maley, Jean. 'Synthèse sur l'histoire de la végétation et du climat en Afrique centrale au cours du quaternaire recent,' in *Peuplements Anciens et Actuels*, eds. Froment and Guffroy: 53–75.

Malheiro, Alexandre. *Chronicas do Bihé*. Lisbon: Livraria Ferreira, 1903.

Matos, José Pedro Gamito de Saldanha. 'Hydraulic-hydrologic Model for the Zambezi River, Using Satellite Data and Artificial Intelligence Techniques.' Unpublished PhD Dissertation, École Polytechique Fédérale de Lausanne, 2014.

McKittrick, Meredith. 'Making Rain, Making Maps: Competing Geographies of Water and Power in Southwestern Africa.' *Journal of African History*, 58, 2 (2017): 187–212.

Mendelsohn, John, and B. Weber. *Atlas e Perfil do Moxico, Angola—An Atlas and Profile of Moxico, Angola*. Windhoek: Raison, 2015.

Mendelsohn, John, and Antonio Martins. 'River Catchments and Development Prospects in South-Eastern Angola.' 2018.

Menezes, Joaquim Antonio de Carvalho e. *Memoria Geografica e Politica das Possessões Portuguezas n'Affrica Occidental, que diz Respeito aos Reinos de Angola, Benguela, e suas dependencias*. Lisbon: Typografia Carvalhense, 1834.

Menezes, Joaquim António de Carvalho e. *Demonstração Geográphica e Política do Territorio Portuguez na Guiné Inferior, que Abrange o Reino de Angola,*

Benguella, e suas Dependencias; Causas da sua Decadencia e Atrasamento, suas Conhecidas Producções, e os Meios que se Podem Applicar para o seu Melhoramento e Utilidade Geral da Nação. Rio de Janeiro: Typ. Clássica de F. A. de Almeida, 1848.

Miller, Joseph C. 'The Significance of Drought, Disease and Famine in the Agriculturally Marginal Zone of West-Central Africa.' *Journal of African History*, 23, 1 (1982): 17–61.

Miller, Joseph C. *Way of Death: Merchant Capitalism and the Angolan Slave Trade, 1730–1830.* London: James Currey, 1988.

Mohammed, Elias A. 'Hydrodynamics Flood Modelling in Barotse Floodplain, Zambia: Effectiveness of Digital Elevation Models.' Unpublished MSc Dissertation: Abba Minch University, Ethiopia, 2015.

Muyunda, Sakwiba. 'Agricultural Change in Sesheke District of Western Zambia, 1899–1964.' Unpublished MA Dissertation: University of Zambia, 2019.

Neto, Maria da Conceição. 'In Town and Out of Town: A Social History of Huambo, Angola, 1902–1961.' Unpublished PhD Dissertation: SOAS University of London, 2012.

Nicholson, Sharon E. 'Historical and Modern Fluctuations of Lakes Tanganyika and Rukwa, and Their Relationship to Rainfall Variability.' *Climatic Change*, 41, 1 (1999): 53–71.

Nicholson, Sharon E. 'The Nature of Rainfall Variability over Africa on Time Scales of Decades to Millennia.' *Global and Planetary Change*, 26 (2000): 137–58.

Nicholson, Sharon E. 'Spatial Teleconnections in African Rainfall: A Comparison of Nineteenth and Twentieth Century Patterns.' *Holocene*, 24, 12 (2014): 1840–48.

Nicholson, Sharon E., Chris Funk, and Andreas H. Fink. 'Rainfall over the African Continent from the Nineteenth Through the Twenty-First Century.' *Global and Planetary Change*, 165 (2018): 114–27.

Omboni, Tito. *Viaggi nell'Africa Occidentale.* Milan: Stabilimento Civelli e Comp, 1845.

Oswell, W.E. *William Cotton Oswell, Hunter and Explorer: The Story of His Life.* London: William Heinemann, 1900.

Palmer, Robin and Neil Parsons, eds. *The Roots of Rural Poverty in Central and Southern Africa.* London: Heinemann, 1977.

Parsons Corporation. 'Final Report: The Water Resources of Barotse Province.' Los Angeles: Unpublished, 1968.

Peires, J.B. *The Dead Will Arise: Nongqawuse and the Great Xhosa Cattle-Killing Movement of 1856–7.* Johannesburg: Ravan Press, 1989.

Pinto, Alexandre de Serpa. *Como eu Atravessei África do Atlántico ao Mar Indico: Viagem de Benguella à contra-costa, atravès regiões desconhecidas.* London: Sampson, Low, Marston, Searle, and Rivington, 1881.

Prins, Gwyn. *The Hidden Hippopotamus: Reappraisal in African History; the Early Colonial Experience in Western Zambia*. Cambridge: Cambridge University Press, 1980.

Reason, Chris J.C. and Sandi Smart. 'Tropical South East Atlantic Warm Events and Associated Rainfall Anomalies over Southern Africa.' *Frontiers in Environmental Science*, 3 (2015).

Reenen, Reenen J. van. *Report of the Drought Investigation Commission of South West Africa, 1924*. Pretoria: Weather Bureau, 1949.

Roberts, A.D. 'A Note on Drought, Flood, Famine, and Pestilence in and Around Zambia.' *African History Seminar*. SOAS, 15 May 1974.

Schulz, Aurel, and August Hammar. *The New Africa: A Journey Up the Chobe and Down the Okavango Rivers; a Record of Exploration and Sport*. London: William Heinemann, 1897.

Selous, Frederick C. *A Hunter's Wanderings in Africa, Being a Narrative of Nine Years Spent Amongst the Game of the Far Interior of South Africa*. London: Richard Bentley & Son, 1881.

Selous, Frederick C. 'Journeys into the Interior of South Central Africa.' *Proceedings of the Royal Geographical Society*, 2nd series, 3, 3 (1881): 169–75.

Shaw, Paul. 'The Desiccation of Lake Ngami: An Historical Perspective.' *Geographical Journal*, 151, 3 (1985): 318–26.

Siiskonen, Harri. *Trade and Socioeconomic Change in Ovamboland*. Helsinki: SHS, 1990.

Silva, Raúl J. Candeias da. 'Subsídios para a história da colonização do Distrito de Moçâmedes.' *Studia* (Lisbon), 32, 371–78; 33, 341–72; 34, 481–534; 35, 421–39; 36, 293–390 (1971–1973).

Silva Porto, António F. Ferreira da. *Silva Porto e Livingstone: Manuscripto de Silva Porto encontrado no seu espólio*. Lisbon: Sociedade de Geographia de Lisboa, 1891.

Silva Porto, António F. Ferreira da. *Silva Porto e a Travessia do Continente Africano*, ed. G. de Sousa Dias. Lisbon: Agência Geral das Colonias, 1938.

Silva Porto, António F. Ferreira da. *Viagens e Apontamentos de um Portuense em África: Excerptos do 'diário' de António Francisco Ferreira da Silva Porto*, ed. G. de Sousa Dias. Lisbon: Agência Geral das Colonias, 1942.

Silva Porto, António F. Ferreira da. *Viagens e Apontamentos de um Portuense em África: Diário de António Francisco Ferreira da Silva Porto*, ed. Maria-Emília Madeira Santos. Coimbra: Biblioteca Geral da Universidade de Coimbra, 1986.

Silva Porto, António F. Ferreira da. *The lands of the Lui: The Upper Zambezi journals of António Francisco Ferreira da Silva Porto, 1847–1884*, ed. Jack Hogan and Ana Rita Amaral. Oxford: Oxford University Press, forthcoming.

Singh, Deepti, Richard Seager, Benjamin I. Cook, Mark Cane, Mingfang Ting, Edward Cook, and Mike Davis. 'Climate and the Global Famine of 1876–78.' *Journal of Climate*, 31, 23 (2018): 9445–67.

Spillmann, Joseph, comp. *Vom Cap zum Sambesi: Die Anfänge der Sambesimission, aus den Tagebüchern des P. Terörde, S.J., und aus den Berichten der andern Missionäre*. Freiburg im Breisgau: Herder'sche Verlagshandlung, 1882.

Stevenson-Hamilton, James. *The Barotseland Journal of James Stevenson-Hamilton, 1898–1899*, ed. J. P. R. Wallis. London: Chatto & Windus, 1953.

Tabler, Edward C. *Trade and Travel in Early Barotseland: The Diaries of George Westbeech, 1885–1888, and Captain Norman MacLeod, 1875–1876; Illustrated with the Sketches of Lieutenant William Fairlie*. London: Chatto & Windus, 1963.

Tams, G. *Visit to the Portuguese Possessions in South-Western Africa*. London: T. C. Newby, 1845.

Therrell, Matthew D., David W. Stahle, Lydia P. Ries, and Herman H. Shugart. 'Tree-Ring Reconstructed Rainfall Variability in Zimbabwe.' *Climate Dynamics*, 26 (2006): 677–85.

Van Horn, Laurel. 'The Agricultural History of Barotseland, 1840–1964,' in *The Roots of Rural Poverty*, eds. Palmer and Parsons: 144–69.

White, Sam, Christian Pfister, and Franz Mauelshagen, eds. *The Palgrave Handbook of Climate History*. London: Palgrave Macmillan, 2018.

Xulu, Nkosinathi G., Hector Chikoree, Mary-Jane M. Bopape, and Nthaduleni S. Nethengwe. 'Climatology of the Mascarene High and Its Influence on Weather and Climate over Southern Africa.' *Climate*, 8, 86 (2020): 1–11.

Zambia. *Totals of Monthly and Annual Rainfall for Selected Stations in Zambia*. Lusaka: Department of Meteorology, 1972.

Zimba, Henry, Banda Kawawa, Anthony Chabala, Wilson Phiri, Peter Selsam, Markus Meinhardt, and Imasiku Nyambe. 'Assessment of Trends in Inundation Extent in the Barotse Floodplain, Upper Zambezi River Basin: A Remote Sensing-Based Approach.' *Journal of Hydrology: Regional Studies*, 15 (2018): 149–70.

CHAPTER 6

Droughts and Political Crisis in Imerina, Madagascar, 1825–1829

Gwyn Campbell

Droughts in Madagascar in 1825–1826 and 1828–1829, hitherto unstudied, were probably part of a wider drought crisis affecting Indian Ocean Africa, and contributed to provoke political crises that indelibly shaped the history of nineteenth-century Madagascar. That of 1825–1826 helped cement the Merina crown's rejection of the 1820 Britanno-Merina Treaty, while that of 1828–1829 confirmed the rupture of the British alliance, and emboldened the Merina to declare suzerainty over the entire

This research was supported by the Social Sciences and Humanities Research Council of Canada.

The original version of this chapter was revised: Incorrect text has been updated on page number 174. The correction to this chapter is available at https://doi.org/10.1007/978-3-030-98198-3_12

G. Campbell (✉)
Indian Ocean World Centre, McGill University, Montreal, QC, Canada
e-mail: gwyn.campbell@mcgill.ca

P. Gooding (ed.), *Droughts, Floods, and Global Climatic Anomalies in the Indian Ocean World*, Palgrave Series in Indian Ocean World Studies, https://doi.org/10.1007/978-3-030-98198-3_6

island, expel the British Resident Agent, reject Western influence, and implement autarky.

I have hitherto explained these processes in largely economic and political terms. The 1820 Treaty emerged through Britain's recognition of Madagascar as an important supplier of provisions to their recently acquired colony of Mauritius. In it, King Radama I (r. 1810–1828) of Imerina agreed to ban slave exports and accept a British Resident agent and missionaries, in return for British recognition of Merina sovereignty over the entire island, and the promise of military aid to realise Merina domination within the island, compensation for the ban on the slave export trade, and technical assistance to promote alternative exports.[1] However, it failed to produce the anticipated benefits for the Merina crown. The slave exports ban, inadequate British compensation for that ban, and failed experiments with 'legitimate exports' reduced royal revenues, while expenditure increased in attempts to conquer the entire island. Moreover, Radama soon feared that the British wished to make Madagascar at best a satellite power, or worse, a colony. In short, it plunged the Merina economy and Radama's rule into crisis.[2]

However, environmental crises, and Malagasy perceptions thereof, need to be brought into the equation, especially in terms of the impact of drought. Riziculture was central to Imerina's economy, cosmology, and politics. Droughts, which significantly impacted the rice harvest, also affected inter-personal relations and Merina relations with the spiritual world. What follows is an overview of the human–environment context underpinning Imperial Imerina in the early nineteenth century, an explanation of the impacts of the 1820s droughts therein, and the wider consequences they had for Merina domestic and foreign policies through an analysis of Merina cosmology.

Environment and Riziculture

Imerina is located in the centre of the central highlands of Madagascar, at an elevation of between 1200 and 1500 metres. It forms an irregular parallelogram of about 18,000 km^2 in area, extending roughly 160 km

[1] British Library (hereafter: BL) Add. 20131 f. 116 *Papers Relating to the Abolition of the Slave Trade in the Mauritius: 1817–1820*, vol. 18 (House of Commons, 1821), 360.

[2] See: Gwyn Campbell, *An Economic History of Imperial Madagascar 1750–1895: The Rise and Fall of an Island Empire* (Cambridge: Cambridge University Press, 2005).

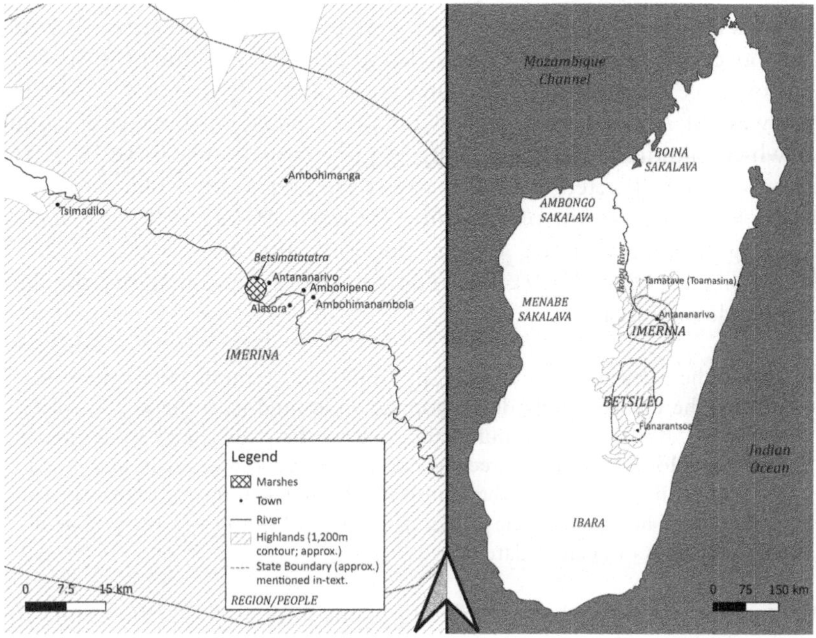

Fig. 6.1 Map of Madagascar and the core of Imerina on the Ikopa River. Drawn by Philip Gooding

north to south, and about 112 km east to west.[3] The region comprises chiefly granite or gneiss rock, which forms the summits of all the hills, overlaid at lower levels by hard red laterite.[4] This made cultivation, and thus subsistence, difficult until marshland riziculture was introduced from the early 1600s, first in Antsihanaka and Betsileo, highland provinces, respectively, to Imerina's north and south, and subsequently in Imerina itself (Fig. 6.1).

In response to adverse climatic conditions from the late seventeenth to early eighteenth centuries, hydraulic riziculture was introduced to bring

[3] James Sibree, 'Imerina, the Central Province of Madagascar, and the Capital, Antananarivo,' *Proceedings of the Royal Geographical Society and Monthly Record of Geography*, 14, 11 (1892), 742; Samuel Pasfield Oliver, *Madagascar: An Historical and Descriptive Account of the Island and Its Former Dependencies* (London: Macmillan, 1886), I, 222–23.

[4] Sibree, 'Imerina,' 742.

hillsides under cultivation, and dikes were built in lowland areas to enable the conversion of marshland into fields to cultivate rice, the most spectacular example of which was the conversion of the Betsimitatatra marshes, just west of Antananarivo. Following the *Tsimiofy*, a seven-year famine in which hundreds perished, King Andriamasinavalona (c.1675–1710) had a dike constructed, several metres deep and wide, flanking the river Ikopa for some 26 km of its course between Alasora and Andriantavy. An English representative of the London Missionary Society (LMS), James Sibree (1836–1929), later commented of Betsimitatatra, the core rice-producing area:

> This is the granary of the capital, and doubtless accounts for its position, and for the comparatively dense population around it to the north, west, and south. But there are innumerable valleys where the slopes are terraced with rice-plots, like great green staircases, where the grain is first sown broadcast, and from which the young plants are taken up and transplanted in the larger fields along the banks of the rivers, and in the beds of small dried-up lakes of ancient date.[5]

Riziculture was limited to valleys and plains until about 1790 when, due to population pressure, and possibly the introduction of more sophisticated techniques from Betsileo, lower concave hillsides were increasingly brought under cultivation.[6] In early 1821, Welsh LMS missionary David Jones (1796–1841) noted of Antananarivo: 'the country around it is very hilly and on every little eminence there [is] one or more villages—The low ground is well cultivated—rice grows luxuriantly in it and it is well

[5] Ibid., 742.

[6] BL Add.18128 Nicolas Mayeur, 'Voyage au pays d'ancove, autrement dit des hovas ou Amboilamba dans l'intérieur des terres, Isle de Madagascar' (1777), 162–63, 171–72; Alfred Grandidier and Guillaume Grandidier, *Histoire Physique, Naturelle et Politique de Madagascar* (Paris: Imprimerie Nationale, 1908), IV, I, 79–80; Guillaume Grandidier, *Histoire Physique, Naturelle et Politique de Madagascar* (Paris: Hachette et Societe d'Editions Geographiques, Maritimes et Coloniales, 1928), IV: *Ethnographie de Madagascar*, IV, 3, 5, 9, 30–31, 40; Charles Robequain, *Madagascar et les bases dispersées de l'union française* (Paris: Presses Universitaires de France, 1958), 275; Archives de l'Académie Malgache (hereafter: AAM) Raombana, *Histoires* (1853), 24, 26–27, 35; François Callet, R.P. *Histoire des Rois. Tantaran ny Andriana*, trans. George S. Chapus and Emmanuel Ratsima (Tananarive: Librarie de Madagascar, 1974), III–IV, 749, 754; Oliver, *Madagascar*, I, 250–51 and II, 53; Raymond Decary, 'La population de Madagascar,' *Bulletin de l'Académie Malgache*, 28 (1947–8), 36–37.

watered by the Ecoupa [Ikopa] river.'[7] Irrigated riziculture necessitated enormous labour investment, French trader Nicolas Mayeur commenting of Imerina in 1785:

> The continual attention, toil and vigilance of the cultivator, his skill in channelling the water necessary to irrigate the soil ... his unshakeable perseverance, his incredible industry ... and the large number of workers permits the cultivation of the greater part of the land; and this labour, although unceasing, and however thankless and laborious, ultimately reaps a reward.[8]

Plateau soil was cultivated with such assiduity that, by 1800, two rice crops, the *vary aloha* (first rice) and *vary vaky ambiaty* (second rice), were annually cultivated. An average Merina family of four to five people, working 330 days a year on one hectare of land, produced from 2000 to 2500 kg of rice, although this was inferior to yields of coastal wet rice and of swidden.[9]

Rice and Rainfall

Soil fertility, temperature, periodic tropical storms, hail, and locust plagues could all impact rice cultivation. However, the most important factor in its production was rainfall. About 2000 L of water is required to cultivate 1 kg of rice.[10] Rice in Malagasy is termed *vary*, 'seed of the water.'[11] As David Griffiths of the LMS (1792–1863), commented, 'It is a proverbial

[7] *Archifdy Llyfrgell Genelaethol Cymru* / Archives of the National Library of Wales (hereafter: ALGC) Jones to Hastie, 23 Oct. 1820, quoted in: Hastie to Griffiths, Port Louis, 18 Feb. 1821, 19157E.

[8] BL Add.18128 Mayeur, 'Voyage au pays d'ancove, par le pays d'ancaye autrement dit des Baizangouzangoux,' related by Dumaine (1785), 224.

[9] BL Add.18128 Mayeur, 'Voyage au pays d'ancove' (1785), 224; National Archives, Kew, London (hereafter: TNA, UK) CO 167/34 James Hastie, 'Diary' (1817), 187; Grandidier, *Histoire* (1928), 3–7, 11; H.M. Dubois, *Monographie des Betsileo (Madagascar)* (Paris: Institut d'ethnologie, 1938), 429; Campbell, *Economic History*, 26.

[10] Maksika Sipa, 'The Grain of Life—Rice Cultivation in Madagascar,' *Mada Magazine*, https://www.madamagazine.com/en/das-korn-des-lebens-reisanbau-auf-madagaskar/ [accessed: 23 Mar. 2021].

[11] Ralph Linton, 'Rice, a Malagasy Tradition,' *American Anthropologist*, 29 (1927), 654.

saying in Madagascar that access to water is the decisive factor in riziculture. Water is required to soften, sow, and plant the rice, to enable it to ripen, to cook it, and also in order to eat it.'[12] First, rice seeds and shoots had to be nurtured in water, for not only did the topsoil harden, crack, and crumble during the dry season, exposing the roots to a quick death in the sun, but plateau soil, devoid of phosphoric acid, chalk, potassium, and nitrogen, was low in fertility. Long-standing marshland contained a residue of fertility, but dry land needed to be assiduously broken, oxygenized, irrigated, and fertilised with humus or human or animal excrement before a reasonable yield could be expected. In May and June, the topsoil was turned with an *angady*, a long-handled iron spade, which produced clods approximately 300 mm square and 130 mm thick. Also, irrigation and drainage channels were repaired, and seed nurseries (*ketsa*) were prepared:

> The seed beds are usually made on hillsides, one above the other. They are surrounded by little walls of earth which keep in the water and also serve as paths when the fields are flooded. The preparation of the beds takes a long time. First little canals are dug across them to dry them thoroughly. Then the earth is cut out in square clods, like bricks, and these are piled up to dry and air.[13]

In September, nurseries were flooded to a height of about 50 mm, and well manured, before being planted with rice seeds. As one Malagasy informant told American anthropologist Ralph Linton:

> The best [fertilizer] is the half rotted straw from the cattle pits. Ashes from the fireplace must be spread on the seed beds and our ancestors also covered them with the sweepings from their dwellings. They kept the sheep and chickens in their houses, so these sweepings were good fertilizer.

[12] David Griffiths, *Hanes Madagascar, neu Grynodeb o Hanes yr Ynys, ei Chynyrch, ei Masnach, ac Ansawdd ei Thrigolion* (Machynlleth: Richard Jones, 1843), 16. As a Malagasy informed Linton: 'For breakfast we have Sosoa, whole rice boiled with a great deal of water. This is also given to sick people. For dinner we have Ampangoro, rice boiled until the water is all gone. We also make rice dumplings, doing up the meal in pieces of banana leaf and boiling it' (Linton, 'Rice, a Malagasy Tradition,' 658).

[13] Linton, 'Rice, a Malagasy Tradition,' 655.

> If nothing better can be gotten the fields are covered with green lily leaves which are dug in when the field is cultivated.[14]

The resulting shoots were carefully tended until they stood 125–150 mm above the water's surface. In October, the main fields were flooded. Water from the River Ikopa was used for the vast Betsimitatatra plain, while rice fields elsewhere relied on water chiefly from springs at the heads of valleys. As André Coppalle, a visiting French painter, noted in 1826:

> Rice is grown only on the plains or mountain sides supplied with a source of water; because this crop needs constant water, the first task of the Merina rice cultivator is digging canals to bring water, and preparing the land to receive it, retain it for as long as it is needed, and to expel it when it becomes harmful.
>
> Many cultivators join forces to dig these canals which are sometimes 15 to 18 miles [24-29km] in length. The possessor of the highest land receives water first; his field is divided into small well-leveled squares surrounded by clay ditches the elevation of which is proportionate to the amount of water needed for the rice. When the ditches are full, the excess water pours over the sides to irrigate the square [portions of land] and the lower fields. Irrigating or drying the fields is sometimes accelerated and regulated by means of small locks; but the work is usually arranged with so much skill, and the height of the ditches so well calculated, that the water pours out so evenly that it is normally absorbed by the time it would otherwise become harmful.[15]

In September and October, such large quantities of water were needed to irrigate rice fields that streams almost dried up, and the level of the Ikopa fell to the point that the river was easily fordable.[16]

In smaller fields, two or three men driving between 20 and 30 cattle, and on larger plots up to 30 men with 50–100 cattle, within a day trampled the broken clay sods into liquid mud (although from around 1800, cattle were barred from the Betsimitatatra because of the damage they

[14] Ibid. See also: James Sibree, 'The Changing Year in Central Madagascar,' *Antananarivo Annual and Madagascar Magazine*, 18 (1894), 214.

[15] André Coppalle, 'Voyage dans l'intérieur de Madagascar et à la capitale du Roi Radama pendant les années 1825 et 1826,' *Bulletin de l'Académie Malgache*, 7–8 (1909–1910), 79.

[16] Sibree, 'Changing Year in Central Madagascar,' 213.

Fig. 6.2 Transplanting rice. James Sibree, *A Naturalist in Madagascar* (London: Seeley, 1915), 112

could inflict upon its complex dike system). Women then replanted in the larger fields every shoot from the nurseries, at the rate of two to three per second, 150–225 mm apart.[17] Sibree commented (Fig. 6.2):

> Towards the beginning of December the earlier crop of rice comes into ear; and should the rains fall as usual during November, the remaining portions of the great rice-plain will be all planted out with the later crop, the whole of the level and its branching valleys presenting an unbroken expanse of green. Of this, the early rice shows distinctly as a darker shade

[17] BL Add.18128 Lescalier, 'Voyage à l'isle de Madagascar' (1792), 320; Grandidier, *Histoire* (1928), 31; Oliver, *Madagascar*, II, 4–5, 53; TNA, UK CO 167/34 Hastie, 'Diary' (1817), 187; Maurice Bloch, *Placing the Dead: Tombs, Ancestral Villages, and Kinship Organization in Madagascar* (London: Seminar Press, 1971), 75–76, 93–94; Campbell, *Economic History*, 25–27; Sibree, 'Changing Year in Central Madagascar,' 210–32.

> of colour, although it will soon begin to turn yellow, as the grain ripens under the steady heat and the plentiful rainfall.[18]

He noted further,

> Generally, both crops of rice the earlier and the later are all cut by the end of April, although in the northern parts of the province it is usually five or six weeks after that date. But if the rains are late, and should happen to be scanty in February and March, harvest work is still going on at the end of May. In fact, owing to there being these two crops of rice, with no very exactly marked division between the two, autumn, in the sense of rice harvest, is going on for about four months, and sometimes longer, as just mentioned, and extends over the later months of summer as well as the two months of autumn or *Fararano* (March and April).[19]

Despite the network of dikes and canals constructed to counter famines that afflicted Imerina from the early to mid-eighteenth century, highland rizicuture continued to be critically dependent on appropriate rainfall and was highly vulnerable to variations in precipitation patterns. Too much or too little rain, at critical junctures in the agricultural year, could do immense damage to rice production. Too much rain could cause rivers to overflow, dikes to collapse, and ruin to crops. Too little rain could stunt the growth of the rice plant, or even kill it.

Preventative Measures Versus Environmental Crises

Merina farmers were highly aware of the risks posed by changing weather and had developed through the generations a number of strategies designed to avert possible adverse environmental impacts on agricultural production, centred on riziculture.

Some of these were practical. Thus, great pains were taken to protect the rice nurseries. Once the newly planted rice seed had sprouted, and the plant reached a height of from 10 to 13 cm, the cultivator commonly planted small branches along the borders of the plot, and covered the rice

[18] James Sibree, *A Naturalist in Madagascar* (London: Seeley, 1915), 92.

[19] Ibid., 103.

plants with a layer of long dry grass or ferns to protect them from the sun during daytime and cold at night.[20]

> Again, in a very few weeks' time the watery covering of the plain is hidden by another green crop, but not of so bright and vivid a tint as the fresh-planted and growing rice. This is the ***kolikoly***, or after-crop, which sprouts from the roots of the old plants. This is much shorter in stalk and smaller in ear than the first crop, and is often worth very little; but if the rains are late, so that there is plenty of moisture, it sometimes yields a fair quantity, but it is said to be rather bitter in taste.[21]

To lessen the risk of over-dependence on rice, the Merina cultivator diversified by also planting hillside plots of manioc, sweet potatoes, yam, sugarcane, and various fruits.[22] However, these could also be greatly damaged adverse environmental events, in particular by severe drought.

Other protective measures involved invoking the spiritual world. Of particular importance in this regard was the role of diviners who could often foretell natural disasters. For example, during Ranavalona's reign (1828–61), the female diviner Rakapila

> could tell in advance years of good or bad harvest. This is how she went about her prediction. She poured water into a jug and into a clay dish, then placed both in a place surrounded by a circular fosse. After three days, she inspected the state of the water. If this foamed and so spilled out of the jug or dish, it signified a propitious year. If, to the contrary, the amount of water in the recipients had decreased, Rakapila warned cultivators saying: "Beware, take all precaution, for it will be a bad year". Then, people asked her if she had any means of remedying the situation, to which she responded: "impossible, because the fate of the year has already crossed three valleys."[23]

Even so, many diviners were believed to be able to avert natural calamities. Thus, Linton was informed that

[20] Sibree, 'Changing Year in Central Madagascar,' 213.

[21] Sibree, *A Naturalist in Madagascar*, 104.

[22] ALGC Jones to Hastie, 23 Oct. 1820, quoted in: Hastie to Griffiths, Port Louis, 18 Feb. 1821, 19157E.

[23] Callet, *Histoire des Rois*, I, 135.

> In former times almost every village had a sorcerer who claimed to be able to keep the locusts away. When a swarm was seen he went out of the village, usually climbing a hill, and stood there without saying anything. As long as he remained standing the locusts would not settle on the rice fields. Every one injured and insulted him, for this was part of the charm. The sorcerer's power over the locusts was inherited from his ancestors. The same sorcerer made charms against the hail. When the storm began he would go out naked except for a loin cloth. He carried a rice pestle with which he beat the wind and rain. Afterwards he leaned the pestle against the wall of the house, close to the southeast corner. He also carried a small round pebble, like a hailstone, in his mouth, and it was believed that in this way he kept the hail in his power. While the rice was growing it was forbidden for the people to eat peanuts or to burn green plants at the fire in the house. If locusts came or hail fell the sorcerers would claim that it was because these rules had been broken.[24]

However, the most widespread preventative measures in early nineteenth-century Imerina was the use of talismans. These fell into two general categories: *ody*, or individual charms, and *sampy*, collective talismans representing the spirit of the ancestors. All were considered essential in offering protection to the rice crop. Raombana, a court official of Ranavalona, commented of the *ody*:

> I do not believe that there are above two or three thousand persons in Madagascar, who has not Idols in their possessions, that is mature people, without reckoning the children, and the women; and a great number of the latter also possess Idols or charms like the men. Some Idols are said to possess power over the Hails, so as to prevent them from falling and destroying the rice growing in the rice plantations; and the persons who possess the above Idols, on perceiving any appearance of hails, goes naked to the front of their houses, menacing the hail, and ordering them not to fall on the growing rice but to fall on the plains, and when it happen, (as it is frequently the case) that the rice are destroyed, they always give out, that some persons has "Manota fady" the Idols, and which consequently has drawn the displeasures of the Idols and caused the destruction of the rice.
>
> Their foolish statements are always believed. Thousands of people are believed to possess Idols or charms who has powers to prevent the due effect of the hails; and when the rice is ripe and brought in, they have a

[24] Linton, 'Rice, a Malagasy Tradition,' 656–57.

> certain part of the rice given to them, on the supposition that it is through the powers of their Idols that their rice were preserved from the Hails.[25]

Sampy were more powerful than *ody*. Each clan possessed its own *sampy*, and by the early nineteenth century, Imerina possessed 'twelve' (a symbolic rather than actual figure) national *sampy*—a development dating to the seventeenth century. Rakelimalaza was the most important national *sampy*, reputedly so designated by Andriamasinavalona.[26] The other chief *sampy* were Ramahavaly, Rafantaka, Manjakatsiroa, Rabehaza, and Ratsimahalahy.[27] These were of particular importance in times of environmental crisis. All were believed to offer protection against illness, but Ramahavaly also protected against drought (by inducing rain), lightning, disease, and crocodiles,[28] and was summoned to perform the *mitiodrano* ceremony whereby sacred water was sprinkled over an assembled population to ensure a successful outcome when the rice crop was ripening.[29] Other *sampy* included Kelimanjaka lanitra ('little but ruler of the heavens') which could preserve the rice crop from hail, and Rakapila ('half dishevelled') which could foresee and provide preventative charms against rice blights and human diseases.[30]

All the supernatural entities and talismans noted above had *fady*, or taboos, associated with them that had to be respected, lest ill-fortune follow. Thus disrespecting the *fady* of a supernatural entity such as the

[25] AAM Raombana, *Annales* (1853), 8.

[26] James Richardson, *A New Malagasy-English Dictionary* (Antananarivo: LMS, 1885), 453; Herbert F. Standing, *The Children of Madagascar* (London: Religious Tract Society, 1887), 178–79; Gwyn Campbell, *David Griffiths and the Missionary 'History of Madagascar'* (Leiden: Brill, 2012), 189.

[27] Gabriel Ferrand, *Les Musulmans à Madagascar et aux iles Comores Pt. I. Les Antaimorona* (Paris: E. Leroux, 1891), 30–31; Callet, *Histoire des Rois*, I, 124; David Griffiths, *Hanes Madagascar, neu Grynodeb o Hanes yr Ynys, ei Chynyrch, ei Masnach, ac Ansawdd ei Thrigolion* (Machynlleth: Richard Jones, 1843), 18.

[28] William Ellis, *History of Madagascar: Comprising also the Progress of the Christian Mission Established in 1818; and an Authentic Account of the Recent Martyrdom of Rafaravavy; and of the Persecution of the Native Christians* (London: Fisher, Son, & Co., 1838), I, 224–25, 402–3, 407–10; Callet, *Histoire des Rois*, 1, 177, 207; R. Valmy, 'Les sampy. Idoles royales,' *Revue de Madagascar*, 27 (1956), 56.

[29] Gwyn Campbell, 'Crisis of Faith and Colonial Conquest. The Impact of Famine and Disease in Late Nineteenth-Century Madagascar,' *Cahiers d'Études Africaines*, 32, 127 (1992), 413.

[30] Ellis, *History of Madagascar*, I, 412–13.

vazimba (reputedly Madagascar's original inhabitants) or *razana* (ancestors), would result in grave misfortune for the guilty individual, or even for the entire community.[31] There were many *fady* connected to rice cultivation, including the *ody havandra*, the four *fady* which protected the crop against hail, and *ody valala* to protect the crop from locusts.

Taboos associated with riziculture took several forms.[32] For example, it was taboo when rice ears were sprouting, to

1. sit in the doorway, lest it impeded the growth of the plant
2. allow grass or straw to lie on the floor, lest it caused disorder in the rice fields
3. crush white stones, lest it induced heavy hail
4. transport stones to build tombs
5. bring into the village specified grasses, vegetables, other plants, and fruits (*zozoro* and *hazondrano*—used to make brooms and mats; *vinda*, *rindra*, and *hisatra*—mats; *harefo*—mats, hats, and baskets; *haravola*—baskets; *ampanga*—tinder; raffia palm branches—poles and rafts; bamboo—poles; *saonjo*; manioc; *anantsinahy*, lemons) lest they cause hail, lightning or other misfortune.

Additionally, when rice was maturing in the rainy season, it was taboo to:

6. make noise near rice fields, lest it induces hail
7. burn green coloured things, as that would induce death to the burner through lightning, cause hail to destroy the crop, or provoke famine
8. walk with a basket upside down on the head
9. catch fish with plant poison, lest it attract lightning
10. prepare a threshing place before the rice has ripened, lest it induce lightning
11. leave white rice on domestic shelves, lest it provoke hail
12. burn rice-chaff, lest it induce it to hail
13. play the Merina kicking game, lest it brings hail

[31] Jørgen Ruud, *Taboo: A Study of Malagasy Customs and Beliefs* (Tananarive: Trano Printy Loterana, 1970), 21.

[32] Campbell, *David Griffiths*, 543; Ellis, *History of Madagascar*, I, 417; Herbert F. Standing, 'Malagasy "Fady",' *Antananarivo Annual & Madagascar Magazine*, 7 (1883), 70–71; Ibid., 66–73.

14. stone someone, lest it bring hail
15. wear dark coloured clothing, lest it attract locusts.

And at harvest time, it was taboo to:

16. use a grass wreath to dampen the fire at night, lest tangled growth in the rice fields render harvesting difficult
17. push stones over a precipice or down a steep hill, lest heavy hail destroys the crop
18. inform anyone of the time the rice be first cut, lest it provoke bad weather
19. sing short snatches of song, lest it induces famine
20. cut bamboo or branches of the raffia palm, lest devastating weather ruin the harvest and induce famine
21. lay out newly cut rice for drying before the entire crop had been cut
22. clean the threshing place of rice before the entire crop had been harvested
23. let water into harvested fields before the entire crop had been cut.

Furthermore, lest it induce lightning, during thunderstorms it was taboo to:

24. run quickly
25. pass another person
26. break the soil
27. sit back to back with someone
28. lie flat on the back
29. whistle
30. pound rice
31. hold an iron or copper object
32. carry a spade home from the rice field in such a way that sunlight gleamed on it.

And more generally, it was taboo to

33. use a rice straw as a flute, lest it induce famine
34. push a rice pestle hard on the ground, lest it cause famine

35. leave a spade in the field at night, lest the owner fall ill and die
36. stir rice while it was being cooked, lest it offend the *sampy* Ramahavaly.

Faditra 'any offering made to avert evil' or *sorona* 'a sacrifice ... to obtain a desired benefit from that to which one prays' could also serve as preventive measures.[33] They were generally made to a *vazimba*, *razana*, or *sampy* whose particular powers were required. For example, at the commencement of the planting season, geese were killed and their heads impaled on sticks near sacred rocks or trees as a sacrifices to the ancestors, in order to ensure the success of the rice crop.[34] The *mpsikidy* (diviner) commonly directed that sacrifices be made at the sites of *vazimba* tombs, considered to be shrines, and generally positioned on top of prominent hills. The *vazimba* was deemed *masina*, or holy, when such a supplication was successful, but *masiaka*, or fierce, when unsuccessful.[35]

Environmental Crisis and ENSO in the 1820s

Despite these measures, global climatic factors contributed to two severely depleted rice harvests during the 1820s. In general terms, Madagascar aligns with eastern Africa's basic climatic zones. These are to some degree decided by latitude.[36] The one major difference in Madagascar's climatic regime is prevailing southeast trade winds that bring all-year-round rainfall to the island's eastern littoral. The Intertropical Convergence Zone (ITCZ) also generally ensures plentiful rainfall for Madagascar's northern reaches. By contrast, the southern and southwestern regions of Madagascar appear to have much the same climate as Southern Africa, which experienced reasonably good rainfall from 1810 to 1819, but severe and extensive drought in 1820–1821, 1825–1827, and 1833–1836. During that period, Natal experienced its worst climatic conditions between 1820

[33] Richardson, *New Malagasy-English Dictionary*, 145, 591.

[34] Lucy A. Jarosz, 'Taboo and Time-Work Experience in Madagascar,' *Geographical Review*, 84, 4 (1994), 440.

[35] Ellis, *History of Madagascar*, I, 84, 424–25.

[36] Gwyn Campbell, 'Environment and Enslavement in Highland Madagascar, 1500–1750: The Case for the Swahili Slave Export Trade Reassessed,' in *Bondage and the Environment in the Indian Ocean World*, ed. Gwyn Campbell (Cham, CH: Palgrave Macmillan, 2018), 52.

and 1824. These adverse events had devastating impacts on agriculture and human and animal life.[37]

If conditions in Mozambique can be taken as a rough indication of climate in the high central plateau of Madagascar where Imerina is located, that region experienced good rainfall until about 1819, followed by a generally arid interval until the mid-nineteenth century. The period 1822 to 1838 was exceptionally arid.[38] Drought of this order is often associated with the El Niño phase of El Niño Southern Oscillation (ENSO). Estimates of ENSO indicate the possibility of medium strength El Niños from 1823 to 1825, 1827 to 1828, and 1831 to 1832 (see Table 6.1). However, from their South American reconstruction, William Quinn and Victor Neal identified a long-term period of apparently anomalous El Niño behaviour from 1812 to 1832,[39] a tendency that is apparent in the reconstruction of rainfall patterns by Sharon E. Nicholson, Amin K. Dezfuli, and Douglas Klotter for several regions in sub-Saharan Africa,[40] as well as in general patterns in the wider Indian Ocean World in these, the final years of the Little Ice Age.[41]

[37] Clare Kelso and Colleen Vogel, 'The Climate of Namaqualand in the Nineteenth Century,' *Climatic Change*, 83 (2007), 357–80; Clare Kelso and Colleen Vogel, 'Diversity to Decline-Livelihood Adaptations of the Namaqua Khoikhoi (1800–1900),' *Global Environmental Change*, 35 (2015), 257–58; Clive Spinage, *African Ecology: Benchmarks and Historical Perspectives* (Berlin: Springer, 2012), 186–88; Rudolf Brázdil, Andrea Kiss, Jürg Luterbacher, David J. Nash, and Ladislava Řezníčková, 'Documentary Data and the Study of Past Droughts: A Global State of the Art,' *Climate of the Past*, 14 (2018), 1930; Dirk Verschuren, 'Decadal and Century-Scale Climate Variability in Tropical Africa During the Past 2000 Years,' in *Past Climate Variability Through Europe and Africa: Developments in Paleoenvironmental Research*, eds. Catherine E. Stickley, Richard W. Battarbee, and Françoise Gasse (Dordrecht: Springer, 2004), 153; M.D.D. Newitt, 'Drought in Mozambique 1823–1831,' *Journal of Southern African Studies*, 15, 1 (1988), 19.

[38] Brázdil et al., 'Documentary Data and the Study of Past Droughts,' 1931; Sharon E. Nicholson, Amin K. Dezfuli, and Douglas Klotter, 'A Two-Century Precipitation Dataset for the Continent of Africa,' *Bulletin of the American Meteorological Society*, 93, 8 (2012), 1227.

[39] Joëlle L. Gergis, 'Reconstructing El Niño-Southern Oscillation: Evidence from Tree-Ring, Coral, Ice and Documentary Palaeoarchives, A.D. 1525–2002' (PhD diss., University of New South Wales, 2006), 219; W. Quinn and V. Neal. 'The Historical Record of El Niño Events,' in *Climate Since A.D. 1500*, eds. R. Bradley and P. Jones (London, Routledge, 1992), 623–48.

[40] Nicholson, Dezfuli and Klotter, 'A Two-Century Precipitation Dataset,' 1227.

[41] Indian Ocean World Centre, McGill University (hereafter: IOWC), 'Appraising Risk': https://www.appraisingrisk.com/ [accessed: 13 Apr. 2021].

Table 6.1 Estimates of ENSO episodes, 1818–1833[42]

Year	*ENSO phase*	*Strength*	*Year*	*ENSO phase*	*Strength*
1818–1819	El Niño	medium	1825	La Niña	medium
1819	La Niña	strong	1827–1828	El Niño	medium
1820	La Niña	strong	1829–1830	El Niño	weak
1820–1821	El Niño	medium	1831–1832	El Niño	medium
1823	La Niña	medium	1832–1833	El Niño	weak
1823–1824	El Niño	medium	1833–1834	El Niño	strong
1824–1825	El Niño	medium			

The highlands of Madagascar, including Imerina, are affected by the southeast trade winds that, rising up the high eastern escarpment, create a rain shadow across the centre of the island, characterised by a strong east to west rain gradient. This impacts Imerina, which otherwise, in common with the west central plains, tends to experience much the same climatic regime as southeastern Africa. Imerina experiences two distinct seasons, hot and humid from November to March and dry from April to November. Average annual rainfall in the region is currently 1084 mm, most of which falls in the austral summer when hailstorms also occur. The region is also occasionally visited by cyclones travelling inland from the northeast coast.[43]

This forms the global climatic context for rainfall patterns in Imerina during the early nineteenth century. According to archival sources, severe droughts occurred in Imerina from September to the end of December in both 1825 and 1828. Such was the lack of rain from September 1825, that in early November, cultivators warned the Merina court that 'drought threatens to destroy the rice crop.'[44] By Christmas Day the

[42] Richard Grove and George Adamson, *El Niño in World History* (London: Palgrave Macmillan, 2018), 147; Joëlle L. Gergis and A.M. Fowler, 'A History of ENSO Events Since A.D. 1525: Implications for Future Climate Change,' *Climatic Change*, 92 (2009), 369.

[43] BL. Add.18137 Hugon, 'Aperçu de mon dernier voyage à ancova de l'an 1808,' 11; TNA, UK CO.167/34 Le Sage, 'Mission to Madagascar' (1816), 102; Oliver, *Madagascar*, I, 450, 465; Grandidier, *Histoire* (1928), 6–8, 30–39; Robequain, *Madagascar*, 58–59; Hubert Deschamps, *Les Migrations Intérieures à Madagascar* (Paris: Berger-Levrault, 1959), 13, 15.

[44] Coppalle, 'Voyage dans l'intérieur de Madagascar,' 65.

rains had still not fully commenced, Coppalle commenting 'The rains, which every year from November to March come regularly every day to water the land and fertilise it, are very late this year, causing people to fear for the crops.'[45] Again, in late 1828, only about 50 mm of rain fell between September and November, while in November and December 1828, some 305 mm of rainfall fell, compared to the current average of 450 mm.[46] By contrast, 423 mm fell in January 1829, compared to the average of about 345 mm, 9 mm in February 1829 compared to an average of almost 300 mm, and 293 mm for March compared to the average of just under 300 mm.[47]

The shortfall in precipitation was accompanied by other calamities. In late 1825, Imerina was visited by plagues of locusts whose

> thick clouds truly occlude the glow of the sun. They seem to take advantage of the small breezes from the northeast to rise above the mountains, from which they rush into the rice plains. The natives drive them away uttering loud cries, and either by taste, or to save themselves from a dangerous enemy, they take them home in large bags that they dry to eat.[48]

Additionally, there is a direct connection between El Niño events, famines, and epidemics of malaria, cholera, and influenza at the close of El Niño episodes.[49] For example, the biggest malarial outbreaks occur a year after an El Niño event, the hypothesis being that reduction in malarial transmission during an El Niño drought reduced human immunity which was further weakened by famine, so that when the rains returned and with it a malarial transmission season, the population was more vulnerable than normal.[50] In Imerina, the stunted growth of the rice meant

[45] Ibid., 69.

[46] TNA, UK CO 167/116 Robert Lyall, 'General Remarks on the Weather in Madagascar, and Chiefly at Its Capital, Tananarivou, from 27 June 1828 Till 1 January 1829': https://weather-and-climate.com/average-monthly-precipitation-Rainfall,antananarivo,Madagascar [accessed: 23 Mar. 2021].

[47] Robert Lyall, 'General Remarks on the Weather in Madagascar, and Chiefly at Its Capital, Tananarivou; with a Meteorological Journal,' *Journal of the Royal Institution of Great Britain*, 1 (1831), 47–56.

[48] Coppalle, 'Voyage dans l'intérieur de Madagascar,' 70.

[49] Gergis and Fowler, 'A History of ENSO Events,' 343–87.

[50] Grove and Adamson, *El Niño in World History*, 165–66.

that harvesting probably extended into May 1826, when an epidemic, almost certainly of malaria, erupted, 'raging in the west of Imerina'[51] before spreading throughout the kingdom, causing high mortality.[52] The disease may well have been brought back to Imerina by one of the six military expeditions Radama launched that year against lowland communities. Conscripts from the traditionally malaria-free highlands also proved highly susceptible to disease (malaria) and starvation (due to both their opponents' raze and burn guerrilla tactics and inadequate army provisioning). In 1821, from 25,000 to 30,000 Merina troops (an estimated 69% of the total) and, in 1828, 9800 troops (98% of the total) perished of malaria and starvation in campaigns against the Sakalava. Altogether, an estimated 50% of Merina army recruits died each year. Thus, large numbers were annually conscripted simply to maintain army numbers.[53]

Moreover, military service comprised just one of a number of aspects of *fanompoana*, or unremunerated forced labour for the state. Originally small scale and honorific, it was expanded by Radama to ensure a cheap docile workforce for Merina military expansion, and industrial production based on the manufacture of army equipment and textiles. For example, Radama formed a British-trained standing army of 13,000 men in 1822 that by 1830 under his successor, Ranavalona I, had increased to 30,000. In its turn, industrial *fanompoana* increased massively from 1825 producing, for example, bayonets, spears, boots, gunpowder, and soap.[54] As highland riziculture, in contrast to that practised on the coast, required a constant and heavy labour input, the forcible transfer, or flight, of peasant labour, led to manpower-depleted villages, fields, and dikes. Consequently, in many areas, the complex infrastructure required to sustain irrigated riziculture, the basis of the local economy, was undermined.[55]

[51] Coppalle, 'Voyage dans l'intérieur de Madagascar,' 70.

[52] Gwyn Campbell, 'Malaria in Precolonial Imerina (Madagascar), 1795–1895,' in *Disease Dispersion and Impact in the Indian Ocean World*, eds. Gwyn Campbell and Eva-Maria Knoll (London: Palgrave Macmillan, 2020), 150.

[53] Gwyn Campbell, *Africa and the Indian Ocean World from Early Times to 1900* (Cambridge: Cambridge University Press, 2019), 221.

[54] Gwyn Campbell, 'The Adoption of Autarky in Imperial Madagascar, 1820–1835,' *Journal of African History*, 28, 3 (1987), 395–411; Ibid., 207, 221.

[55] Campbell, *Africa and the IOW*, 242.

Cultural factors also played a role in exacerbating the effects of drought. This related specifically to the number of days considered to be *fady* (taboo), on which no work was done. Some such taboo days were assigned to individuals after consultation with a *mpsikidy*. Others were universally applicable. These included the taboo days associated with *sampy*. For example, Saturday was *fady* for Ratsisimba 'the incorruptible,' who 'is not malicious; but if you do her ill, she will pay you back in kind.'[56] Saturday was also designated a *fady* day by Rakapila. As stated in the *Tantara*, a compendium of Merina royal traditions:

> Rice should not be sown or replanted on that day. Should some stubborn person break this ban, his rice would be eaten by weevils, destroyed by hail, or submerged in a flood—the waters of which would cause the grain and stalk to perish, and thus the harvest to fail. When such catastrophes became known, everyone said: "famine will break out throughout the Tsimandilo territory for the rice entrusted to the protection of Rakapila has perished in the flood". Ramatoa Rakapila had in effect taken it upon herself to that extent to watch over the lives and fields of the inhabitants of Tsimadilo. Should all ensure that the Saturday taboo was respected, the ears of the first crop of rice [in Tsimadilo] showed before those of any other crops, and constituted the "first fruit" offered to the sovereign. The cultivators of Ambohimanambolo and those of Ramatoa Rakapila were the first to carry the "first fruits" of their harvest to the sovereign. In seeing them pass, people said; "This year's crop will be abundant, for Ramatoa Rakaplia's rice is already en-route to the capital".[57]

Lucy Jarosz argues that *fady* days could offer the opportunity on those days for cultivators to engage in other vital tasks, such as local trading.[58] However, in terms of riziculture, a 2011-study found that two or more days a week were *fady* for 18% of the population of agricultural households, resulting overall in 5% lower agricultural productivity.[59]

56 Callet, *Histoire des Rois*, I, 219. See also: William Ellis and Joseph John Freeman, *Madagascar and Its Martyrs: A Book for the Young* (London: John Snow, 1842), 28–29; J.J. Freeman and David Johns, *A Narrative of the Persecution of the Christians in Madagascar* (London: John Snow, 1840), 100.

57 Callet, *Histoire des Rois*, I, 135.

58 Jarosz, 'Taboo and Time-Work Experience in Madagascar,' 448.

59 David Stifel, Marcel Fafchamps, and Bart Minten, 'Taboos, Agriculture and Poverty,' *Journal of Development Studies*, 47, 10 (2011), 1456.

This issue was critical after the death of Radama, when an official twelve-month period of mourning was announced from August 1828 during which 'no one should ride a horse, nor be carried in a chair… none should weave silk, nor make pottery, nor work in the precious metals, nor manufacture sugar… none should work as carpenters, nor write, nor plait hats.'[60] Traditionally, 'Working in the rice-grounds was to be permitted, lest by any neglect a famine should arise'[61] but Ranavalona ignored this when in December 1828 she ordered the reopening of mission schools that were used as institutions for the recruitment of soldiers. At this time, 15,000 young men, who would otherwise have been trying to salvage what remained of the rice crop, were immediately drafted into the army.[62] Fortunately for other regions of Imerina, teachers refused the order to reopen country schools on 10 February 1829, arguing that they could only do so in May, at the close of the rainy season, for fear of malaria.[63] Griffiths states that the period of mourning effectively ended in April 1829,[64] possibly in recognition of the manpower needed to harvest as much as possible of a damaged rice crop, although Robert Lyall (1790–1831), the British Agent at the Merina Court in 1828–1829, claimed that the period of mourning for Radama ended officially on 27 May 1829 in Antananarivo and on 5 June 1829 in Tamatave.[65]

Curative Measures

In cases of severe national crises, such as in late 1825 and late 1828, when ordinary preventative rituals had failed, curative measures needed to be implemented. These invariably involved the sovereign, the most

[60] Ellis, *History of Madagascar*, II, 398. See also: Griffiths, *Hanes Madagascar*, 46; Standing, 'Malagasy "Fady",' 73–74.

[61] Ellis, *History of Madagascar*, II, 399.

[62] Council for World Missions/London Missionary Society (hereafter: CWM/LMS) Madagascar Incoming Letters (hereafter: MIL) Bx.3 F.2 J.C Minute Book of the Madagascar Mission, 1828; CWM/LMS MIL Bx.3 F.2 J.A Jones to Hankey, 29 July 1829; Ibid., 416.

[63] CWM/LMS MIL Bx.3 F.2 J.C Minute Book of the Madagascar Mission, 1828; Ellis, *History of Madagascar*, II, 416.

[64] Griffiths, *Hanes Madagascar*, 48.

[65] CWM/LMS MIL Bx.3 F.2 J.A. Extracts of the Minutes of the Madagascar Mission (4 May–8 July 1829); TNA, UK CO 167/116 Lyall to Colville, Tamatave, 9 June 1829.

powerful living representative of the ancestors. Sovereigns were considered to be gods. Thus, Radama responded to a question from James Hastie (1786–1826), British resident to the Merina capital in 1820–1826, as to why, during a thunderstorm, he fired some cannon: 'We are answering one another—both of us are gods. God above is speaking by his thunder and lightning, and I am replying by my powder and cannon.'[66] Rice was so essential to the Malagasy that they also referred to it as 'God' (*Vary Andriamaninitra*: 'Rice is a god').[67] According to tradition, when temporal sovereigns married, God's daughters stole rice and presented it to them as wedding gifts, and the famed Merina king Andrianampoinimerina consistently linked the sacred character of rice to the crown.[68]

Thus, in an environmental crisis, the people turned to the crown. In December 1825, Coppalle commented that such was the extent of the drought that year that a delegation of small farmers travelled to Antananarivo to beg Radama to intercede on their behalf. For that reason, the entire court made a pilgrimage to the tomb of Radama's father, Andrianampoinimerina, in Ambohimanga, to ask that he induce the rains.[69] This inevitably involved offering a sacrifice to the former king. As noted in the *Tantara*,

> when the kingdom was hit by misfortune, an epidemic, late rains etc., the sovereign sacrificed a *volavita* cow to ask the ancestors for rain or dispel the epidemic… Then we roast the rump of the cow at the holy-house, to its east, to anoint the stone standing there… And [as] the *volavita* animal—is a cow, a beast with little fat, we immediately sacrifice a *malaza* bullock with the *volavita* cow: because the *malaza* bullock is big and very fat.[70]

In extreme circumstances, the *mpsikidy* ordained that a child be sacrificed to the gods in order to stop an epidemic, famine, or other major

[66] Quoted in: William Burder, *Religious Ceremonies and Customs; or, the Forms of Worship Practised by the Several Nations of the Known World, from the Earliest Records to the Present Time; on the Basis of the… Work of Bernard Picart. To Which Is Added, a Brief View of Minor Sects Which Exist at the Present Day* (London: T. Tegg, 1841), 564.

[67] Griffiths, *Hanes Madagascar*, 20; Linton, 'Rice, a Malagasy Tradition,' 654.

[68] Jarosz, 'Taboo and Time-Work Experience in Madagascar,' 443.

[69] Coppalle, 'Voyage dans l'intérieur de Madagascar,' 69.

[70] Callet, *Histoire des Rois*, I, 255.

misfortune. This occurred, for example, in Ranavalona's reign. However, once a child had been selected and delivered to the palace to be sacrificed, the queen ordered that a chicken be substituted for the child who then became known as *tsy maty manota* ('who cannot be put to death for a fault').[71]

These curative measures reinforced the view that droughts were attributable to powerful malignant forces that resisted supplications and sacrifices to the ancestors. Consequently, in 1825 and 1828 (as in other years), it became critical for the Merina first to identify, and then nullify, the causal agent. The most malign agent of misfortune was a *mpamosavy*, a sorcerer or witch in possession of *ody mahery*, powerful unknown charms that were difficult to counter.[72] Foreigners were the chief suspects. The three main categories of outsiders in early nineteenth-century Imerina were slaves, non-Merina Malagasy, and foreigners. The customary way to detect and eradicate *mpamosavy* was through the application of the *tangena* poison test, although the *tangena* was not foolproof, as some *mpamosavy* possessed charms potent enough to counter it.[73] The *sampy* Rakelimalaza, described in 1823 as 'a little bag that contains some roots, wrapped up in bits of red cloth, and a chain of silver and shells huddled together and fastened to the end of a black stick,'[74] was traditionally believed to have the powers to detect *mpamosavy*. Radama allegedly belittled Rakelimalaza and did not immediately utilise it to detect the malign agent responsible for the 1825 drought.[75] However, in May 1826, at the close of a disastrous rice-growing season, when an epidemic erupted, he summoned the *sampy*

[71] Ibid., 316–17.

[72] Ruud, *Taboo*, 13.

[73] Griffiths, *Hanes Madagascar*, 129.

[74] Charles Theodore Hilsenberg and Wenceslaus Bojer, 'A Sketch of the Province of Emerina, in the Island of Madagascar, and of the Huwa, Its Inhabitants; Written During a Year's Residence,' in *Botanical Miscellany; Containing Figures and Descriptions of Such Plants as Recommend Themselves by Their Novelty, Rarity, or History, or by the Uses to Which They Are Applied in the Arts, in Medicine, and in Domestic Economy Together with Occasional Botanical Notes and Information*, ed. William Jackson Hooker (London: John Murray, 1833), III, 257–58.

[75] Campbell, *David Griffiths*, 189.

Ramahavaly, famed for its protective powers against enemy charms and its ability to combat illness and disease.[76]

The king may also have resorted to Ramahavaly to identify the agent responsible for the spate of adverse environmental factors affecting Imerina. As foreigners, notably those representing European political powers, were believed to harbour the most potent and potentially damaging of powers, suspicion would naturally have fallen upon Hastie, the British Agent to the Merina court. According to traditional beliefs, such fears might have been confirmed when from May 1826, at precisely the time that the epidemic erupted and the powers of Ramahavaly were invoked, Hastie suffered a series of accidents that precipitated his death. On 19 May, while sleepwalking aboard a ship sailing from Tamatave to Port Louis, he fell down the hatchway, lost consciousness and 'sustained considerable injury.'[77] Upon arrival in Mauritius, he received medical attention, but it was feared he might die. After recovering somewhat, Hastie decided to accompany the newly arrived missionary, David Johns (né Jones; 1796–1843), and two missionary artisans, James Cameron (1799–1875) and John Cummins (1805–1872), to Antananarivo. They sailed from Port Louis aboard *HMS Wizard* and reached Tamatave on 2 August, where Hastie experienced another severe fall aboard ship. He remained ill on the coast for some time. He then participated in a duck-shooting expedition during which he severely injured his right hand. On the evening of 4 September, he was within 13 km of Antananarivo when his bearers slipped, throwing Hastie from his palanquin, so that he fell on his head, again injuring himself. On 25 September, he experienced considerable pain in his side, his liver became enflamed, and his tongue and face swelled up. On 15 October, his condition further deteriorated, and he died on 18 October.[78]

Following Hastie's death, Radama explicitly rejected the British alliance established by the 1820 Britanno-Merina Treaty that established, amongst other things, free trade for British merchants trading in Madagascar. On 25 October, exactly a week after Hastie died, he signed a five-year exclusive commercial contract, operative from 1 January 1827, with Louis

[76] Coppalle, 'Voyage dans l'intérieur de Madagascar,' 70; Ellis, *History of Madagascar*, I, 224–25, 407–8.

[77] Quoted in: Campbell, *David Griffiths*, 685.

[78] Coppalle, 'Voyage dans l'intérieur de Madagascar,' 3, fn.3; Ellis, *History of Madagascar*, II, 371–74.

Blancard, representing the Mauritian firm, Blancard & Cie. The Blancard 'treaty' stipulated that, in return for annual payments of $30,000 in 1827 and $40,000 a year thereafter, cash loans on demand, and the use of Blancard's fleet when required, Radama authorised Blancard to purchase 4000 bullocks annually at $6 each, and duty-free imports and exports from all east coast ports, and 25% in other ports. The exception, important in the context of the agricultural crisis, was for the export of rice which carried a 100% duty.[79] A generally 25% duty was levied on imports and exports carried by other foreign ships which were restricted to twelve east coast ports.[80] Radama followed this, in March 1827, with a warning to the governor of Mauritius that if Hastie's brother did not arrive soon, as was envisaged, to take up the planned sugar enterprise, he would need to look for other partners.[81] The Blancard 'treaty' served notice to the Mauritian authorities that Radama was serious in his aim of breaking free of British tutelage and was roundly condemned on Mauritius, which was heavily dependent for provisions upon Madagascar.[82]

Radama died in mid-1828 at the same time as Robert Lyall, Hastie's replacement as British Agent, reached Antananarivo. Between then and the end of the year, Imerina experienced almost no rainfall, precipitating a crisis analogous to that of 1825–1826, and a similar search for the *mpamosavy* responsible. Suspicion naturally fell on Lyall, and he was from the start closely surveyed. Over the months following his arrival, local suspicions appeared confirmed by the scientific activities of Lyall, who was a qualified surgeon and ardent naturalist, mineralist, astrologer, and meteorologist who chronicled all his collections and observations.[83] His role as a doctor aroused probably the most suspicion. As naturalists Charles Hilsenberg (1802–1824) and Wenceslas Bojer (1795–1856) had noted during a visit to Madagascar in 1823, 'The peoples of Madagascar ...

[79] National Archives, Mauritius (hereafter: NAM) HB 4 'Convention passé entre Sa Majesté Radama, Souverain de Madagascar, et le Sr. Louis Blancard, agissant au nom de M.M. Blancard & Co., Négociant de Maurice,' Tananarive, 25 Oct. 1826. See also: CWM/LMS MIL Bx.2 F.3 J.D Jones and Griffiths to [LMS], 9 Nov. 1826.

[80] NAM HB 4 Radama I, 'Proclamation,' 25 Oct. 1826.

[81] NAM HB 4 Radama I to Cole, 9 Mar. 1827.

[82] NAM HB 4 Dupuy frères to Cole, 13 Dec. 1826.

[83] See: Gwyn Campbell, *The Travels of Robert Lyall, 1789–1831: Scottish Surgeon, Naturalist and British Agent to the Court of Madagascar* (Cham, CH: Palgrave Macmillan, 2021).

accord the title of sorcerer (ambamousavou) [*mpamosavy*] to those who distribute alleged remedies.'[84]

In the context of the severe drought that continued until the close of 1828, Lyall's persistence in distributing western medicines aroused acute anxiety at court. The rains eventually started on 22 November 1828, well after normal, but contrary to the usual pattern of afternoon downpours, they were light and occurred haphazardly. Many at court became convinced that Lyall was to blame.[85] On 29 November 1828, Ranavalona informed Lyall that she no longer felt bound by the British treaty and would not recognise his position as British agent,[86] and by early 1829 a growing consensus had formed amongst the queen's advisors that he should be expelled.[87]

The drought was broken by rains in January 1829, but such heavy rain fell in the first days of March that by the 8th day of that month the lowlands were flooded, for which Lyall was widely blamed.[88] On 29 March, after he had ridden a horse 'through ignorance or carelessness'[89] into the home village of a *sampy* where horses were taboo (probably Ambohimanambola, home of the national *sampy*, Rakelimalaza), he was arrested for sorcery and taken into custody in the village of Ambohipeno, 11 km to the east of Antananarivo. The guardians (*vadin-tany*) of Rakelimalaza were members of the powerful Tsimiamboholahy clan,[90] and

[84] NAM HB 3 Hilsenberg and Boyer to Burke, 24 Oct. 1823.

[85] Lyall, 'Journal,' 2 Nov. 1827, in *Le Journal de Robert Lyall*, eds. Georges-Sully Chapus and Gustave Mondain (Tananarive: Imprimerie Officielle, 1954), 82; Lyall, 'General Remarks on the Weather,' 47–56; A Resident [Edward Baker], *Madagascar Past and Present: With Consideration as to the Political and Commercial Interests of Great Britain and France; and as to the Progress of Christian Civilisation* (London: R. Bentley, 1847), 49.

[86] AAM Raomabana, *Annales* (1853), 181; Ellis, *History of Madagascar*, II, 417–18.

[87] TNA, UK CO 167/116 Lyall, 'Journal,' 15 Jan. 1829, 21 Feb. 1829.

[88] Lyall, 'Journal,' 2 Nov. 1827, in *Journal de Robert Lyall*, eds. Chapus and Mondain, 82; Lyall, 'General Remarks on the Weather,' 47–56; Resident, *Madagascar, Past and Present*, 49.

[89] Richard Lovett, *The History of the London Missionary Society 1795–1895* (London: Henry Frowde, 1899), II, 687. Rakelimalaza's many taboos included horses. See: Griffiths, *Hanes Madagascar*, 18; Callet, *Histoire des Rois*, I, 186.

[90] Ellis, *History of Madagascar*, II, 418.

in the *sampy's* name identified Lyall as a warlock.[91] However, rather than utilise Rakelimalaza to condemn Lyall, the court decided to employ Ramahavaly, which Ranavalona had fetched from its home village to reside in the palace. Ramahavaly, the *sampy* of revenge, was famed for its protective powers during war, or against enemy charms.[92] Lyall was subsequently expelled, leaving Madagascar for Mauritius in May 1829.[93]

Postscript

Lyall's expulsion failed to prevent a particularly harsh wave of adverse environmental events and disease outbreaks that continued until 1834. In 1831, a malaria epidemic erupted in northern and western Imerina,[94] and in 1832, the rains commenced early, in late October, accompanied by such storms that sorcery was suspected.[95] In the wet season of 1833–1834, Imerina experienced the highest rainfall in living memory, the rice harvest was ruined, and in 1834 a malaria epidemic erupted.[96] In some cases, local Merina were blamed. Epidemics in early 1830 caused Ranavalona to accuse some 60 Merina, ordinary subjects and members of the elite, of being *mpamosavy*. They were subjected to the *tangena* poison test, and many died as a result.[97] However, in the main, suspicion

[91] AAM Raombana, *Annales* (1853), 249–51, 254, 445; Régis Rajemisa-Raolison, *Dictionnaire Historique et Géographique de Madagascar* (Fianarantsoa: Librarie Ambozontany, 1966), 283; Georges-Sully Chapus and Gustave Mondain, *Un Homme d'état Malgache. Rainilaiarivony* (Paris: Diloutremer, 1953).

[92] TNA, UK CO 167/116 Lyall, 'Journal,' 1 Oct. 1828; Callet, *Histoire des Rois*, 211–12; Ellis, *History of Madagascar*, I, 224–25, 407–8, II, 419; Kari Mason, 'Customs and Habits of the Merina Tribe of Madagascar' (Unpublished MS: Leicester, 1959), 22; Rajemisa-Raolison, *Dictionnaire Historique et Géographique*, 163.

[93] Ellis, *History of Madagascar*, II, 420–21.

[94] Campbell, 'Malaria in Precolonial Imerina (Madagascar), 1795–1895,' 150.

[95] ALGC 19157 Johns to Thomas Philip, 26 Nov. 1832.

[96] NAM HB 9 Freeman to Dick, 13 Oct. 1834; Thomas Trotter Matthews, *Thirty Years in Madagascar* (London: Religious Tract Society, 1904), 100; Charles Moss, A Pioneer in Madagascar. *Joseph Pearse of the L.M.S.* (London: Headley Bros, c1913), 142, 164–65.

[97] Edward Baker (1831), quoted in: Joannes Chatin, 'Recherches pour server à l'histoire botanique, chimique et physiologique du tanguin de Madagascar' (Thèse: École Supérieure de Pharmacie de Paris 1873), 10. See also the claims in: George L. Robb 'The Ordeal Poisons of Madagascar and Africa,' *Botanical Museum Leaflets, Harvard University*, 17, 10 (1957), 272.

fell upon non-Merina Malagasy and foreigners.[98] Thus the storms and torrential downpours of October 1832 were blamed by the Merina court on a group of 70 people from the 'south' of the island who had recently arrived in Antananarivo to pledge allegiance to Ranavalona. Johns noted that

> these poor wretches were accused of having charms to draw down the [thunder bolts] and it was soon reported, that they pretended to come up to submit to the queen, while in fact, they came up to draw down the lightenings with their cçharms and by that means to kill the queen and the people. 9 of them were put in a rice hole and boiling water was poured upon them till they died. All the rest took the *Tangena* (ordeal) 38 of them died the rest got over the poison.[99]

However, as indicated earlier, even greater suspicion fell on non-Malagasy. Indeed, when the Sultan of Oman despatched Hamisy as his ambassador to the Merina court in 1833, Ranavalona would not permit him to enter Imerina lest he bewitched her.[100]

Environmental crises thus contributed to a general economic crisis in the 1820s to 1830s. The Merina crown blamed both environmental and wider economic woes on foreigners, most notably in the 1820s on Hastie and Lyall, the appointed British political agents to the Merina court. The death of Hastie and expulsion of Lyall cleared the way for the Merina crown to disengage from the 1820 Britanno-Merina Treaty, which had established informal British hegemony in Imerina and free trade privileges for British merchants. Immediately after the death of Hastie, in October 1826, Radama imposed higher taxes on all foreign traders, and through import substitution, inaugurated efforts to stimulate domestic industry. The expulsion of Lyall in 1829 by Ranavalona, Radama's successor, opened the way for the implementation of a fully autarkic policy. This was based on the military conquest of the entire island and exploitation of its human and natural resources, and the founding of a modern industrial centre based chiefly upon the production of armaments.

[98] ALGC 19157 Johns to Thomas Philip, 26 Nov. 1832.

[99] Ibid.

[100] AAM Raomabana, *Annales* (1853), 330.

Conclusion

This chapter is the first to interpret the rupture of the 1820 Britanno-Merina alliance in the context of human–environment interaction, rather than in purely human terms. In 1825–6 and 1828–9, Imerina, the central province of Madagascar, experienced environmental crises, notably severe droughts that were probably part of a wider drought crisis affecting Indian Ocean Africa. The droughts and accompanying events, such as locust plagues and epidemics, contributed to provoke a political crisis that led the Merina crown to reject the 1820 treaty and British pretensions to political hegemony, and emboldened it to expel the British Resident Agent, declare suzerainty over the entire island, and implement autarky—events that indelibly shaped the history of nineteenth-century Madagascar.

Bibliography

Archival Sources

Archives de l'Académie Malgache, Antananarivo, Madagascar (AAM).

Archifdy Llyfrgell Genelaethol Cymru/Archives of the National Library of Wales, Aberystwyth, Wales (ALGC).

British Library, London, UK (BL).

Council for World Missions/London Missionary Society, SOAS, London, UK (CWM/LMS).

National Archives, Mauritius (NAM).

The National Archives, Kew, London (TNA, UK).

Published and Unpublished Sources

A Resident [Edward Baker]. *Madagascar Past and Present: With Consideration as to the Political and Commercial Interests of Great Britain and France; and as to the Progress of Christian Civilisation*. London: R. Bentley, 1847.

Bloch, Maurice. *Placing the Dead: Tombs, Ancestral Villages, and Kinship Organization in Madagascar*. London: Seminar Press, 1971.

Bradley, R., and P. Jones, eds. *Climate Since A.D. 1500*. London: Routledge, 1992.

Brázdil, Rudolf, Andrea Kiss, Jürg Luterbacher, David J. Nash, and Ladislava Řezníčková. 'Documentary Data and the Study of Past Droughts: A Global State of the Art.' *Climate of the Past*, 14 (2018): 1915–60.

Burder, William. *Religious Ceremonies and Customs; or, the Forms of Worship Practised by the Several Nations of the Known World, from the Earliest Records to the Present Time; on the Basis of the... Work of Bernard Picart. To Which Is Added, a Brief View of Minor Sects Which Exist at the Present Day*. London: T. Tegg, 1841.

Callet, François. *Histoire des Rois. Tantaran ny Andriana*, trans. George S. Chapus and Emmanuel Ratsimba. 5 Vols. Vols. I–IV: Tananarive: Editions de la Librairie de Madagascar, 1974. Vol. V: Antananarivo: Academie Malgache, 1978.

Campbell, Gwyn. 'The Adoption of Autarky in Imperial Madagascar, 1820-1835.' *Journal of African History*, 28, 3 (1987): 395–411.

Campbell, Gwyn. 'Crisis of Faith and Colonial Conquest: The Impact of Famine and Disease in Late Nineteenth-Century Madagascar.' *Cahiers d'Études Africaines*, 32, 127 (1992): 409–53.

Campbell, Gwyn. *An Economic History of Imperial Madagascar 1750-1895: The Rise and Fall of an Island Empire*. Cambridge: Cambridge University Press, 2005.

Campbell, Gwyn. *David Griffiths and the Missionary 'History of Madagascar.'* Leiden: Brill, 2012.

Campbell, Gwyn, ed. *Bondage and the Environment in the Indian Ocean World*. Basingstoke, Hampshire: Palgrave Macmillan, 2018.

Campbell, Gwyn. 'Environment and Enslavement in Highland Madagascar, 1500–1750: The Case for the Swahili Slave Export Trade Reassessed,' in *Bondage and the Environment in the Indian Ocean World*, ed. Gwyn Campbell. Cham, CH: Palgrave Macmillan, 2018: 49–78.

Campbell, Gwyn. *Africa and the Indian Ocean World from Early Times to 1900*. Cambridge: Cambridge University Press, 2019.

Campbell, Gwyn. 'Malaria in Precolonial Imerina (Madagascar), 1795–1895,' in *Disease Dispersion and Impact in the Indian Ocean World*, eds. Gwyn Campbell and Eva-Maria Knoll. London: Palgrave Macmillan, 2020.

Campbell, Gwyn. *The Travels of Robert Lyall, 1789–1831: Scottish Surgeon, Naturalist and British Agent to the Court of Madagascar*. Cham: CH: Palgrave Macmillan, 2021.

Campbell, Gwyn, and Eva-Maria Knoll, eds. *Disease Dispersion and Impact in the Indian Ocean World*. London: Palgrave Macmillan, 2020.

Chapus, Georges-Sully, and Gustave Mondain. *Un Homme d'état Malgache. Rainilaiarivony*. Paris: Diloutremer, 1953.

Chapus, Georges-Sully, and Gustave Mondain, eds. *Le Journal de Robert Lyall*. Tananarive: Imprimerie Officielle, 1954.

Coppalle, André. 'Voyage dans l'intérieur de Madagascar et à la capitale du Roi Radama pendant les années 1825 et 1826.' *Bulletin de l'Académie Malgache*, 7-8 (1909–1910): 7–10.

Deschamps, Hubert. *Les Migrations Intérieures à Madagascar*. Paris: Berger-Levrault, 1959.
Dubois, H.M. *Monographie des Betsileo (Madagascar)*. Paris: Institut d'ethnologie, 1938.
Ellis, William. *History of Madagascar. Comprising also the Progress of the Christian Mission Established in 1818; and an Authentic Account of the Recent Martyrdom of Rafaravavy; and of the Persecution of the Native Christians*. 2 Vols. London: Fisher, Son, & Co., 1838.
Ellis, William, and Joseph John Freeman. *Madagascar and Its Martyrs: A Book for the Young*. London: John Snow, 1842.
Ferrand, Gabriel. *Les Musulmans à Madagascar et aux iles Comores Pt. I. Les Antaimorona*. Paris: E. Leroux, 1891.
Freeman, J.J., and David Johns. *A Narrative of the Persecution of the Christians in Madagascar*. London: John Snow, 1840.
Gergis, Joëlle L. 'Reconstructing El Niño-Southern Oscillation: Evidence from Tree-Ring, Coral, Ice and Documentary Palaeoarchives, A.D. 1525–2002.' Unpublished PhD dissertation, University of New South Wales, 2006.
Gergis, Joëlle L., and A.M. Fowler. 'A History of ENSO Events Since A.D. 1525: Implications for Future Climate Change.' *Climatic Change*, 92 (2009): 343–87.
Grandidier, Guillaume. *Histoire Physique, Naturelle et Politique de Madagascar*. Paris: Hachette et Societe d'Editions Geographiques, Maritimes et Coloniales, 1928.
Grandidier, Alfred, and Guillaume Grandidier. *Histoire Physique, Naturelle et Politique de Madagascar*. Paris: Imprimerie Nationale, 1908.
Griffiths, David. *Hanes Madagascar, neu Grynodeb o Hanes yr Ynys, ei Chynyrch, ei Masnach, ac Ansawdd ei Thrigolion*. Machynlleth: Richard Jones, 1843.
Grove, Richard, and George Adamson. *El Niño in World History*. London: Palgrave Macmillan, 2018.
Hilsenberg, Charles Theodore, and Wenceslaus Bojer. 'A Sketch of the Province of Emerina, in the Island of Madagascar, and of the Huwa, Its Inhabitants; Written During a Year's Residence,' in *Botanical Miscellany*, III, ed. Hooker: 246–75.
Hooker, William Jackson, ed. *Botanical Miscellany: Containing Figures and Descriptions of Such Plants as Recommend Themselves by Their Novelty, Rarity, or History, or by the Uses to Which They Are Applied in the Arts, in Medicine, and in Domestic Economy Together with Occasional Botanical Notes and Information*. 3 Vols. London: John Murray, 1833.
Jarosz, Lucy A. 'Taboo and Time-Work Experience in Madagascar.' *Geographical Review*, 84, 4 (1994): 439–50.
Kelso, Clare, and Colleen Vogel. 'The Climate of Namaqualand in the Nineteenth Century.' *Climatic Change*, 83 (2007): 357–80.

Kelso, Clare, and Colleen Vogel. 'Diversity to Decline-Livelihood Adaptations of the Namaqua Khoikhoi (1800–1900).' *Global Environmental Change*, 35 (2015): 254–68.

Linton, Ralph. 'Rice, a Malagasy Tradition.' *American Anthropologist*, 29 (1927): 654–60.

Lovett, Richard. *The History of the London Missionary Society 1795–1895*. 2 Vols. London: Henry Frowde, 1899.

Lyall, Robert. 'General Remarks on the Weather in Madagascar, and Chiefly at Its Capital, Tananarivou; with a Meteorological Journal.' *Journal of the Royal Institution of Great Britain*, 1 (1831): 47–56.

Mason, Kari. 'Customs and Habits of the Merina Tribe of Madagascar.' Unpublished MS: Leicester, 1959.

Matthews, Thomas Trotter. *Thirty Years in Madagascar*. London: Religious Tract Society, 1904.

Moss, Charles. *A Pioneer in Madagascar: Joseph Pearse of the L.M.S.* London: Headley Bros, c1913.

Newitt, M.D.D. 'Drought in Mozambique 1823-1831.' *Journal of Southern African Studies*, 15, 1 (1988): 15–35.

Nicholson, Sharon E., Amin K. Dezfuli, and Douglas Klotter. 'A Two-Century Precipitation Dataset for the Continent of Africa.' *Bulletin of the American Meteorological Society*, 9, 8 (2012): 1219–31.

Oliver, Samuel Pasfield. *Madagascar: An Historical and Descriptive Account of the Island and Its Former Dependencies*. 2 Vols. London: Macmillan, 1886.

Quinn, W., and V. Neal. 'The Historical Record of El Niño Events,' in *Climate Since A.D. 1500*, eds. Bradley and Jones: 623–48.

Rajemisa-Raolison, Régis. *Dictionnaire Historique et Géographique de Madagascar*. Fianarantsoa: Librarie Ambozontany, 1966.

Richardson, James. *A New Malagasy-English Dictionary*. Antananarivo: LMS, 1885.

Robb, George L. 'The Ordeal Poisons of Madagascar and Africa.' *Botanical Museum Leaflets, Harvard University*, 17, 10 (1957): 265–316.

Robequain, Charles. *Madagascar et les bases dispersées de l'union française*. Paris: Presses Universitaires de France, 1958.

Ruud, Jørgen. *Taboo: A Study of Malagasy Customs and Beliefs*. Tananarive: Trano Printy Loterana, 1970.

Sibree, James. 'Imerina, the Central Province of Madagascar, and the Capital, Antananarìvo.' *Proceedings of the Royal Geographical Society and Monthly Record of Geography*, 14, 11 (1892): 737–53.

Sibree, James. 'The Changing Year in Central Madagascar.' *Antananarivo Annual and Madagascar Magazine*, 18 (1894): 210–32.

Sibree, James. *A Naturalist in Madagascar*. London: Seeley, 1915.

Sipa, Maksika. 'The Grain of Life—Rice Cultivation in Madagascar.' *Mada Magazine*: https://www.madamagazine.com/en/das-korn-des-lebens-reisanbau-auf-madagaskar/ [Accessed: 23 Mar. 2021].

Spinage, Clive. *African Ecology: Benchmarks and Historical Perspectives*. Berlin: Springer, 2012.

Standing, Herbert F. 'Malagasy "Fady".' *Antananarivo Annual & Madagascar Magazine*, 7 (1883): 253–265.

Standing, Herbert F. *The Children of Madagascar*. London: Religious Tract Society, 1887.

Stickley, Catherine E., Richard W. Battarbee, and Françoise Gasse, eds. *Past Climate Variability Through Europe and Africa: Developments in Paleoenvironmental Research*. Dordrecht: Springer, 2004.

Stifel, David, Marcel Fafchamps, and Bart Minten. 'Taboos, Agriculture and Poverty.' *Journal of Development Studies*, 47, 10 (2011): 1455–81.

Valmy, R. 'Les sampy. Idoles royales.' *Revue de Madagascar*, 27 (1956): 56–64.

Verschuren, Dirk. 'Decadal and Century-Scale Climate Variability in Tropical Africa During the Past 2000 Years,' in *Past Climate Variability*, eds. Stickley, Battarbee and Gasse: 139–58.

CHAPTER 7

The Great Ilocos Flood of 1867

James Francis Warren

From 25 to 27 September 1867 a great typhoon swept over the Philippines, hitting Northern Luzon hardest, especially Ilocos Sur and the area around its provincial capital of Vigan. The typhoon's torrential rains caused the Abra River and its tributaries to overflow with chest-deep floodwaters. Worst hit were the areas surrounding the provincial capital of Vigan, a still predominately rural region dotted with small villages and rice fields. The main road leading from Vigan through the Abra Valley and flood plain became impassable with the water rising above 12 feet (4 meters) along stretches of the small highway.

The typhoon left the neighbouring provinces of Ilocos Sur and Cagayan with hundreds of thousands of pesos worth of crop and livestock damage. The provincial governor reported from a shattered capital building in Vigan to the governor general in distant Manila that thousands of hectares of rice that were almost ready to harvest were destroyed by the floodwaters and storm. Among the settlements in Abra Province,

J. F. Warren (✉)
Asia Research Centre, Murdoch, WA, Australia
e-mail: j.warren@murdoch.edu.au

P. Gooding (ed.), *Droughts, Floods, and Global Climatic Anomalies in the Indian Ocean World*, Palgrave Series in Indian Ocean World Studies, https://doi.org/10.1007/978-3-030-98198-3_7

known for its fertile valley and flood plain, the *barrio* (village) of San Julian suffered the most losses. This paper looks closely at this cyclonic storm, which caused the Abra River to overflow, sending floodwaters cascading over its banks.

The huge storm-related flood in the Northern Philippines was associated with the effects of a strong positive El Niño Southern Oscillation (ENSO) event. This investigation of the 1867 Ilocos flood is concerned with two general questions: First, why did this flood calamity occur at such magnitude; and secondly, how could disruption on this scale change a community, and transform colonial disaster policy and practice? I explore why particular groups and individuals, differentiated by gender, age, status, and class, were more vulnerable than others in such a disaster. Certain of these characteristics affected both how and why some people were more severely affected than others, and how they were able to respond.

The El Niño of 1867

The destructive effect of the September 1867 typhoon on the province of Ilocos Sur, especially along the Abra River flood plain, was frightening to behold. Except for the onset of the global warming occurring since the end of the Little Ice Age in c.1840, caused by the industrial revolution and deforestation of the earth, El Niño is the most important recurring event affecting climatic change in the Philippines. Here, we need to think beyond mere event-based processes and adjustment, and acknowledge that ENSO and typhoons have multiple synergistically chain-related impacts, including a great flood, food insecurity, and disease dispersion, resulting from the multi-faceted interplay between climates, local environmental factors, and societal vulnerability.[1]

El Niño, a seasonal warming of the Pacific Ocean that disrupts normal weather patterns, refers specifically to changes in the flow of warm-and cold-water currents in the Pacific. It is directly linked to water-based cascading hazards, and the cause of the greatest annual oscillations in global climate, shifting currents, trade-winds, rain-bearing systems, and storm fronts in the Philippines and the Indo-Pacific world. This major

[1] John Hay, David Easterling, Kristie L. Ebi, Akio Kitoh, and Martin Parry, 'Conclusion to the Special Issue: Observed and Projected Changes in Weather and Climate Extremes,' *Weather and Climate Extremes*, 11 (2016), 103–5.

climatic anomaly presently occurs every three to five years.[2] The decade of the 1860s has been identified as a particularly active El Niño period with events occurring in 1860, 1862, 1864, then followed by an extended ENSO from 1865 to 1869.[3] Collectively, these El Niños marked the beginning of a 60-year period in which El Niño events were particularly strong. Other notable El Niños in this context include those of 1877–1878, 1888–1890, 1896, 1902, 1911–1915, 1918, and 1925.[4]

ENSO's effects on tropical storms in the western Pacific are still under investigation. The latest research suggests that ENSO anomalies affect tropical storms' genesis location, duration, tracks, and frequency, and that the linkages between these phenomena vary depending on time of year. Year-round, El Niño is associated with an eastward displacement of the average location of tropical cyclone genesis in the western Pacific Ocean and China Seas, thus contributing to their making landfall in the Philippines and other parts of the eastern IOW less frequently.[5] However, in July–September, the frequency of typhoons—the strongest type of tropical cyclone—making landfall and the number of days they endure in the Philippines is above average.[6] Thus, a positive ENSO anomaly may have created the climatic context in which a particularly strong typhoon was more likely to make landfall in the Philippines in September 1867.

By the time it did so, many ecosystems in Ilocos Sur had already been damaged or destroyed through colonial policies, setting the stage for an escalation of cascading disasters caused by cyclonic storms. This was despite a commonly held belief at the time that floods were a 'natural' hazard. The idea that this flood could have been exacerbated by human

[2] Brian Fagan, *Floods, Famines and Emperors: El Niño and the Fate of Civilizations* (New York: Basic Books, 1999); Michael Glantz, *Currents of Change: El Niño's Impact on Climate Variability and Society* (Cambridge: Cambridge University Press, 1996).

[3] Ross Couper-Johnston, *El Niño: The Weather Phenomenon That Changed the World* (London: Coronet, 2001), 13. See also: Chapter by Clarence-Smith, this volume.

[4] See, for example: Mike Davis, *Late Victorian Holocausts: El Niño Famines and the Making of the Third World* (London: Verso, 2001); Richard Grove and John Chappell (eds.), *El Niño: History and Crisis* (Cambridge: White Horse Press, 2000); Richard Grove and George Adamson, *El Niño in World History* (London: Palgrave Macmillan, 2018), 93–104; Chapters by Clarence-Smith, Gooding, Ventura, and Williamson, this volume.

[5] James B. Elsner and Kam-biu Liu, 'Examining the ENSO-Typhoon Hypothesis,' *Climate Research*, 25 (2003), 43–54.

[6] Irenea L. Corporal-Lodangco, Lance M. Leslie, and Peter J. Lamb, 'Impacts of ENSO on Philippine Tropical Cyclone Activity,' *Journal of Climate*, 29 (2016), 1877–97.

activity, such as the unplanned growth and development taking place on the Abra plains, was not yet recognised or fully understood by the Spanish colonial authorities. They were bent on blaming rampaging nature for the disaster and returning to business as usual, rather than identifying in the aftermath of the flood the key socio-economic issues that affected people's lives, livelihoods, and coping mechanisms (Fig. 7.1).[7]

The Initial Setting: Manila and Its Environs

On 25 September 1867, strong winds and heavy downpours buffeted Manila as the typhoon entered Philippine waters near the capital. The colonial government paid heed to the serious situation unfolding in Manila, but further flood calamities in distant Northern Luzon followed over the course of the next several days, and the authorities in the capital failed to respond rapidly to the situation. The initial lack of communication and slow recovery response was due to the tyranny of distance, destruction, and damage to most roads and bridges connecting Northern Luzon with the capital, and a bureaucracy facing immediate recovery challenges in that same capital.

The tropical storm flooded the fields and suburbs of Manila but did not develop into a full-blown typhoon.[8] But flooding in the capital was bad enough to cause a lack of resources in certain parts of the city. Consequently, an official edict was proclaimed, in Spanish and Tagalog, soliciting assistance for the flooded neighbourhoods. The wealthy and the religious orders responded to the governor general's exhortation to provide comfort and aid to flood victims, but 'not without being exposed themselves to obvious personal danger.'[9] Afterwards, on the evening of 26 September, the barometer began to rise, and the floodwaters receded within 48 hours. Officials on the spot readily understood with a sense of relief that the reported impacts of the flood in the capital were far less disastrous than what had been originally feared. Even so, the governor general's detailed report to the King mentioned the destruction suffered

[7] Ma. Theresa M. Alders, 'Floodwaters of Death: Vulnerability and Disaster in Ormoc City, Philippines: Assessing the 1991 Flood and Twenty Years of Recovery' (Unpublished PhD diss., Murdoch University, 2017), 2.

[8] Grove and Adamson, *El Niño in World History*, pp. 96–97.

[9] 'Official Report: Ministry of Overseas Affairs,' *La Esperanza* (19 Dec. 1867).

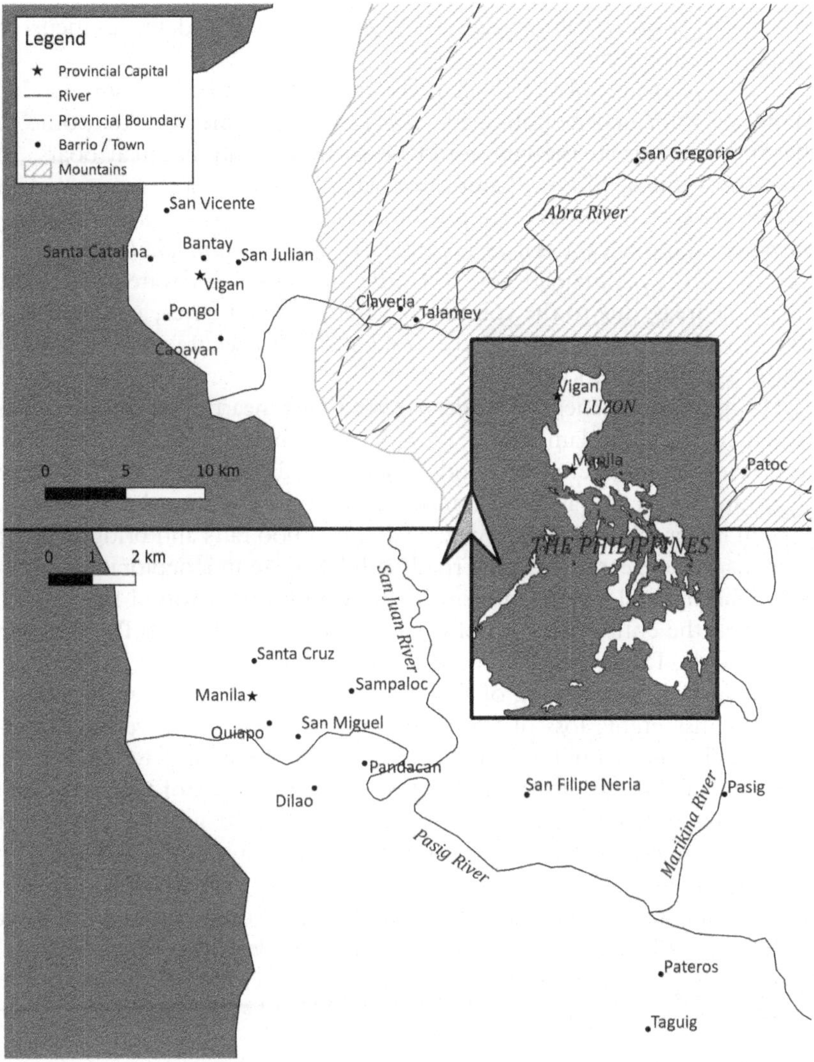

Fig. 7.1 Map of the Philippines, with close-ups of the lower Abra River and environs (above) and Manila (below). Drawn by Philip Gooding

to Manila's tobacco factories and warehouses, as well as damages incurred to this important crop.

The overflowing of the Pasig River and the estuaries that surrounded Manila transformed neighbouring areas into an immense lake. Some of the busier streets in the city could be traversed only in small boats. In the *barrios* of San Miguel, Sampaloc, Quiapo, and Santa Cruz, municipal patrols constantly moved around the flooded areas on *bancas* (double-outrigger dugout canoes), as members of the *guardia civil*, the public security force, maintained law and order in waist-high waters. In some low-lying areas, due to the volume of water in the estuaries, assistance was also provided to save endangered residents or to secure their makeshift homes against the floodwaters.

The governor general's report stressed that nearby provincial towns presented the same lamentable picture, with the level of floodwaters in crop fields varying between 1 and 2 meters, and consequently preventing communication; all travel and transport had to be conducted using *bancas* or even larger sailing vessels. All cane and bamboo rafts and bridges in the rivers had disappeared, and the roads still left were in a deplorable condition. The tropical storm, in terms of food insecurity around the capital, destroyed the entire rice crop of the towns of Pasig, Taguig, Pateros, San Felipe Neria, Dilao, and Pandacan.[10]

The sudden overflowing of all the rivers and streams in Central Luzon caused considerable loss of livestock, and the winds destroyed innumerable small homes built of cane and bamboo belonging to poor rural folk; generally the well-off families lost only the rooves of the wood and stone houses that had often been built on elevated ground to protect them against floods. The report stated that fruit trees were uprooted in large numbers, while the coffee and cocoa crops near Manila were also largely destroyed. It was against this backdrop of desolation and disaster in Manila and its environs that some *gobernadorcillos* (town officials) from Ilocos Sur arrived in the capital on 27 September to report on the horrendous scale of the calamity in parts of Northern Luzon.

[10] 'Philippine Islands: Overseas,' *La Espana* (17 Dec. 1867), 1–2.

The Distant Setting: The Abra River Valley and Flood Plain

Between 1565 and 1815, the Spanish Philippines was utterly dependent on the *situado*, a financial administrative grant in aid, and the Manila galleons which carried trade items from China and other parts of Asia via the Philippines and Mexico to the West. The chronic sinking of these Manila galleons in typhoons exposed the vulnerability and economic dependence of the colony. The galleon trade ended in 1815, and the Spanish colonial government opened the Philippines to foreign trade and monocrop agricultural development—tobacco, abaca, sugar, and coconuts.[11]

In the late-eighteenth and nineteenth centuries, this monocrop agriculture developed apace. Lowland forests disappeared to be replaced with monocrop farming, populations increased, and pioneering settlements encroached on upland watersheds. The great rivers of the Philippines that flow into the sea remained largely untamed because major flood infrastructure programs for construction, rehabilitation, and improvement projects were not prioritised until the 1880s. The disastrous effects of flooding and typhoon damage led to severe loss of income, famine, and disease because the pioneer farmers no longer produced sufficient food crops. In the Ilocos provinces, periodic storm-related floods were unleashed by the Abra River.

The Abra is one of the largest and longest rivers in Luzon. Rising in the Cordillera Central in Northern Benguet, it flows through the Ilocos Mountains with several other rivers. Near the town of Vigan, the provincial capital of Ilocos Sur, it forms a flood plain that is 64 kilometres long at its lower end. Historically, population densities in Ilocos Sur exceeded the colonial average. While the Ilocos coast was one of the most densely populated regions in the nineteenth-century Philippines, the mountainous province of Abra had a relatively sparse population that was largely concentrated upon the narrow flood plain of the Abra River.[12]

[11] Ed. C. de Jesus, *The Tobacco Monopoly in the Philippines: Bureaucratic Enterprise and Social Change, 1766–1880* (Quezon City: Ateneo de Manila University Press, 1980), 22–46; John A. Larkin, *The Pampangans: Colonial Society in a Philippine Province* (Quezon City: New Day Publishers, 1993), 41–62.

[12] Frederick L. Wernstedt and Joseph E. Spencer, *The Philippine Island World: A Physical, Cultural and Regional Geography* (Berkeley, CA: University of California Press, 1967), 332–33.

Each rainy season, lasting from June to November, typhoons brought renewed threats of floods. People living along the banks of the Abra River, or nearby, paid close attention to its flowing water because of the life-giving and death-dealing potential it claimed upon their daily lives. Across the centuries, it has served or destroyed generations of Ilocanos and their livelihoods. The memories of the settlers living nearby can only partially reveal the struggles to contain its turbulent waters as well as the disillusionment and pain felt at failed efforts. What has this river, located in remote north-western Luzon annually accomplished during its long, frequently unrestrained passage to the sea? First and foremost, the Abra helped provide food for the pioneer farmer's table. It also became the primary conduit via which to convey recently introduced monocrops, like tobacco, to market towns and coastal ports. In the nineteenth century, these crops were cultivated by peasant farmers at the behest of the Spanish colonial government.

The Flood Event

The 'great flood' for the nineteenth-century inhabitants of the Ilocos region occurred between 25 and 27 September 1867. The typhoon sent so much water rushing down the steep hills of the Abra Valley that it formed a fast-moving wall of uprooted trees, boulders, dead animals, houses, and human corpses, that were swept along by the torrent as the riverbanks collapsed under the deluge and pressure. Once the Abra breached its banks, due to its rising height, speed, and violent force, the raging river bulldozed everything in its path, swamping buildings and drowning people *en masse*. This extraordinary flood, when the surging waters reached a height of 25 meters above its normal level, killed 1800 persons and many thousands more farm animals.[13] The chronicle of Santa Maria, Ilocos Sur, called it 'the greatest flood ever experienced in this town,'[14] while in the chronicle of Santa Domingo, Ilocos Sur, it was simply called the 'great flood.'[15] But it was the local historian of Tayum who commented on the vast scale of death and destruction,

[13] Miguel Selga, *Charts of Remarkable Typhoons in the Philippines, 1902–1934, Catalogue of Typhoons 1348–1934* (Manila: Bureau of Printing, 1935), 35.

[14] Archives of the Manila Observatory (hereafter: AMO), Selga collection, Bx.10, It. 37 'Floods,' 25–27 Sep. 1867, 14.

[15] Ibid.

stating succinctly that a great many people had died in the province of Abra because of an extremely strong typhoon and deadly flood.[16]

The politico-military governor's report concerning the catastrophe in the districts of Abra and Ilocos Sur, belatedly sent from the stricken province, begins with a detailed description of a change in the weather and the sudden arrival of the typhoon, based on eyewitness accounts.[17] On the afternoon of 23 September, a weak southwest storm front began to prevail, bringing some rain showers by evening. The wind picked up and grew colder as the rains increased, but without arousing undue fear of a major storm. Then, on the 24th, the winds dropped, and the rainfall diminished. This weather then remained constant until 8 o'clock in the evening, when the wind worsened and the rain became heavy, raising fears in Vigan, and on the flood plain, that a *baguio*, or typhoon, was threatening. However, these fears were mistakenly allayed because, although the force of the wind progressively increased, and the rain was torrential, the storm had remained in the same space with only a slight shift in direction from the south to southwest.

But, between 2 and 3 o'clock in the morning, just before dawn on the 25th, the wind suddenly changed direction to the west, constantly gaining strength in the process, and remaining in this powerful state for more than two hours; and it was during this time of transition while still very dark that the flood occurred, inundating the entire plains of the Abra district. While the local inhabitants slept, the flood caused a significant loss of life and goods. When startled awake, the terrified people perceived the dangerous situation, and some of them realised that their sole hope of salvation was to be found in nearby trees and tall bushes. Most individuals were still tired, dazed, and frightened about the safety of their children. In that moment, desperate fathers and mothers forgot about their own personal safety to assist those 'so precious to their hearts.'[18] However, the pitch-black night prevented many from seeing an extended hand of help. The rampaging force of the river currents and this general lack of organised assistance caused panic and numerous deaths. At midday on the 25th, the storm began to wane and the floodwaters receded, but not fast

[16] AMO Selga collection, Bx.10, It. 47 'Records of floods in the Philippines, 1691–1911,' 14.

[17] 'Philippine Islands: Overseas,' 1–2.

[18] Ibid.

enough for desperately needed aid to be administered to those who were fortunate enough to still be alive—even if displaced and in precarious circumstances.

It was not until 1 October, a full week later, that the beleaguered political-military governor of Ilocos Sur was able to report to Manila, from the provincial capital of Vigan, on the scale of the unfolding catastrophe. He stated that due to torrential typhoon rains, a great flood had taken place, one larger than any recollected in living memory. The water level, he wrote, was 5.5 meters higher than during any of other large floods of the nineteenth century. As a result, Vigan proper had almost entirely disappeared under the floodwaters, except for the administrative centre of the city located on high ground, which included the governor's headquarters. Of nearby farmlands, the governor ominously wrote, 'There is nothing useful left.'[19] At this stage, the exact number of residents that were either dead or missing was still unknown—the river dragged away many, while many others were buried under layers of mud and debris deposited by the rushing waters inundating the landscape, by the shifting sand in other places, and by the piles of broken trees and foliage that created chokepoints in the river.[20] In any case, by the time of this first report, from among a population of around 10,000 in Vigan, more than 600 corpses had already been interred.

Outside of Vigan, the Abra River, as it overflowed at its mouth, extended for over 10 kilometres, inundating the nearby villages of Bantay, San Vincente, Santa Catalina, Santa, and Caoayan with 2 meters of water. The floodwater overspread into the interiors of churches and destroyed entire *barrios* in the process—among them San Julian, which was one of the smaller communities that suffered the most damage. The first news sent to Manila concerned the number of cadavers found upon undertaking initial investigations, the number of people missing in these towns, as well as the countless people saved by the concerted efforts of the *gobernadorcillos, subalternos* (the *gobernadorcillos'* local subordinates) and *cuadrilleros* (municipal policemen). The governor noted the local people had been caught off guard and were surprised, if not stunned, by the extraordinary height of the floodwaters. He foreshadowed in his initial report that he expected there would be far more bodies discovered 'as

[19] Ibid.

[20] Ibid.

the radius of the search area for casualties extended further and further away from Vigan, as well as taking into account the height of the flood levels in particular areas.'[21]

In Patoc, only crops had been lost, but in neighbouring Talamey, apart from goods, a 'grievous' number of people had also perished. The armed forces stationed there sought refuge in the ceiling and rafters, over 4.6 meters above the floor, of their barracks. But, despite their desperate efforts, only around half a metre separated them from the rising flood-waters. They stayed crouched in the ceiling rafters—wet and fearful—for two days with nothing to eat except for a few husks of roasted corn. The well-built barracks were in danger of collapsing, while all the houses in the town had been destroyed. So, as soon as the floodwaters began to recede, the trapped troops climbed down from the rafters and fled to the nearby *barrio* of Claveria.[22] In neighbouring San Gregorio, the governor recorded the tragic obliteration of the town, with the total loss of all goods and the deaths of a number of unidentified individuals. The rest of the inhabitants survived by fleeing to the nearby mountains where, as a consequence, they suffered hunger and starvation.[23]

The rice fields that were ready to harvest were almost entirely lost; some were smothered in sand, others covered with a mixture of sand and stones, and still others were submerged by a thick layer of river mud full of debris. The stocks of the previous year's maize harvest and the current stores of rice were damaged or destroyed in towns throughout the flood plain and valley. The desperate authorities in Vigan stressed the relief measures the battered populace needed to take to stave off famine. But the sowing of nutritious food crops was near impossible in such deplorable conditions, and on such short notice. The problem of planting the proposed emergency crops was hampered, not only by the devastated state of the fields, a lack of farm implements, and a scarcity of draft animals, but also by a lack of labour required to do the farm work, as most flood survivors were considered physically and emotionally unfit to undertake manual labour for quite some time. Many able-bodied men had already perished due to the 30 traumatic hours they had suffered through in agony, without eating or sleeping, while enduring the onslaught of

[21] 'Official Report: Ministry of Overseas Affairs,' 2–3.

[22] Ibid.

[23] Ibid.

extraordinary weather. Others had been sorely frightened, fearful that the weak trees sheltering them would not be able to withstand the strength of the extreme winds or the swift pace of the river's currents before the storm subsided and the floodwaters receded. Many were psychologically overwhelmed and in pain from the harrowing family scenes they had witnessed.[24] Indeed, each individual case provided a powerful emotional reason to cause the survivors to feel shattered, which contributed to extreme illness and nervous breakdowns.[25]

The governor lamented the fact that it was not yet possible to either traverse or properly survey, even at a distance, the plains because of the enormous layer of slime deposited on them by floodwaters, as well as the serious hazards to travel posed by fallen trees, damaged houses, and the foliage and waste piled up by the flood. This huge, entangled mass of debris obstructed communication and travel to and from Vigan. To further complicate matters for the local authorities, most of the roads and all bridges were destroyed, making it impossible to travel from one place to another by horseback, and only with great difficulty on foot. In the aftermath of the flood, the journey on foot between Talamey and Vigan, a distance of only around 16.5 kilometres, took twelve hours to complete.[26]

In such disastrous circumstances, it soon became apparent that Abra Province had suffered a sudden drastic decline in its population size and economy. In this state of emergency, to avoid the possibility of social-political collapse, the political-military governor knew he had to repair the main roads. However, he could not devote his immediate attention to this crucial task due to the lack of labour and building supplies. He also recognised that it was equally important to attend to the critical needs of the innumerable families spread across the plains who were reduced to begging for help, as there were very few individuals who could deliver the aid required to those so desperately in need.[27]

In Vigan itself, the needy and suffering received clothes, blankets, and food, and they were given as much personal aid as possible thanks to the generosity and assistance of some wealthy individuals and the charity of religious organisations. On that same day in the capital (27 September),

[24] Ibid.

[25] 'Philippine Islands: Overseas,' 1–2.

[26] Ibid.

[27] Ibid.

the Spanish authorities with local assistance also began the gruesome job of gathering together the corpses closest to the capital. Nearly 300 bodies were collected and, late in the day, the *gobernadorcillos* of Tayum and Pidigan, having overcome incredible obstacles, reported to the governor the scale of the damage also inflicted upon their towns, as well as the estimated number of victims. The extent of the flood was such that it would 'never fade from the memory of those who had survived it; [it was] a flood that had no equal in their recorded history.'[28]

Reconstruction, Recovery, and Trauma

In late September 1867, the survivors of the flood—the vast majority homeless—now faced new challenges. Some of the first to arrive in Vigan were women, children, the indigent, and elderly, who were the most vulnerable and who came from the poorest sectors of society. Running water and hygiene were major concerns, as were a lack of clothing and blankets for the children. Cholera, typhoid, pneumonia, and diarrhoea had exacted a heavy toll in the aftermath of previous flood events.[29] Ominously, desperate survivors washed and relieved themselves in filthy water. Others drank from polluted wells and pools of contaminated rainwater.[30] Local officials thus moved quickly to prevent outbreaks of flood-related diseases after the rancid waters had receded from fields, village streets, and town plazas. When the floodwaters had subsided, corpses and dead livestock constituted a further and grave public health problem, as did the logistical problems of travel on damaged or destroyed roads.

To further complicate matters, locusts appeared in massive numbers in the aftermath of the flood, destroying the remains of the rice crop.[31] Colonial officials on the spot in Ilocos Sur feared that the wholesale destruction of these rice fields and of tobacco seedbeds would inevitably

[28] 'Official Report: Ministry of Overseas Affairs,' 2–3.

[29] See: Linda A. Newsom, *Conquest and Pestilence in the Early Spanish Philippines* (Honolulu: University of Hawaii Press, 2009); Dean C. Worcester, *A History of Asiatic Cholera in the Philippine Islands* (Manila: Bureau of Printing, 1908), 3–15; Ken De Bevoise, *Agents of Apocalypse: Epidemic Disease in the Colonial Philippines* (Princeton: Princeton University Press, 1995), 164–84.

[30] 'Official Report: Ministry of Overseas Affairs,' 2–3.

[31] 'Philippine Islands: Overseas,' 1–2.

lead to destitution, food shortages, and famine. Thousands of families had been displaced and left homeless; many survivors, by late January, were experiencing hunger and urgent requests were again made for additional financial assistance and relief aid.

A first measure post-typhoon was the digging of mass graves. On 26 September, the first cadavers began to be gathered—those found close to Vigan. The number reached 173 on that day, but had reached 600 by the next, with hundreds still missing.[32] Authorities began preparing mass burials to minimise health risks as the death toll rapidly escalated. Rescue and relief efforts outside the capital were being spearheaded by government troops and able-bodied survivors determined to make sure that not a single corpse, as far as possible, would be left unburied. They were doing this gruesome work for public health purposes. The bodies were decomposing and there was no place where they could be stored—not in a still enclosed building, and not in a church. Many of the Vigan dead were just piled up and laid outside partially gutted buildings.

Those that went in search of the cadavers found great numbers who perished violently wedged in the branches of uprooted trees.[33] A lot of cadavers and dead livestock of all types were scattered along the entire course of the river and as far as the shores of the sea. Some carabao and a number of bulls and cows had been pasturing in nearby mountains on the night of the flood. On the coast, some distance from Pongol, the governor with two companions found the whole beach strewn with all kinds of drowned animals, but they had to retreat due to the foul stench of bloated decomposing carcasses.[34] The burial teams, including all male survivors mustered from Vigan, organised by *gobernadorcillos, subalternos*, and *cuadrilleros*, worked around the clock for five days gathering cadavers and burying them, and burning animals. In one of the worst-hit *barrios* of San Julian, a provisional cemetery was prepared for the internment of those bodies that could not be brought to the town cemetery. By 28 September, 380 cadavers had already been buried in this race against time to stave off a possible epidemic, and 440 persons remained missing: a total of 820 victims. However, the governor believed that the losses in lives would eventually exceed more than 2000. He stressed, several days later,

[32] 'Official Report: Ministry of Overseas Affairs,' 2–3.

[33] Ibid.

[34] 'Philippine Islands: Overseas,' 1–2.

in an urgent letter to Manila, that more than 600 corpses had now been recovered, and that over 950 houses and 6500 animals had disappeared in the rushing waters.[35] He emphasised that the scores of people who had been swept far away by the river and the dead animals would be buried and burned in other localities, some distance away from the proposed cemeteries.[36]

It is important to stress here that, in normal circumstances, more people died from infectious diseases annually than as a direct consequence of floods and typhoons.[37] Thus, the threat of cholera caused Spanish soldiers and local volunteers to hastily collect the decaying corpses, piling them up for mass burials and cremations, contrary to traditional Roman Catholic burial procedures and practices. The huge number of cadavers and the innumerable dead animals—both great and small—and their state of putrefaction compelled the authorities to quickly order the preparation of makeshift cemeteries in convenient spots where the rotting corpses could be interred. They also indicated other places where uncontaminated bodies were to be buried to prevent the spread of cholera, and to defend public health.[38] The number of people who drowned was soon estimated at more than a thousand, and as the number of animals rose to more than several thousand, a majority of which had not yet been burned, it was clear that this operation was going to last well into October. The governor noted so many animals were in a state of advanced decomposition that the stench of the foul air made it difficult to breathe in places.[39] However, despite the establishment of provisional cemeteries and the designation of places where countless dead animals were to be either buried or cremated, the medical practitioners of the colony believed for a brief time that public health had been compromised.[40]

[35] Selga, *Charts of Remarkable Typhoons*, 35.

[36] 'Philippine Islands: Overseas,' 1–2.

[37] See: Norman G. Owen, *Death and Disease in Southeast Asia: Explorations in Social, Medical and Demographic History* (Singapore: Oxford University Press, 1987), 8–16; Luis Dery, *Pestilence in the Philippines: A Social History of the Filipino People, 1571–1800* (Quezon City: New Day Publishers, 2006), 57–144; Newsom, *Conquest and Pestilence*, 16–52.

[38] 'Philippine Islands: Overseas,' 1–2.

[39] Ibid.

[40] 'Official Report: Ministry of Overseas Affairs,' 2–3.

The Newspapers

In newspaper accounts of the time, there is a wealth of information about the great Ilocos flood of 1867, increased exposure and vulnerability due to the event, and colonial disaster relief. The large number of newspapers that followed the flood disaster became an integral part of this unfolding tragic historical event and how it was imagined. Philippine newspapers with daily circulation in Manila and provincial capitals included *Diaro de Manila* and *Gaceta de Manila*, while Spanish broadsheets, that covered the story on the peninsular between November 1867 and January 1868, included *La Esperanza*, *La Espana*, *La Epoca* and *El Imparcial*. These records fortunately, for historians, still exist. They remind the Philippines, but especially the people of the Abra Valley, about the flood's calamitous past. Additionally, the news articles also bear witness to a past marked by the increasing intensity and scale of cyclonic storms over the course of the nineteenth and twentieth centuries, a phenomenon which is related to the effects of global warming.[41]

The September 1867 flood tragedy in Ilocos Sur occurred only four years after the great Manila earthquake of 3 June 1863, one of the most dramatic and violent catastrophes in the city's history.[42] In late September 1867, ruined buildings still dotted the landscape, including the celebrated cathedral. While constantly viewing scenes of shattered buildings remained a disturbing reality in the everyday life of Manila's populace, the capital's consciousness was once again preoccupied with confronting the damage inflicted by a large natural hazard on 24 September 1867. However, at this point, Manila's residents were not aware yet of the extent of the flood disaster that had rocked Northern Luzon, especially in Abra Province, until newspapers began to run breaking stories on the scale of the tragedy. But they soon grew accustomed over the next several months to reading newspaper articles and published official accounts about the Northern catastrophe, its flooded environs, and its displaced people. News articles concerning the crisis began to appear on the front

[41] See: A. Henderson-Sellers, H. Zhang, G. Berz, K. Emanuel, W. Gray, C. Lansea, G. Holland, J. Lighthill, S.-L. Shieh, P. Webster, and K. McGuffie, 'Tropical Cyclones and Global Climate Change: A Post-IPCC Assessment,' *Bulletin of the American Meteorological Society*, 79 (1998), 19–28.

[42] See: Susana Maria Ramirez Martin, *El Terremoto de Manila de 1863 Medidas Politicas y Economicas* (Madrid: Consejo Superior De Investigaciones Cientificas, 2006), 37–56.

pages of Manila's broadsheets just three days after the remarkable storm struck the capital.

During the second half of the nineteenth century, when most rural Filipinos under colonial rule lived in difficult social and economic circumstances, the sensitive news reporter sometimes could not get it all in 'at-a-glance' in late September 1867. For a historian attempting to reconstruct such a disastrous event, these published flood accounts still do not do full justice to its enormous impact on the society and environment. One is left astonished at the difference between reading the big picture reports of the calamity in the press, as opposed to the official eyewitness accounts of those who were there on the spot, and bore witness to the actual extent of the physical damage and loss of life and property, such as, for example, the archived testimony surrounding the remarkable deeds of Valentina Mendoza (see below).

The official reports reaching Manila slowly provided a treatment of this unprecedented flood event. The embattled governor general of the Philippines systematically released extracts from these reports to the Press. Thus, as a historical source, these newspaper accounts devote attention to the concrete and specific, and, as a journalistic piece of historical reconstruction, they shed light on the great flood and its victims, largely inarticulate and rarely documented as individuals, while eschewing the sensational. The governor general believed full-scale press coverage of this catastrophic event 'as it happened' would have a crucial positive effect on the thinking and generosity of the donor public in the Philippines and Spain.

While the spate of articles could not restore to Filipinos in the capital and countryside the damage to life and property inflicted by the workings of the flood, they did graphically depict the crucial role typhoons played in the life and history of the colony at both national and local levels. These articles vividly captured the plight of the Spanish governor general and peasant farmers of Ilocos Sur who were both inextricably bound together as product and fate by extreme weather. The articles also drew specific attention to the precarious lives Ilocano farmers led in the Abra Valley in the second half of the nineteenth century.

The economic and political costs of this remarkable typhoon proved immensely troubling for Spanish colonial officials. Hence, newspapers, especially in Manila, in the years following this event continued to publish the reports of governor generals and local officials about particular natural disasters, in order to provide victims with a voice and moral support. The strategy also helped to raise additional funds for emergency relief. It was against this background that the beleaguered Manila government, in an effort to respond to the disturbing scale of the flood accounts appearing in the news, launched a public subscription in the colony, and appealed to the Spanish Crown to do likewise in Spain, in order to meet the disaster relief needs of the unfortunate people of Ilocos Sur for family assistance and reconstruction of their houses.

The Spanish Government and people of Spain learned about the extreme typhoon and floods in Ilocos Sur and elsewhere on Luzon from belated newspaper accounts in mid-December 1867. The news articles were based on extracts from the official correspondence of the civil governor of the Philippines. His reports about the unfolding flood catastrophe had been unduly delayed because of extreme weather. Steamships transporting official mail via the Cape of Good Hope had to temporarily cancel their scheduled departures. One valiant mail boat attempted to leave Manila Bay despite the impending storm, but it was forced to turn back and to seek shelter at Cavite shortly before the typhoon crossed over Manila. For this reason, detailed news of the significant disruption that this devastating El Niño-related storm had caused, particularly in Northern Luzon, did not reach Spain and take hold of the public imagination there for nearly two months—arriving sometime in November.[43]

On 17 December 1867, extracts of official letters from the Philippines published in Spanish newspapers, with news that only covered events up to 23 October, revealed news of the terrible disaster in Ilocos Sur to the Spanish public. These early reports and newspaper accounts, however, stressed that the government in Manila did not have any further news about these tragic events in Ilocos Sur at that time. But it was believed, based on the limited evidence at hand, that the typhoon had caused considerable damage and personal loss. These early incomplete disaster accounts all ended abruptly, simply stating the head of the beleaguered province, the politico-military governor in Vigan, was trying his best to respond to the scale of the damage and loss suffered in his province. As Spanish newspaper accounts began to appear from this point onwards,

[43] National Archives of the Philippines (hereafter: NAP), folios S244–S245 Don Fernando de Santa Coloma to the King of Spain, 15 June 1868, in 'Valentina Mendosa Provides for Flood Victims,' Varia Provincias-Ilocos Sur, 1796–1898.

Spain's public belatedly learned about the damaging extent of the great flood, causing widespread food insecurity and a sudden dramatic loss of agricultural production.

There was also critical concern expressed in the Spanish press in November 1867 over the government's lack of contingency planning in a colony where floods occurred annually, but especially during the typhoon season. Thus, papers like *La Esperanza* called for rapid deployment of funds, food, and medicine on behalf of the survivors located in Ilocos Sur. Additionally, some articles portrayed the resilience and courage of individual Filipinos and Spaniards in the midst of the horrendous impacts of the storm and flood. On 31 December 1867, *La Esperanza* received a batch of periodicals from Manila that contained news about the disaster up until 7 November, which recounted the situation in the capital. The typhoon had left parts of Manila submerged, but the articles mentioned the efforts being made to return to business as usual. *La Esperanza* reported that the authorities in Manila were selectively repairing major buildings damaged by the typhoon, and that the public subscription, to aid the 'misfortunate' who had lost virtually everything they possessed in the storm and flood, had reached 120,000 pesos. It was also noted that the inmates of Manila's prison continued to provide 'indispensable services,' leaving the prison daily to repair the streets and small channels damaged by the flood.[44]

The same day, 31 December 1867, *La Espana* published, under the column heading 'Overseas,' a detailed, albeit moving, account of the 'misfortunes' that had occurred in Ilocos Sur, due to the September flood: Dead inhabitants: 1145; Lost ships: 5 schooners, 14 *pontines,* 16 *pancos,* 7 *bancas,* 10 *paraos* and 12 *lanchas* and *botes*; Livestock: 1141 carabaos, 1809 cows, 1270 horses, 730 pigs, 349 rams and goats.[45] The article noted that the citizens of Vigan, although located on high ground, had suffered great personal tragedy, while Santa Catalina was the village that had lost the most livestock. According to the article, based on the report of the political-military governor of Ilocos Sur, the subscription raised in Manila to help alleviate these calamities had already produced satisfactory benefits. Thousands of dollars (*duros*) had already been sent to the 'poor souls' and a large amount of clothes had been handmade for them by the

[44] 'Philippine Islands: Overseas,' *La Esperanza* (31 Dec. 1867).

[45] 'Overseas,' *La Espana* (31 Dec. 1867), 2.

women of Manila, and even by girls mustered into service from Manila's municipal school.

The stricken political-military governor stressed the horrific nature of what had occurred, and felt it was senseless to discuss everything in detail. Instead, in his report, he asked readers to imagine the thousands of people, caught by surprise, who had clambered up trees in the dead of night to seek refuge, calling out for aid. But it was impossible to reach them. He then stated he could not find appropriate words to praise the merits of the parents and mothers who managed to save their children, nor could he conceive of how those parents managed to clamber up into nearby trees under such extreme duress with their children, many of whom were still infants. He considered it a miracle stating, 'providence had enabled this to occur in many of these cases,' and then provided the following example:

> A woman had just given birth to a son the day before the flood: this baby arrived, and then the husband carried the newborn child and climbed to the top of a cane tree. His wife then followed him. She grabbed hold of another branch, and then her sister-in-law was entrusted with carrying the other little child of that marriage, and she also then climbed up the tree. The wind forced the tree downward, swaying back and forth and submerging them momentarily in the rushing water. The branches rose up again and these unfortunate ones surfaced. These forced 'baths' were repeated over the thirty hours it took for the flood waters to recede: none of them ate, the newborn could not even suckle his mother's breast during the entire ordeal. All lived, my friend: nothing is impossible through Divine Providence.[46]

It is no wonder that the governor general lamented, shortly after news of the flood tragedy in Ilocos Sur gradually emerged, that there had been an excessive number of calamities—great and small—in the colony in recent years. He also realised in this situation the newspapers were the most powerful and persuasive entity at his disposal, in order to shape public opinion to support a general subscription to remedy the flood calamity

[46] Ibid.

the colony now faced. The Spanish Treasury likewise published such official reports from the Philippines in the *Gaceta de Madrid*, describing the terrible nature of this typhoon and flood, for similar reasons.[47]

The General Committee for Aid for the Philippines and Puerto Rico

The typhoon, flood, and hurricane accounts published in the newspapers of Manila (and later in Madrid) in the weeks and months following the disaster in Ilocos Sur contributed to the formation of the General Committee for Aid for the Philippines and Puerto Rico—the latter place also experiencing one of the most destructive hurricanes in its history in 1867.[48] The royal decree of 10 December, which ordered the opening of a general subscription in Spain and its overseas colonies with the full support of the King, had the stated purpose of obtaining voluntary contributions from businesses, government organisations, religious institutions, and the Spanish population.[49] The King convened the committee to aid the at-risk inhabitants of the Philippines and Puerto Rico on 14 December, taking full advantage of the prestige associated with his high office, in order to ensure the agreement of the clergy and religious orders in making voluntary contributions to the appeal, as well as helping to raise funds in their parishes and bishoprics.[50]

The *Gaceta de Madrid* that day (10 December) contained a notice of the prominent persons whom the King had appointed to the committee, *Junta de socorros*, that would oversee the donations destined for the Philippines and Puerto Rico. He appointed the Cardinal Archbishop of Toledo to the position of vice president, with the members being: the Duke of Medinaceli, the Marquis of Socorro, the Marquis of Perales, the Patriarch of India, the Marquis of Novaliches, the Marquis of Manzanedo, the Marquis of O'Gavan, and the Count of Govenache. The secretary was Don Salvador de Albacete y Albert. The committee, convened in

[47] 'Official Report: General Committee for Aid for the Philippines and Puerto Rico,' *La Esperanza* (20 Dec. 1867), 2.

[48] Orlando Pérez, *Notes on the Tropical Cyclones of Puerto Rico, 1508–1970* (Pre-printed) (Report. National Weather Service), 16: https://www.aoml.noaa.gov/hrd/data_sub/perez_11_20.pdf [accessed: 16 Apr. 2021].

[49] 'Overseas,' 2.

[50] 'Official Report: General Committee for Aid,' 2.

one of the salons of the Royal Palace, then set out the formal arrangements that would govern the appointment of the sub-committee for each province, area, and parish. At the province level, the committee comprised the governor, a provincial deputy, a member of the clergy appointed by the prelate, a provincial consultant, the *regidor sindico* (union councillor), and a leading taxpayer to be chosen by the city council. Local committees in the area were composed of the mayor, the senior parish priest, a *regidor*, and a large taxpayer also chosen by the city council. In the parishes the committee was to be formed by a parish priest, a member of the city council, and two residents chosen by the council.[51]

The King also released superior orders so that the funds collected could be deposited in the Caja General de Depositos (central bank) and its branches without undue difficulty. The Crown placed their principal confidence in the church from whom it hoped to receive considerable assistance; it could do no less than trust the clergy in such difficult times, relying upon them as the main support for its charitable aspirations. The King and Queen looked upon the reverend priests and clergy as determined protectors of the faith and fervent apostles of all charitable works. The religious were charged not only with accepting cash contributions irrespective of the amount, but also those donations that were in-kind: in the case of the latter, they were to be sold immediately by the parish council at current prices, and the profits were to be deposited in Madrid in the *Caja General de Depositos*.

The *junta* created by the Royal Decree of 10 December wasted no time in fulfilling one of its initial duties when it directly requested all Spaniards, who 'felt in their hearts a desire to do good,' to urgently assist 'the unfortunate inhabitants of the Philippine Islands and Puerto Rico, victims of the hurricanes, floods and earthquakes that recently occurred.'[52] The official report of the General Committee, published in *La Esperanza* on 20 December 1867, stressed the leading role and compassionate zeal of the King and Queen as always being the leaders in 'drying the tears of their faithful subjects and consoling them in their misfortunes with all kinds of benefits.'[53] News readers learned the royal couple was doing everything possible to alleviate the suffering of their

[51] 'Official Report: Ministry of Overseas Affairs,' 2–3.

[52] 'Official Report: General Committee for Aid,' 2.

[53] Ibid.

loyal subjects and overcome the tyranny of distance separating them from one another. They were using the most rapid means of communication—steamboats and telegraph—to transmit to those beleaguered colonies news from Madrid, regarding what decisions had been recently resolved on behalf of their subjects, due to the strong concerns of their King and Queen.[54]

By mid-January 1868, the Royal Orders promulgated by the Ministry of Overseas Affairs were already being implemented in the Philippines. The colonial authorities, in order to provide relief from the havoc caused in Ilocos Sur, quickly authorised the extension of the use of donations and local funds in areas where it was absolutely essential, especially on the Abra River Flood Plain. The government in Manila also recommended, wherever possible, the gainful employment of local labourers on flood reconstruction projects. It gave assurances that there would be proper payment of bills of wages, and mandated the inclusion, in preparation of the next budget, of a provision for public calamities.[55]

Worthy of Commendation

The governor general's local subscription initiated in Manila to help alleviate the calamity in distant Ilocos Sur produced satisfactory results by late December 1867. By then, funeral honours had been held for those who died in the flood. On that occasion, the congregation who gathered in the partially destroyed cathedral in Vigan, canonically known as the Metropolitan Cathedral of St. Paul the Apostle, was in extreme mourning: the mass was solemn.[56] The governor general's report to the Crown mentions the solemnity of the occasion, as well as several prominent local men from Vigan, whose flood deeds were considered above and beyond the call of duty. They were singled out for commendation, including prominent local figures such as Don Teodoro Reyes and a former captain, Don Marcos Alegre, who—crossing the raging river in boats loaded with provisions—were able to save 'many unfortunates, making them deserving of the blessings of the whole town.'[57]

[54] Ibid.

[55] *El Imparcial* (13 Jan. 1868), 2.

[56] 'Philippine Islands: Overseas,' 2.

[57] 'Official Report: Ministry of Overseas Affairs,' 2–3.

However, while in the midst of compiling such accounts of courage and sacrifice to report to the Crown, the governor general had not yet stumbled across the illuminating story of a remarkable female first responder from the flood ravaged barrio of St. Julian in the Abra River Valley. With local tales circulating about the incredibly level-headed and courageous Valentina Mendoza, the faraway Spanish Government in Manila was to learn, not only what a capable woman of some means could do as a self-appointed community leader in this flood crisis, but the widely held Spanish idea that women were worse under pressure in such a situation than men was simply proven wrong. The colonial authorities in the capital belatedly learned that this indigenous woman (Tinguian), the wife of a wealthy Chinese mestizo trader, had saved over two hundred local lives at the height of the flood.

At the crack of dawn, she stood staring out one of the windows of her large finely built wood frame house that was raised high off the ground with post-and-beam joinery. It stood well above the rushing waters. She felt, in that moment, an urgent need to help save her neighbours, who were trapped in their small homes of cane and nipa or who clung desperately to branches in nearby trees—trees that were at the mercy of the raging floodwaters and in danger of being dragged in the direction of the sea. Mendoza, observing the unfolding tragedy from her window, immediately sent rafts, crewed by her servants, to help those in need of assistance, especially children and the elderly. They were brought across to her large safe house, where she opened her wardrobe and trunks, distributing all her clothes, in order to cover the naked and freezing bodies of the survivors. With foresight she ordered staff to light a fire in the fireplace, using her wooden furniture as fuel, to warm up the hundreds of freezing, exhausted, and frightened people who were now in her house. During the ordeal, she fed them and emptied her pantry in the process. As floodwaters receded, Mendoza recognised that the recovery process was not going to happen quickly. Consequently, when the rice in her pantry ran out, she sent her servants and a number of able-bodied displaced men to Vigan to purchase several *cavans* of rice to distribute among her neighbours. Subsequently, she continued to provide a point of trust amidst the trauma and destruction, allowing some families who had lost their homes and possessions to stay in her house long after the floodwaters receded.[58]

[58] NAP folios S203–S245 'Valentina Mendosa Provides for Flood Victims,' Varias Provincias-Ilocos Sur, 1796–1898.

In May 1868, Don Mariano Sales, the *gobernadorcillo* of the mestizo community of Vigan, handed a remarkable document, certified by the parish priest of Vigan, to Don Fernando de Santa Coloma, Naval Captain of the Port, to forward to Manila. The document certified that Valentina Mendoza was a middle-class woman of the wider Vigan community residing in the *barrio* of St. Julian, which had been particularly hard hit by the typhoon and ensuing flood. Described in the report as pious and charitable, Valentina Mendoza had opened her home and heart to her neighbours in their hour of greatest need and undoubtedly saved the lives of scores of people. Her selfless actions earned her the admiration and gratitude of the entire community in Vigan. In an effort to honour the character and sacrifices made by this brave woman, Don Fernando de Santa Coloma, in his covering letter to the Crown, dated June 15, summed up the praiseworthy and meritorious efforts employed by this 'noble woman' to save her neighbours from a grave danger in the following manner:

> Finally, in conclusion, I do not ask anything on behalf of Valentina Mendoza, but rather trust the munificence of Your Majesty and the Queen, as well as Your Illustrious Government that knows how to recognise with gratitude the real worth of so many philanthropic acts and the remarkable character of selflessness and self-denial displayed during the horrific days of the storm and flood that was experienced in this Province, and that you will know how best to reward the noble and kind Valentina Mendoza.[59]

The Flood of Tears

Little has been written from a historical standpoint about the social and psychological impacts of the wholesale destruction of Philippine towns and communities caused by cyclonic storms and floods under Spanish rule. Within a short space of time, often in less than an hour or in the worst-case scenario, simply a matter of minutes due to the speed and height of a flood, an individual, family, or even a community's identity and entire way of life could suddenly be swept away by raging floodwaters. In the case of the Abra River Valley in September 1867, there was no

[59] NAP folios S244–S245 Don Fernando de Santa Coloma to the King of Spain, 15 June 1868 in 'Valentina Mendosa Provides for Flood Victims,' Varias Provincias-Ilocos Sur, 1796–1898.

warning. Terrified people woke up in the pitch black in waste-high water. Focusing on such singular, albeit terrifying, moments enables the historian to describe and analyse at close range the nature of vulnerability and fear, and of coping and resilience. In the midst of widespread trauma and loss, many individuals found strength and hope when it really mattered. Such an investigation, by its very nature, is concerned with collapse and the loss of social cohesion, personal possessions, and life itself. It is also concerned with the political and moral economy of death as, in this case, the stricken residents of Ilocos Sur attempted to negotiate the outcomes of colonial relief initiatives, and their prospects and failures.[60]

Archival documents, concerning the impacts of the flood on survivors, reveal different perceptible effects and insights. Houses in Vigan were either flattened or partially blown apart during the typhoon. But afterwards, people could still re-enter the remains of their damaged homes to reclaim personal belongings. This situation stood in marked contrast to the nearby Abra River Valley, where the impact of the flood crumpled, submerged, and swept homes away, and then mingled, spread, and buried the debris over vast distances. When the floodwaters receded, survivors traversed a damp, muddy, contaminated landscape marked by utter devastation. There was nothing recognisable left except a heap of debris where their houses and fields had once stood.

The official reports of 1867 floods in Ilocos Sur that appeared in Spanish newspapers depicted a devastated landscape akin to a war zone, a place of total destruction and death, where traumatised survivors searched in vain for their loved ones and personal mementos. Shocked children watched in disbelief as a parent or relative picked over the 'furniture of self' in the piles of rubble that had once been their home. For some individuals in this precarious situation, the distress of having lost everything, from both a personal and material standpoint, nearly drove them to breaking point.[61]

Some survivors could not cope. For many, the memory of the fury of the extreme typhoon and great flood remained to haunt them. There was never again any sense of security for some in Vigan or those living along the banks of the Abra River after 1867, who now truly understood that

[60] Kai T. Erikson, *Everything in Its Path: Destruction of Community in the Buffalo Creek Flood* (New York: Simon and Schuster, 1976).

[61] Ibid., 6.

mass death and destruction were an inevitable fact of life in the typhoon season in the context of Spanish colonial rule. This horrific flood disaster changed many persons' perceptions of the world around them and sapped the confidence of some Ilocanos in the natural world and social order. They now kept an even closer watch on the changing character of the weather and their environment and the places that shaped their outlook and made them who they were; the places where nature and the weather were big, and they were small.[62]

Some survivors were undoubtedly tormented by recurring nightmares. In the following decades, in the public mind, the rapid rate at which the Abra River suddenly rose due to subsequent typhoon-related rainfall in October 1871 and 1881, and, the resultant destruction and loss of life, created a widespread dread of flooding and storm fatigue. In 1934 Fr. Miguel Selga, Director of the Manila Observatory, described this generalized feeling of apprehension and timidity as *tifonitis* and acquired 'pathological state owing to nervous overstimulation produced by the frequency or extraordinary intensity of typhoons.'[63]

Conclusion

In this chapter I have discussed a huge typhoon-related flood that devastated Ilocos Sur, Northern Philippines, underpinned by a strong El Niño event. I have argued how a natural hazard like the great flood of 25–27 September 1867 can illuminate the crucial relationship between big weather and national and local development under Spanish rule in the Philippines. I also explored why particular groups and individuals differentiated by gender, age, status, and class are more vulnerable than others in a disaster like this major flood.

This extraordinary flood on the Abra River, with surging waters reaching a height of 25 meters (82 feet) above normal level, killed approximately 1800 people and many thousand more farm animals; many of the

[62] Geraldine Brooks, *The Idea of Home: Boyer Lectures 2011* (Sydney: HarperCollins, 2011), 28.

[63] Greg Bankoff, *Cultures of Disaster: Society and Natural Hazard in the Philippines* (London and New York: Routledge Curzon, 2003), 17. See also: F.J. Aguilar Jr., 'Disasters as Contingent Events: Volcanic Eruptions, State Advisories, and Public Participation in the Twentieth-Century Philippines,' *Philippine Studies: Historical and Ethnographic Viewpoints*, 64, 3–4 (2016), 593–624. For more on the Taal eruption, see: Chapter by Ventura, this volume.

bodies were never recovered by the understaffed and ill-equipped Spanish forces. The survivors, many of whom were homeless and destitute, had to face the threat of cholera and other lethal waterborne diseases, despite mass burials and cremations. In September 1867, no one had anticipated the record amount of sudden precipitation that Ilocos Sur and the Abra River Valley were to receive. While there were limited contingency plans in place for key stretches of the coast, little disaster planning was centred on Vigan's inland flood plain before the great inundation of 1867.

This ENSO-related typhoon of 1867 stands out as memorable, in both the local collective memory and the periodical literature of the time. The wind caused considerable damage, but the subsequent flooding caused far more devastation. The people of Abra Province had experienced flooding before but nothing quite like this. The rainfall fell heavily on the nearby mountains and this flowed down to the river system and the Abra had quickly overflowed.

The 1867 flood in distant Ilocos Sur was officially declared a national calamity. The colonial government launched a public subscription in the Philippines and Spain to meet the disaster relief needs of the people of Ilocos Sur for family assistance and reconstruction of their homes. The hands of authorities in Manila, meanwhile, were tied from tackling storm damage closer to the capital. They were bogged down with typhoon-related infrastructural problems and challenges in Manila and its environs, and they could do nothing in practical terms to rapidly improve the condition of the desperate people in coastal Ilocos Sur and the peripheral province of Abra. It is set against this background that I reconstruct this flood story based upon first-person accounts, including of the bravery and skill of Valentina Mendoza. Her hands were not tied, practically speaking, and she was able to address the unfoldingly tragic situation on the spot. She realised the speed with which the flood was occurring and the fact that it would be deadly. Mendoza's quick thinking, decisive deeds, and compassionate concern for the well-being of her less fortunate neighbours saved hundreds of lives. She became, in the process, a living symbol of courage, strength, and hope in the besieged *barrio* of St. Julian, where she resided in her large, well-built house. That house became a safe haven in a collapsing local world.

Certain characteristics of groups and individuals affect their vulnerability to natural hazards and their ability to mitigate against them, as exemplified by Valentina Mendoza's remarkable response to the threat the flood posed to her less fortunate neighbours. She demonstrated that there was nothing more empowering than people themselves and community connection in such a crisis. Nevertheless, it was also through social memory associated with this event that individual and collective trauma were fused across time. In certain small communities dotting the Abra River Valley, like the *barrio* of St. Julian, there developed a more permanent sense of collective trauma, if not fatalism, towards what was considered elsewhere in the archipelago only a temporary setback and a transitional state of post-traumatic stress, due to chronic typhoon-related flooding.[64]

Bibliography

Archival Sources

Archives of the Manila Observatory (AMO), Ateneo de Manila University, Quezon City, The Philippines.

National Archives of the Philippines (NAP), Manilla, The Philippines.

Newspapers

El Imparcial.

La Espana.

La Esperanza.

Published and Unpublished Sources

Aguilar Jr., F.J. 'Disasters as Contingent Events: Volcanic Eruptions, State Advisories, and Public Participation in the Twentieth-Century Philippines.' *Philippine Studies: Historical and Ethnographic Viewpoints*, 64, 3–4 (2016): 593–624.

Alders, Ma. Theresa M. 'Floodwaters of Death: Vulnerability and Disaster in Ormoc City, Philippines: Assessing the 1991 Flood and Twenty Years of Recovery.' Unpublished PhD dissertation, Murdoch University, 2017.

Bankoff, Greg. *Cultures of Disaster: Society and Natural Hazard in the Philippines*. London and New York: Routledge Curzon, 2003.

Bevoise, Ken De. *Agents of Apocalypse: Epidemic Disease in the Colonial Philippines*. Princeton: Princeton University Press, 1995.

[64] Erikson, *Everything in Its Path*, 163–64, 185.

Brooks, Geraldine. *The Idea of Home: Boyer Lectures 2011*. Sydney: HarperCollins, 2011.

Corporal-Lodangco, Irenea L., Lance M. Leslie, and Peter J. Lamb. 'Impacts of ENSO on Philippine Tropical Cyclone Activity.' *Journal of Climate*, 29 (2016): 1877–97.

Couper-Johnston, Ross. *El Niño: The Weather Phenomenon That Changed the World*. London: Coronet, 2001.

Davis, Mike. *Late Victorian Holocausts: El Niño Famines and the Making of the Third World*. London: Verso, 2001.

Dery, Luis. *Pestilence in the Philippines: A Social History of the Filipino People, 1571–1800*. Quezon City: New Day Publishers, 2006.

de Jesus, Ed. C. *The Tobacco Monopoly in the Philippines: Bureaucratic Enterprise and Social Change, 1766–1880*. Quezon City: Ateneo de Manila University Press, 1980.

Elsner, James B., and Kam-biu Liu. 'Examining the ENSO-Typhoon Hypothesis.' *Climate Research*, 25 (2003): 43–54.

Erikson, Kai T. *Everything in Its Path: Destruction of Community in the Buffalo Creek Flood*. New York: Simon and Schuster, 1976.

Fagan, Brian. *Floods, Famines and Emperors: El Niño and the Fate of Civilizations*. New York: Basic Books, 1999.

Glantz, Michael. *Currents of Change: El Niño's Impact on Climate Variability and Society*. Cambridge: Cambridge University Press, 1996.

Grove, Richard, and John Chappell, eds. *El Niño: History and Crisis*. Cambridge: White Horse Press, 2000.

Grove, Richard, and George Adamson. *El Niño in World History*. London: Palgrave Macmillan, 2018.

Hay, John, David Easterling, Kristie L. Ebi, Akio Kitoh, and Martin Parry. 'Conclusion to the Special Issue: Observed and Projected Changes in Weather and Climate Extremes.' *Weather and Climate Extremes*, 11 (2016): 103–5.

Henderson-Sellers, A., H. Zhang, G. Berz, K. Emanuel, W. Gray, C. Lansea, G. Holland, J. Lighthill, S.-L. Shieh, P. Webster, and K. McGuffie. 'Tropical Cyclones and Global Climate Change: A Post-IPCC Assessment.' *Bulletin of the American Meteorological Society*, 79 (1998): 19–28.

Larkin, John A. *The Pampangans: Colonial Society in a Philippine Province*. Quezon City: New Day Publishers, 1993.

Martin, Susana Maria Ramirez. *El Terremoto de Manila de 1863 Medidas Politicas y Economicas*. Madrid: Consejo Superior De Investigaciones Cientificas, 2006.

Newsom, Linda A. *Conquest and Pestilence in the Early Spanish Philippines*. Honolulu: University of Hawaii Press, 2009.

Owen, Norman G. *Death and Disease in Southeast Asia: Explorations in Social, Medical and Demographic History*. Singapore: Oxford University Press, 1987.

Selga, Miguel. *Charts of Remarkable Typhoons in the Philippines, 1902–1934, Catalogue of Typhoons 1348–1934*. Manila: Bureau of Printing, 1935.

Wernstedt, Frederick L., and Joseph E. Spencer. *The Philippine Island World: A Physical, Cultural and Regional Geography*. Berkeley, CA: University of California Press, 1967.

Worcester, Dean C. *A History of Asiatic Cholera in the Philippine Islands*. Manila: Bureau of Printing, 1908.

CHAPTER 8

El Niño and the Human–Environment Nexus: Drought and Vulnerability in Singapore, 1877–1911

Fiona Williamson

The historiography of the Indian Ocean World (IOW) has long reflected how climatic forces have shaped the region's civilisations and economies. As Philip Gooding rightly notes in the introduction to this volume, these

[1] Sunil Amrith, *Unruly Waters: How Mountain Rivers and Monsoons have Shaped South Asia's History* (London: Penguin, 2018); Edward A. Alpers, *The Indian Ocean in World History* (Oxford: Oxford University Press, 2014); Peter Boomgaard, *A World of Water: Rain, Rivers and Seas in Southeast Asian Histories* (Singapore: NUS Press, 2007); Richard Hall and John Stravinsky, *Empires of the Monsoon: A History of the Indian Ocean and its Invaders* (London: Harper Collins, 1996); K.N. Chaudhuri, *Trade and Civilisation in the Indian Ocean: An Economic History from the Rise of Islam to 1750* (Cambridge: Cambridge University Press, 1985).

F. Williamson (✉)
Singapore Management
University, Singapore, Singapore
e-mail: fwilliamson@smu.edu.sg

P. Gooding (ed.), *Droughts, Floods, and Global Climatic Anomalies in the Indian Ocean World*, Palgrave Series in Indian Ocean World Studies, https://doi.org/10.1007/978-3-030-98198-3_8

begin with histories of monsoon and trade interaction before moving towards environmental and atmospheric histories of the IOW.[1] However, such readings have not (or are not), necessarily conveying how shorter-term or multi-scale climatic fluctuations may have impacted society, environment, or culture beyond perhaps some scholarship undertaken on extreme events.[2] Yet, recent science has shown that many atmospheric phenomena—such as the monsoon, one of the most complex atmospheric mechanisms to affect the IOW—are not stable.[3] What this modern awareness brings is a greater appreciation of the impacts of atmospheric dynamics on society and culture at different stages of our past, which includes (with direct relevance for our purposes here) the periods of variation and amplitude in the tropical Pacific and El Niño Southern Oscillation (ENSO).[4]

In the late 1990s and early 2000s, climatic and environmental histories of the IOW began to be read through socio-cultural perspectives on impact and change.[5] Championed by geographers as well as historians—Mike Hulme one of the best known of the former—such readings have helped form a deeper appreciation of climatic factors that have influenced past societies.[6] Concurrently, advances in historical climatology and in the recovery of weather narratives and climatic data for the region have

[2] Greg Bankoff and Joseph Christensen (eds.), *Natural Hazards and Peoples in the Indian Ocean World: Bordering on Danger* (New York: Palgrave Macmillan, 2016); James F. Warren, 'Weather, History and Empire: The Typhoon Factor and the Manila Galleon Trade, 1565–1815,' in *Anthony Reid and the Study of the Southeast Asian Past*, eds. Geoff Wade and Li Tana (Singapore: Institute of Southeast Asian Studies, 2012), 183–220.

[3] Yen Li Loo, Lawal Billa, and Ajit Singh, 'Effect of Climate Change on Seasonal Monsoon in Asia and Its Impact on the Variability of Monsoon Rainfall in Southeast Asia,' *Geoscience Frontiers*, 6, 6 (2015), 817–23.

[4] Rosanne D'Arrigo, Edward R. Cook, Rob J. Wilson, Rob Allan, and Michael E. Mann, 'On the Variability of ENSO Over the Past Six Centuries,' *Geophysical Research Letters*, 32, L03711 (2005), 1–4; Michael E. Mann, R.S. Bradley, and M.K. Hughes, 'Long-term Variability in the ENSO and Associated Teleconnections,' in *ENSO: Multiscale Variability and Global and Regional Impacts*, eds. H.E. Diaz and V. Markgraf (Cambridge: Cambridge University Press, 2000), 357–412.

[5] Dipesh Chakrabarty, 'The Climate of History: Four Theses,' *Critical Inquiry*, 35, 2 (2009), 197–222; Richard Grove, *Ecology, Climate and Empire: Colonialism and Global Environmental History, 1400–1940* (Cambridge: White Horse Press, 1997); Richard Grove and J. Chappell, *El Nino: History and Crisis: Studies from the Asia–Pacific Region* (Cambridge: White Horse Press, 2000).

[6] Mike Hulme, *Weathered: Cultures of Climate* (London: Sage, 2017).

enabled historians to better integrate scientific data into their work. This has enabled a deeper understanding of the global teleconnections that created shared experiences across time and space. For example, it is now commonly understood that fluctuations in precipitation are connected to ENSO phases, the Madden–Julian Oscillation (MJO), the Indian Ocean Dipole (IOD) and Pacific Decadal Variability as much as seasonal cycles and monsoon systems.[7] During El Niño events, for instance, tropical warming displaces major rainfall-producing systems and interacts with the Indo-Asian monsoon.[8] As the Asian monsoon system affects more than half the world's population, fluctuation and uncertainty have major potential to affect historical human experiences. Climate can thus be conceptualised as an aspect of *habitus*, assimilated into the totality of learned behaviours and intrinsic to the organising, unconscious structures governing the rhythms of everyday life.

While scientists grapple with the intricacies of understanding ENSO or IOD stability (especially as we further complicate things with Anthropogenic climate change), historians can see how these forcings have shaped our past. The 1860s–1920s, for example, saw particularly strong ENSO fluctuations and teleconnections across vast regions.[9] This brought unusually strong weathers resulting in floods, droughts and famines in places as far afield as India, America, England, Africa, Brazil, Australia and Russia, with population displacement and millions of deaths globally.[10] The fundamentality of climate in affecting lives and transforming environments, especially for those in developing worlds or with predominantly rural economies, should thus not be underestimated. Climate reaches across national borders and political structures, binding disparate cultures in shared experiences that transcend nationhood. As a frame through which to view the past, therefore, a climatic perspective can

[7] Edward R. Cook, Kevin J. Anchukaitis, Brendan M. Buckley, Rosanne D. D'Arrigo, Gordon C. Jacoby, and William E. Wright, 'Asian Monsoon Failure and Megadrought During the Last Millennium,' *Science*, 328, 5977 (2010), 486.

[8] Allan, 'ENSO and Climatic Variability,' 4–5.

[9] Rob Allan, 'ENSO and Climatic Variability in the Past 150 Years,' in *ENSO: Multiscale Variability*, eds. Diaz and Markgraf, 3–55.

[10] Ibid., 36–41. For an overview, see: Mike Davis, *Late Victorian Holocausts: El Niño famines and the Making of the Third World* (London: Verso, 2001).

offer a supplementary methodological approach to global or connected history.[11]

This chapter brings a climatic perspective to Singaporean history by exploring the El Niño inspired droughts of 1877, 1902 and 1911. Located within the shifting perimeters of the humid Intertropical Convergence Zone (ITCZ), where trade winds from the northern and southern hemispheres meet, Singapore's normative climate is tropical with high humidity and ample rainfall. When warmer than normal waters develop over the Eastern Tropical Pacific along the coastal regions of South America, however, the trade winds become weaker and the El Niño comes to Southeast Asia. Each of 1877, 1902 and 1911 has been identified by scientists as having experienced strong El Niño events globally[12]; most likely as part of protracted episodes.[13] The significance of protracted events is not simply the duration but the potential for stronger impacts than the 'classical' El Niño event.[14] This was certainly the case in Singapore, when the town experienced some of its worst known droughts since the British had established a settlement in 1819.[15] Scientific studies that have recreated El Niño through evidence from Sea Surface Temperature (SST), Niño-3 and Niño-3.4 indices, the Southern Oscillation Index (SOI) or drought indices have often focussed on the 1877 event as one of the strongest to have occurred in the instrumental record period.[16]

[11] Sanjay Subrahmanyam, 'Connected Histories: Notes Towards a Reconfiguration of Early Modern Eurasia,' *Modern Asian Studies*, 31, 3 (1997), 735–62.

[12] Joëlle L. Gergis and Anthony M. Fowler, 'A History of ENSO Events Since AD 1525: Implications for Future Climate Change,' *Climatic Change*, 92, 3–4 (2009), 368; W.H. Quinn, D.O. Zoff, K.S. Short, and R.T.W. Kuo Yang, 'Historical Trends and Statistics of the Southern Oscillation, El Nino, and Indonesian droughts,' *Fishery Bulletin*, 76, 2 (1978), 672–3.

[13] Gergis and Fowler, 'ENSO Events,' 375.

[14] Robert J. Allan, Joëlle Gergis, and Rosanne D'Arrigo, 'Placing the AD 2014–2016 "Protracted" El Niño into a Long-term Term Context,' *The Holocene*, 30, 1 (2020), 103.

[15] A drought is today defined as more than 15 consecutive days with less than 1 mm of rainfall registered at a climate station.

[16] A small sample of this extensive literature includes: Boyin Huang, Michelle L'Heureux, Zeng-Zhen Hu, Xungang Yin, and Huai-Min Zhang, 'How Significant was the 1877/78 El Nino?,' *Journal of Climate*, 33, 11 (2020), 4855; Deepti Singh, Richard Seager, Benjamin I. Cook, Mark Cane, Mingfang Ting, Edward Cook, and Mike Davis, 'Climate and the Global Famine of 1876–78,' *Journal of Climate*, 31, 23, (2018), 9445–67; J.M. Lough, K.D. Anderson, and T.P. Hughes, 'Increasing Thermal Stress for Tropical Coral Reefs: 1871–2017,' *Scientific Reports*, 8, 6079 (2018), 1–8; Patricio

Yet, this article contends that while the 1877 event was indeed incredibly severe, the *impacts* of the event on Singapore were not necessarily worse than those resulting from the slightly weaker events that took place in 1902 and in 1911 especially. There is a huge literature that explores vulnerability across the social and physical sciences and humanities from which this chapter draws, especially that which uses history as a way into exploring relationships of vulnerability and resilience.[17] These factors are the crux of the argument, which shows how human action created the circumstances that led to disaster, over and above the weather. Thus, this chapter will look to the ground, as well as to the atmosphere, to unpack the nexus of nature-inspired versus human-induced vulnerability to drought within the context of colonial urbanisation in this port town.

While the main narrative focuses on Singapore, the town ought not to be considered in isolation. Malaya, neighbouring Borneo, and Indonesia suffered during each of these events and Singapore was not immune to the fluctuations affecting the regionally interlinked economies, especially in the realm of staple food security.[18] In addition, the droughts inspired new scientific questions at regional and even global scales, as a wealth of evidence became available due to the increasingly connected nature of scientific institutions, scientific literature, and communications systems across the colonial IOW and the globe. By exploring Singapore through these three droughts, this chapter aims to connect with the others in

Aceituno, Maríadel del Rosario Prieto, María Eugenia Solari, Alejandra Martínez, Germán Poveda, and Mark Falvey, 'The 1877–1878 El Niño Episode: Associated Impacts in South America,' *Climatic Change*, 92 (2009), 389–416; C.F. Ropelewski and P.D. Jones, 'An Extension of the Tahiti-Darwin Southern Oscillation Index,' *Monthly Weather Review*, 115 (1987), 2161–65.

[17] Paul Erdkamp, Joseph G. Manning, and Koenraad Verboven (eds.), *Climate Change and Ancient Societies in Europe and the Near East: Diversity in Collapse and Resilience* (Cham, CH: Palgrave Macmillan, 2020); Robert Wasson, Arupjyoti Saikia, Priya Bansal, and Chong Joon Chuah, 'Flood Mitigation, Climate Change Adaption, and Technological Lock-in in Assam,' *Ecology, Economy and Society—The INSEE Journal*, 3, 2 (2020), 83–1–4; I. Kelman, J.C. Gaillard, James Lewis, and Jessica Mercer, 'Learning from the History of Disaster: Vulnerability and Resilience Research and Practice for Climate Change,' *Natural Hazards*, 82 (2016), 129–43; Andrea Janku, Gerrit J. Schenk, and Franz Mauelshagen, (eds.), *Historical Disasters in Context: Science, Religion and Politics* (New York: Routledge, 2012); Uwe Lübken and Christof Mauch, 'Uncertain Environments: Natural Hazards, Risk and Insurance in Historical Perspective,' *Environment and History*, 17 (2011), 1–12.

[18] 'Sarawak,' *The Straits Times* (13 Oct. 1877), 2; 'Java News,' *Singapore Daily Times* (14 Dec. 1877), 3. See also: Chapters by Gooding and Ventura, this volume.

the collection to show how, despite the regional and national differences in governance and in culture of each site of exploration, the experience of climate-induced environmental disaster can provide a shared narrative across the IOW.

The sources used in reconstructing the events of 1877, 1902, and 1911 are varied and derive from many years of interdisciplinary archival research on the climatic history of Singapore and the Malaysian peninsula in the libraries and archives of Kuala Lumpur, Singapore, London, and Cambridge. They comprise a multiplicity of documentary evidence, of the history of meteorology and colonial science; weather records; official records on urban development and land-use change; hydraulic engineering and flood and drought mitigation; official medical statistics and narrative accounts of weather extremes, often to be found in the contemporary press. The weather records themselves are derived from what are commonly called the archives of societies, a term now frequently appropriated to explain observational weather records made by human hands, as opposed to the archives of nature (i.e., proxy records, such as tree-rings).[19] In the Straits Settlements, these are piecemeal for much of the nineteenth century and have had to be recovered and re-connected akin to a jigsaw puzzle, albeit one with significant gaps in the chronology and omissions in detail more useful to a climate scientist, such as exact locations or instruments used during their recording. The weather records are problematic in many ways, especially if we look for the precision required for scientific analysis, with question marks over the skill of the observer—often an amateur or a non-scientist in the pre-1880s period—and the quality of the instruments. By the period of this study, observations had been formalised under the Medical Department and the majority of the weather data used here is drawn from the main meteorological observatory at Kandang Kerbau Hospital, today the site of Singapore's Mass Rapid Transit (MRT) station 'Little India' on Tekka Lane. For a historian, the problem of precision is less significant but, of course, little can be understood of the human experience of drought from rainfall records. Hence the need to place the records in context with a range of sources from social, scientific, medical, and planning perspectives.

[19] Sam White, Christian Pfister, and Francis Mauelshagan, 'Archives of Nature and Archives of Societies,' in *The Palgrave Handbook of Climate History*, eds. Sam White, Christian Pfister, and Francis Mauelshagen (London: Palgrave Macmillan, 2018), 27–36.

Making Sense of the Human-Climate-Environment Nexus

Between 1819 and the First World War, three droughts stand out in Singapore's history: in 1877, 1902 and 1911.[20] Each event was inspired by the El Niño phase of the Southern Oscillation and compounded by factors on the ground, including socio-economic dislocation and municipal planning. The socio-economic narratives attending each event are discussed below and they suggest that although the 1877 event was the most severe, climatically speaking, the impacts in 1911 were in fact worse. Looking at the annual rainfall for each drought year for one main urban station at Kandang Kerbau Hospital—one of the few urban meteorological stations to have been in operation during each drought year—gives some insight into this. The annual total rainfall for each year reads at 65.14 ins (1629 mm) for 1877; 77.52 ins (1969 mm) for 1902; and 88.24 ins (2241 mm) for 1911 respectively. To put this into perspective, a modern average annual rainfall (1981–2010)[21] is just over 85 ins (2159 mm), thus only 1877 saw much lower rainfall overall than what might be considered normal. Drought severity maps generated by the Monsoon Asia Drought Atlas (MADA)[22] of reconstructed Palmer Drought Severity Index (PDSI) for 1877 and an instrumental PDSI for 1902 and 1911 also show 1877 to have been a more severe drought (Fig. 8.1).

However, annual rainfall figures can be misleading, the density and spread of rainfall across the year can be more significant. The 1911 drought began in February, with little respite until June when enough rain fell to allow the authorities to resume the normal 12-hour daily water supply.[23] But it reared its head again in July and did not tail off until November. The 1877 drought had seen a similar trajectory with the normal dry season extending across the inter-monsoonal period and, with the exception of a respite in mid-June, failure of the remaining July to

[20] 1819 was the year that East India Company representative Stamford Raffles negotiated a settlement treaty with Temenggong Abdul Rahman to establish a British settlement at Singapore.

[21] http://www.weather.gov.sg/climate-climate-of-singapore/ [Accessed: 15 Apr. 2021].

[22] Cook et al., 'Asian monsoon failure,' 486–9.

[23] Straits Settlements Government Gazette (hereafter: SSGG) Municipal Progress Report for May 1911 (28 July 1911), 1117.

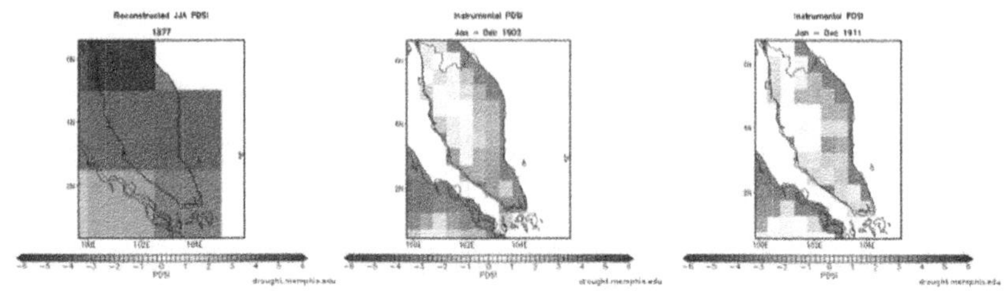

Fig. 8.1 Monsoon Asia Drought Atlas (MADA) reconstructed Palmer Drought Severity Index (PDSI) for 1877 and Instrumental PDSI for 1902 and 1911. http://www.weather.gov.sg/climate-climate-of-singapore/ [Accessed: 15 Apr. 2021]

September southwest monsoon.[24] The 1902 drought, however, only saw its driest periods at the end of the year in October and November, months that do not normally bring the heaviest monsoonal rains.[25] A modern study based on the Standardized Precipitation Evapotranspiration Index (SPEI) dataset is also useful for considering 1902 and 1911 together. This study took into account multi-scale characteristics (including intensity, duration and area) for the evaluation of different types of drought, and is often used for agricultural drought monitoring. While 1877 was not included in the study's scope, it did suggest that 1911 saw more widespread severe drought in Asia than 1902.[26] Building from this, drought is not considered a result of meteorological or atmospheric phenomena alone but a series of complex interactions that also encompass land use, changes to natural water storage and soil moisture, and human water management.[27] Current-day explanations of the causes of urban

[24] Annual Abstract of Meteorological Observations for Kandang Kerbau Hospital, Singapore, 1877.

[25] Annual Abstract of Meteorological Observations for Kandang Kerbau Hospital, Singapore, 1902.

[26] Qianfeng Wang, Jianjun Wu, Tianjie Lei, Bin He, Zhitao Wu, Ming Liu, Xinyu Mo, Guangpo Geng, Xiaohan Li, Hongkui Zhou, and Dachuan Liu, 'Temporal-spatial Characteristics of Severe Drought Events and Their Impact on Agriculture on a Global Scale,' *Quaternary International*, 349 (2014), 15. Data for 1877 are not available.

[27] Rudolf Brázdil, Andrea Kiss, Jürg Luterbacher, David J. Nash, and Ladislava Rezníckova, 'Documentary Data and the Study of Past Droughts: A Global State of the Art,' *Climate of the Past*, 14 (2018), 1916.

vulnerability to drought, for example, explain how an amalgam of hydrological and environmental factors combine with demographic sensitivity and socio-economic adaptive capacities to impact the scale of a disaster.[28] Considering this when looking at the available information for each of our three past droughts, we can uncover several reasons why Singapore's 1911 event might have resulted in more extreme social stress than the previous two.

Each drought realised a burning need to improve the town's water supply. During the 1877 drought, urban inhabitants had largely relied on wells for their water supply; several of which had been built due to the philanthropic donation of Syed Ali Al Junied in the late 1840s and early 1850s.[29] Tanks at Fort Canning and at the port provided additional reserves for dry periods or for combatting fires, but they were still not sufficient for the population during times of stress. However, the town's population was still small, and some people had access to natural water sources. A new reservoir was under development and it was completed in the last month of the year, albeit not early enough to lessen the worst months of drought.[30] Going forward, however, the reservoir, though not without its limitations, did provide a twelve-hour daily piped water to different areas of the town and even some domestic and public buildings.[31] The following year, a further service reservoir was opened at Mount Emily, serving the eastern part of the town and another new facility opened in 1898 at Pearl's Hill (Fig. 8.2).[32]

By 1902, many domestic dwellings had a piped water supply, although many people also relied on standpipes and wells, the latter derived from

[28] C. Joon Chuah, Beatrice H. Ho, and Winston T.L. Chow, 'Trans-boundary Variations of Urban Drought Vulnerability and Its Impact on Water Resource Management in Singapore and Johor, Malaysia,' *Environmental Research Letters*, 13 (2018), 074, 011; Nick Brooks, W. Neil Adger, and P. Mick Kelly, 'The Determinants of Vulnerability and Adaptive Capacity at the National Level and the Implications for Adaptation,' *Global Environmental Change*, 15, 2 (2005), 151–63.

[29] C.B. Buckley, *An Anecdotal History of Old Times in Singapore* (Singapore: Fraser and Neave, 1902), II, 504, 547–8.

[30] Fiona Williamson, 'Responding to Extremes: Managing Urban Water Scarcity in the Late Nineteenth-century Straits Settlements,' *Water History*, 12, 3 (2020), 251–63.

[31] Brenda Yeoh, 'Urban Sanitation, Health and Water Supply in Late Nineteenth and Early Twentieth Century Colonial Singapore' *Southeast Asia Research*, 1, 2 (1993), 147.

[32] Brenda Yeoh, *Contesting Space in Colonial Singapore: Power Relations and the Urban Built Environment* (Singapore: NUS Press, 2003), 207.

Fig. 8.2 (Edited section of) Map of the Island of Singapore and its Dependencies, 1911, War Office (London), 1916. Courtesy of Bibliothèque nationale de France. Original digital map available at: http://catalogue.bnf.fr/ark:/12148/cb407342553 [Accessed: 15 Apr. 2021]

surface water. For example, a contemporary survey of 72 houses in China Street found that 25 per cent of houses still utilised well-water.[33] These were easily contaminated by high tides, floods, and droughts. The authorities recognised this and had closed more than 1904 wells between 1897 and 1901, but the problem was replacing these water sources with efficient and safe supplies. To put this into perspective, when in 1901 the municipality commissioned a survey of remaining wells within town limits, they found that out of 3877 wells, only 609 contained potable water.[34] As well as safety issues, the supply was rapidly falling behind demand.

[33] *The Straits Times* (4 Sep. 1902), 4.

[34] Cited in: James F. Warren, *Rickshaw Coolie: A People's History of Singapore, 1880–1940* (Singapore: Singapore University Press, 2003), 261.

In 1900, the Municipal Engineer had already warned that an increase of 1.5 million gallons per day would be required—up from the current 4 million—over the next decade.[35] In response, a new scheme was created and implemented in phases across the first decade of the new century. Its two main elements were to channel water from the Kallang River watershed and to build a new reservoir, the latter designed by the Municipal Engineer Robert Peirce. This, when completed in 1910, was expected to increase capacity by 3.5 million gallons daily.[36]

In 1911 then, it might be assumed that these schemes would have created a more resilient Singapore. This was not the case. In the nine years between 1902 and 1911—the period during which proposals for the new supply had been enacted—residents had increased by more than 60,000, far more than had been anticipated. This necessitated the creation of new suburbs which demanded in turn new pipelines and water capacity. Newly reclaimed land on the coast had not yet been fully outfitted with an efficient supply.[37] Consumption had also been driven up by commercial expansion, with the newly developed Keppel Docks (then known as New Harbour and fully completed in 1886) piling pressure on the existing water supply. Between 1901 and 1911, water usage had increased from 4.5 million to 5.9 million gallons (slightly over the Municipal Engineer's original estimate), but the worst problem was that the 5.9 million figure was unrepresentative of actual need: the supply was kept deliberately low due to a gap between supply and demand.[38] Hence why, when the rains failed in 1911, the town reserves were not sufficient for the purpose. At the same time, the proposed Kallang River Water works, though partially operational during the 1911 drought, were not fully completed until March 1912. Ironically, work on the scheme actually had to be suspended during the 1902 drought because the water supply failed and the future of the development—which took a further ten years to complete—was in doubt for a period of time due to unexpected costs.[39] This brings

[35] SSGG, Administrative Report of the Singapore Municipality 1901, Appendix M: Minutes by the Municipal Engineer, 28 Nov. 1900.

[36] Yeoh, 'Urban Sanitation,' 149.

[37] Lim Tin Seng, 'Land from Sand: Singapore's Reclamation Story,' *BibioAsia*, 13, 1 (2017), 16–23.

[38] 'Singapore's Water,' *The Straits Times* (27 Mar. 1912), 7.

[39] 'Municipal Commission,' *The Straits Times* (22 Nov. 1902), 5.

us to another key issue: cost. Improvement and extension schemes were immensely expensive, and the structural alterations required to adapt domestic dwellings to piped water were complicated as well as costly. There was also an issue as to from where water should be ascertained in the first place on an island with limited natural fresh water supplies. The clearest possibility—to syphon water from nearby Johore on the mainland peninsula—was a major political, engineering, and financial headache for the Straits Settlements' governors.[40] With deficits in civic improvement and rapid in-migration the norm, municipal authorities had failed to cater even for 'normal' conditions.

Big Picture, Local Impacts: The Socio-Economic Effects of El Niño

This brings us to the second issue, of the level socio-economic adaptive capacity within the town. By the time of the first major drought, in 1877, the port of Singapore was the administrative centre of the Straits Settlements and had developed into a bustling township of around 96,000 inhabitants driven by an influx of (mostly male) migrant labour.[41] Coming out of the customary February dry period with little rain in view, by mid-April 1877 inhabitants were beginning to worry. 'The scarcity of water is now most keenly felt throughout the whole town' claimed one newspaper columnist, and the little that was left was increasingly dirty and undrinkable.[42] This does not come as a surprise when we realise that in 1877, most inhabitants relied on an antiquated system of wells, tanks, and small reservoirs situated in and around the town centre for their water, despite some investment in improving the clean water infrastructure by the municipal government.[43] Piped water to domestic dwellings was a privilege only for the wealthy, and the system of ceramic underground piping for water and for drainage—then commonplace in many

[40] National Archives, Kew, London (TNA, UK) CO273/309, Despatch 266, Straits Settlements, Anderson to Lyttelton, 20 June 1905.

[41] Swee-Hock Saw, 'Population Trends in Singapore, 1819–1967,' *Journal of Southeast Asian History*, 10, 1 (1969), 41.

[42] *Straits Times Overland Journal* (19 Apr. 1877), 11.

[43] Yeoh, *Contesting Space*, 175–212; Yeoh, 'Urban Sanitation,' 143–72.

British cities—had not been adopted in the Straits Settlements.[44] Indeed, it was not long before cholera was being reported in many of the poorer suburbs and *kampong* (villages) across Singapore and other towns on the peninsula—including Georgetown and Malacca—although reports were confused and contradictory. The press noted several anonymous sources that claimed at least twenty deaths a day amongst indigent Singapore 'natives,'[45] yet they were not being reported officially 'on account of the trouble it would get them into.'[46] Others refuted this news as an exaggeration, quoting government figures of 28 deaths in 24 days and even blaming non-European communities for their own illness, citing their apparent choice of contaminated well-water over and above the government sanctioned water available at the Municipal standpipes.[47]

This blame narrative surfaced again in October when eight cholera afflicted Chinese labourers living in shared quarters on River Valley Road were hospitalised and there was an outbreak at the Chinese Immigration depot.[48] It is difficult to assess the reality of the situation, however, with statistics of cholera cases and deaths hard to come by. The annual medical report—though detailed in many ways—does not list cholera as a separate category, only mentioning it specifically if there had been unusual epidemic outbreaks, and there was a particular lack of detail for 1877.[49] On the other hand, a poor state of public health in general was noted, especially amongst the poorest inhabitants and prison inmates. The 1877 Criminal Prison Report, for instance, stated how the health of incoming prisoners was not as good as normal, citing want and disease induced by the long drought as the chief cause.[50] This is corroborated by complaints

44 Warren, *Rickshaw Coolie*, 262.

45 Anon, 'News of the Fortnight,' *Straits Times Overland Journal* (28 Apr. 1877), 7; Nemo, 'Variorum,' *The Straits Times* (28 Apr. 1877), 4.

46 Anon, 'Tuesday 1 May,' *The Straits Times* (5 May 1877).

47 Warning Voice, 'Cholera,' *The Straits Times* (14 Apr. 1877), 4; Anon, 'Fortnight's Summary,' *Straits Times Overland Journal* (28 Apr. 1877), 1.

48 *Straits Times Overland Journal* (18 Oct. 1877), 17.

49 The limited reports for 1877 might be attributed to the illness and absence of Principle Chief Medical Officer (PCMO) for the Straits Settlements, H. L. Randall, during May and June, his role not filled until Acting PCMO Thomas Irvine Rowell took over on 1 July.

50 W.R. Gray, 'Criminal Prison Report 1877,' published in: *Singapore Daily Times* (9 July 1878), 3.

by local inhabitants about drought-induced brackish drains, foul smells, dried-up watercourses, and the unbearable 'great heat.'[51] Singapore was not a healthy place to be in 1877.

Disease and poor public health were also serious preoccupations during the droughts of 1902 and 1911, despite the more ready availability of piped water by the early twentieth century. In 1902, Singapore's population had risen significantly to 581,219 and the annual medical report reveals that these inhabitants' health was generally poor. The number of smallpox cases had been high (though not as high as in Penang) and cholera was exceptionally bad with 842 cases, of which 759 had proved fatal.[52] In 1911, smallpox and malaria both saw a spike over June and July, with cholera alone resulting in 479 hospital admissions and 340 deaths.[53] Although lower than 1902, the number of admissions represented a twofold increase in cases from 1901, leading contemporaries to ponder a connection between the drought and waterborne disease.[54]

Impacts also manifested in the costs of piped water and of food. During 1877, the cost of water rose to 5 cents a bucket, causing 'much distress in consequence amongst the poor,' and in 1902, the increased availability of domestic piped water had hastened the introduction of water metres to prevent wastage—the cost of installation and maintenance chargeable to residents.[55] In terms of food, Singapore was not self-sustainable. While vegetables, pork, and chicken were grown or reared locally, plantation farming across the island was mostly for commercial export, dominated by sugar, coconut, and gambier, the latter used in medicines, dyeing, and tanning. These crops were increasingly phased out in the twentieth century as rubber and pineapple took over plantation capacities. The island relied on the mainland peninsula for rice

[51] Anon, 'The Stamford Road Stream,' *Straits Times Overland Journal* (12 May 1877), 9; Anon, 'Victoria Street Drain,' *Straits Times Overland Journal* (21 July 1877), 9.

[52] Wellcome Trust (hereafter WT): WA28.JM2 S89 1902–7, *The Straits Settlements Medical Report for 1902*, 1.

[53] SSGG Hospitals Report 1911; Hospitals Return Annex Z17. Given the propensity for many inhabitants to avoid hospital, one might assume that the actual numbers of cases were higher than reported.

[54] Anon, 'One Less Per Day,' *The Singapore Free Press and Mercantile Advertiser* (3 June 1911), 7.

[55] Buckley, *An Anecdotal History*, II, 737; Yeoh, 'Urban Sanitation,' 158; Anon, 'The Water Famine,' *Singapore Free Press and Mercantile Advertiser* (11 Sep. 1902).

and fruit, and on the region as a whole for rice; importing from China, Indonesia, Cambodia, Vietnam, Thailand, India, and Hong Kong.[56] The 1877 drought brought a temporary but massive shift from Thai suppliers to British Burmese sources in 1878 to manage the shortfall.[57] Likewise, in 1902, the shortfall was managed by shifting supply from Thailand and by now—Vietnam—to Burma.[58] The total value of imports and exports in general also decreased by 3.75 per cent and 1.75 per cent respectively in sterling. For merchandise, this equated to around a 10 per cent change in dollar value, as opposed to an approximate 1% change the previous year—the shift attributed to 'keenness of competition.'[59] Interestingly, one of the causes for this competition was cited as major conflagrations in Pontianak, Kalimantan and Surabaya, Java, which, while not attributed to dry conditions in the report, were areas also hard hit by drought.[60]

In 1911, imports of general merchandise likewise saw an increase in the cost of around 10 per cent from the previous year, with decreases in volume apparent in rice in particular. Homegrown rice harvests were limited, with only one district—at Beruas in Perak—securing a good yield. Exports declined in volume across several staples, with gambier, tapioca, and preserved pineapple all affected.[61] The press reported how, due to rice harvest failures in Thailand, the price of rice available locally had more than doubled, a fact that was expected to hit the 'coolie [labouring] classes' the most due to their reliance on rice as a staple food source.[62] The price increase was especially apparent in July, August, and September as the drought impacts really began to show. In other areas of the Straits Settlements to the north, the correspondent for Bukit Mertajam wrote

[56] SSGG 1877 Rice imports into Singapore 1876, 285.

[57] SSGG 1877 Appendix U11, 145; SSGG 1878 Appendix U17, 150.

[58] SSGG 1903 Report on trade 1902, Appendix, 27.

[59] SSGG 1903 Report on the trade of the Straits Settlements for the year 1902, Appendix, 25.

[60] S. Robert Aiken, 'Runaway Fires, Smoke-haze Pollution, and Unnatural Disasters in Indonesia,' *Geographical Review*, 94, 1 (2004), 60; 'Sourabaya revisited,' *The Straits Times* (25 Sep. 1902), 2.

[61] SSGG Oct. 1912, Trade 1911, Appendix, 12, 14, 16. It should be noted however that gambier and tapioca were already on a decline, with many estates brought over to the more profitable para rubber.

[62] 'Price of Rice,' *The Singapore Free Press and Mercantile Advertiser* (11 Sep. 1911), 4.

Table 8.1 Import and Export values of rice in Straits Dollars $ per picul before, during, and after each drought. This table summarises import and export costs and reveals the stark contrast with the years proceeding and after each drought

Year	*Import Prices per picul*	*Export Prices per picul*
1877	2.22	2.229
1878	2.77	3.21
1879	2.23	2.95
1902	4.13	4.44
1903	5.07	5.39
1904	4.03	4.37
1911	4.83	4.95
1912	5.57	6.01
1913	4.71	5.10

how, owing to the drought, many central Province Wellesley rice fields had dried up with an estimated 40% drop in output for the year. On top of this, the coconut crop was 50% less, and even para rubber did not escape the ravages of the drought.[63] Once again, the government turned to Burma to make up the shortfall (Table 8.1).

The droughts also showed up stark differences in social vulnerabilities, as poorer inhabitants struggled to manage this further assault on day-to-day living. Nineteenth and early twentieth-century Singapore presents a classic example of a highly stratified colonial society. Despite British dominance politically, in 1877 and 1911 the British and European population only made up 2% and 5% of the population respectively.[64] In terms of social class, society's most well-to-do (as judged by people of all origins who paid rates of more than 40 rupees) equated to only around 385 people in 1877.[65] While it is hard to tell exactly what the social stratification of the remaining population was, the arrival of around 200,000 immigrants annually by the early 1900s meant that the bulk of the population was comprised of refugees and *coolies*, the latter being skilled and unskilled labourers.[66] Not unsurprisingly, this created a situation where poverty characterised the circumstances for many inhabitants,

[63] *The Straits Times* (27 Jan. 1912), 9.

[64] TNA, UK CO277/11, f.109. Return of the population of Singapore (2 Apr. 1877); Saw, 'Population trends,' 41.

[65] SSGG 2 Nov. 1877, 741–3.

[66] W.J. Simpson, *Report on the Sanitary Condition of Singapore* (London: Waterlow, 1907), 6.

leaving large sections of the population vulnerable to the fluctuating prices of staple foodstuffs and water. With some irony, the less well-off also suffered the added burden of water rates levied for improving municipal water supplies while benefitting little from the better water that this provided.[67] That the municipal government knew of this was well-documented, though efforts to improve the status quo were slow and ineffective.[68]

Water shortages also exacerbated pre-existing social tensions. In 1877, for instance, Europeans criticised non-Europeans for using the heat as an excuse to get drunk rather than work, and tempers frayed as people wasted hours queuing under the burning sun to fill buckets of water at standpipes.[69] Officials reported that minor crimes had increased due to a 'general depression … occasioned by the long drought,' though the more likely explanation was increased poverty.[70] As the hours that the water supply was switched on dwindled, competition increased. In 1902, just as enforcement was being suggested to ensure that everyone had fair access to the supply, a violent fight involving stones, sticks, and buckets broke out between rival Chinese and Malay gangs at a hydrant at Victoria Bridge.[71] The authorities came down hard on the men after one of their number was hospitalised and they were arrested for rioting.[72] As the drought continued into October, police were frequently drafted in to guard the standpipes and the municipal government was put under pressure to re-open old wells that had previously been closed for reasons of health and safety. In most cases, the wells were deemed too insanitary to restore, but some inhabitants chose to take matters into their own hands, tearing down boards and coverings.[73] In 1911, the situation intensified. Violence erupted in dramatic fashion with four-hundred rickshaw coolies

[67] Warren, *Rickshaw Coolie,* 5, 261.

[68] *Straits Times Overland Journal* (19 Apr. 1877), 4.

[69] Nemo, 'Variorum,' *Singapore Daily Times* (2 May 1877), 3; 'The Water Supply,' *The Straits Times* (22 Dec. 1877), 6.

[70] SSGG, 5 July 1878, Medical Report for the Prisons, 1043.

[71] *The Straits Times* (30 Aug. 1902), 4; 'Water Famine,' *The Straits Times* (3 Sep. 1902), 5.

[72] 'The Water Famine,' *The Singapore Free Press and Mercantile Advertiser* (11 Sep. 1902), 164.

[73] *The Straits Times* (4 Sep. 1902), 4; 'Municipal Meeting,' *The Straits Times* (28 Feb. 1903), 5.

fighting over a standpipe in Queen's Street in March, lashing out at one another with sticks and staves.[74] Elsewhere on the peninsula, railway coolies went on strike as the water supply was only switched on for four hours a day, at exactly the same time as they were engaged in loading and unloading cargo at Port Swettenham.[75] Of course, it was not the richer inhabitants who had to queue for hours at the street standpipes. They had the luxury of piped water, or servants to collect water, and they had the resources to purchase the bottled water, which had become so popular in the early twentieth century.[76]

Local Impacts, Big Picture: Did Localised Calamity Help Inspire Scientific Change?

It is no coincidence that interest in meteorology—especially in rainfall—increased after 1877 in the Straits Settlements under the auspices of the Medical Department. The event had inspired renewed thinking about the causes of drought, linking the lack of rain to man-made environmental changes and deforestation on the peninsula. This fear led the local government to allocate large areas of rural Singapore as forest reserves in 1882.[77] Studies on the subject were undertaken by colonial officers in the medical and survey departments, many contributing to locally based journals of scientific interest, including the *Journal of the Indian Archipelago and Eastern Asia* and the *Journal of the Straits Branch of the Royal Asiatic Society*. Interest in establishing more, and better, weather records continued to grow, especially because of their relevance to agricultural productivity and for countering both surfeit and deficiency of

[74] 'Effects of the Weather,' *The Singapore Free Press and Mercantile Advertiser* (11 Mar. 1911), 7.

[75] *The Straits Times* (11 Sep. 1911), 6. With some irony, the regional impacts of the drought may have led to more coolies arriving in Singapore than normal just when the drought started in earnest, as they fled drought-related hardship (and political conflict) in other countries, especially China. See: *The Straits Times* (24 Mar. 1911), 6.

[76] *The Straits Times* (6 Sep. 1902), 4.

[77] SSGG, J.F.A. McNair, 'Report by the Colonial Engineer on the Timber Forests in the Malayan Peninsula, 21 June 1879' (3 Oct. 1879), 893–903; National Archives of Singapore (hereafter: NAS), Nathanial Cantley, 'Map of the Island of Singapore. Annexure to Report on the Forests of the Straits Settlements' (1882): Media image no. 20050000974-0093_TM000020_000028_TM.

water.[78] The two-pronged question of how frequent and how severe, was tantamount to drafting schemes for future urban water management policy.

Such thinking was also borne out of a regional and global trend towards increasing the number of long-term weather records made across the British empire, and to making simultaneous observational sets available across the colonies to study such phenomena. When the series of droughts occurred across large swathes of the globe from Brazil to Australia during 1876–1878, it was clear to many that the 'coincidence' could not be ignored.[79] Officers responsible for meteorology within the British Empire, such as Henry Blanford, head of the newly established Indian Meteorological Department (IMD), and Government Astronomers in Australia, including Charles Todd, had utilised synchronous time-series weather observations from different regional stations, as well as international news reports, to deduce that the 'condition of excessive pressure prevailed over not only the Indo-Malayan region and Eastern Australia, but also the greater part if not whole of Asia, probably the whole of Australia and the South Indian Ocean.'[80] Todd concluded that 'there can be little or no doubt that severe droughts occur as a rule simultaneously' referring to reports from India, Singapore and Batavia.[81] News of severe droughts from governmental reports and meteorological outposts from elsewhere across the wider region, including Africa, China, and the Philippines, also pointed to similar conclusions.

The sense that droughts—and thus climate—might be linked across wide areas grew over the ensuing decades. Correspondence between the

[78] C.C. James, *Drainage Problems of the East* (Bombay: Times of India Office, 1906), 224. This was also the case in other British colonies. After an especially severe drought in 1890–91, Hong Kong's Surveyor General requested the Hong Kong Observatory furnish a study of almost forty years of rainfall records to establish patterns for predicting potential future droughts.

[79] See also: Chapter by Gooding, this volume.

[80] Henry F. Blanford, 'On the Barometric See-saw between Russia and India in the Sunspot Cycle,' *Nature*, 21 (1880), 477.

[81] Charles Todd (with H.C. Russel and R.L.J. Ellery), 'The Meteorologist: Droughts in Australia,' *The Australasian* (29 Dec. 1888), 40. See also: Ruth Morgan, 'Prophecy and Prediction: Forecasting Drought and Famine in British India and the Australian Colonies,' *Global Environment*, 13, 1 (2020), 95–132; Richard Grove and George Adamson, *El Niño in World History* (Basingstoke: Palgrave, 2018).

British Association for the Advancement of Science (BAAS) and the Secretary of State for the Colonies in the early 1900s for instance, suggests that the scientific world was keen to invest time and resources into deeper analysis of the several decades long time-series observations. In particular, it was noted how such a study would yield dramatic economic benefits for the British lands of the IOW by enabling a better understanding of monsoon rainfall patterns.[82] Thus, in 1905, the BAAS requested that the British government provide a scientific staff specifically to study the:

> General meteorological conditions which affect the weather in the several British Dominions, and in particular to promote the formulation of meteorological laws, and to apply them to explain and ultimately to anticipate the occurrence of abnormal seasons.[83]

The proposed method was to unite and analyse instrumental records made at sea and on land across the whole IOW to investigate the meteorology of large oceanic areas in relation to their adjacent land mass; the underlying premise that the climatic conditions of India, Australia, South Africa, East Africa, and Egypt were closely related to the Indian Ocean. This, it was hoped, would result in better seasonal predictions and a greater understanding of the conditions generating favourable and unfavourable seasons in India, the droughts of Australia and South Africa, and the relation of the weather of the Mediterranean to Indian cold weather anomalies.[84] As Martin Mahony points out, however, many of these grand plans amounted to nought, as an official imperial meteorological office to co-ordinate these activities failed to materialise; the British

[82] Archive of the British Association for the Advancement of Science (hereafter ABAAS): Papers of Committees, 1896–1912, ff.225r-226r, 225r&v. 'Memorandum on a Proposal for Dealing with Meteorological Questions Affecting the British Dominions beyond the Seas,' 2 June 1905.

[83] ABAAS: Papers of Committees, 1896–1912, ff.223r-4v. 'Draft Memorandum, in further explanation of the proposal for dealing with Meteorology of the Colonies and Dependencies, for the Information of the Secretary of State,' 2 June 1905.

[84] Ibid, f.223v.

Meteorological Office and government relying instead on voluntary individual initiatives by enthusiasts, including some of the first unofficial gatherings of imperial meteorologists from 1909.[85]

In the Straits Settlements too, while scientific journals and reports on the matter were widely available in the colony, the emphasis was even less on thinking about regional atmospheric connections, than local concerns.[86] This may be because the colony did not have a dedicated meteorological service until 1929, in contrast to, say, India (1875) or Hong Kong (1884).[87] Neither did the Straits Settlements have a meteorological champion like Blanford or Todd. Government was largely preoccupied with statistical studies correlating local land-use change or public health with rainfall, not the bigger picture. Indeed, the idea that forest loss was a primary cause of drought dominated local scientific thought well into the twentieth century, and it became connected to the associated narrative of deforestation and soil erosion by the 1930s.[88] It was even considered by some that the long-term weather records were not well-made, as the colony lacked a specialist, trained staff.[89] Only a

[85] These gatherings were the forerunner of the Conference of Empire Meteorologists which became a formal entity in 1919 with the first meeting of the Conference of Meteorologists of British Dominions: Martin Mahony, 'For an Empire of 'All Types of Climate': Meteorology as an Imperial Science,' *Journal of Historical Geography*, 51 (2016), 32.

[86] An interesting article was published in Singapore in 1878, drawing connections between oceanic currents, weather, and global teleconnections. See: 'Extract from an American Paper,' *Singapore Daily Times* (16 Jan. 1878), 3. The colony also received copies of scientific journals and books. In 1911 alone, the Raffles Museums and Library received 100 new texts on subjects of scientific interest, some donated from worldwide museums and institutions including the Smithsonian Institution, Bureau of Science, Manila and the Geological Institute, Mexico: Supplement to the SSGG 23 February 1912, No. 4., 'Annual Report on the Raffles Museum and Library for the Year 1911,' 3.

[87] NAK CO 273/541 ff.9r-13r 'Memorandum on a pamphlet entitled "A Meteorological Department for Malaya" by Sir George Maxwell and Herbert C. Robinson,' (1927), 3.

[88] 'Influence of Forests on Drought,' *The Singapore Free Press and Mercantile Advertiser* (13 Oct. 1911), 9. On soil erosion, see: Fiona Williamson, 'Malaya's "Greatest Menace"? Slow Onset Disasters and the Politics of the environment c. 1920–1950,' *International Review of Environmental History*, 4, 2 (2018), 45–68; ABAAS Sectional Transactions—E.E.P. Stebbing, 'The Encroaching Sahara: Increasing Aridity in West Africa,' Printed Material for the Annual Meetings held in Leicester, Aberdeen and Norwich, eds. O.J.R. Howarth, P.W. Jewson et al. (1935–36), 54.

[89] 'The Agricultural Bulletin,' *The Straits Times* (28 Sep. 1911), 6.

few lone voices bucked the trend, arguing for natural cycles that were bound to repeat and renew, or that sunspot activity might be the reason behind climatic variation.[90]

Conclusion

The Straits Settlements were something of an outlier in the British Empire's journey towards improving meteorological knowledge. Under-provisioned in comparison to Australia, India and Hong Kong, and relying on piecemeal services co-ordinated by the Medical Department, they lacked a dedicated research agenda or facility into the 1920s.[91] In some ways this reflects the British government's lackadaisical attitude towards the science prior to the First World War, which changed rapidly as the strategic and military applications of meteorology for aviation became clear. The situation in the Straits Settlements also reflected the British government's positioning on funding the colony for much of the nineteenth century, preferring the local municipal council to raise its own funds for grand schemes, or to rely on philanthropy, rather than granting loans for major works. The two strands connected in the lack of provision for science, as well as for water works. Arguably though, the lack of scientific knowledge about tropical climates within the ITCZ played a role in shaping these attitudes. It was generally considered that Singapore and the Malayan peninsula did not suffer extreme weathers and, while it was known that the peninsula had two monsoon seasons, they were considered mild in comparison to the rest of the IOW. Thus, the events of 1877 and beyond took the colony by surprise. Drought, and mitigation for drought, were little considered until the last decades of the nineteenth century, and likewise there was little pull to invest in weather science in a country that appeared to have generally abundant rainfall. Nonetheless, while the Straits Settlements lagged behind, arguably, the narrative and practical shift to understanding climate as a teleconnected system was slow elsewhere in British colonial Asia too. While pioneering research was

[90] 'Malacca,' *The Singapore Free Press and Mercantile Advertiser* (26 Mar. 1895), 10; A.M. Skinner, 'Straits Meteorology,' *Journal of the Straits Branch of the Royal Asiatic Society*, 12 (1883), 245–55.

[91] See also: Fiona Williamson, 'Weathering the British Empire: Meteorological Research in the Early Nineteenth-century Straits Settlements,' *The British Journal for the History of Science*, 48, 3 (2015), 475–92.

being undertaken—Henry Blanford and the slightly later IMD scientists, Rai Bahadur Hem Raj and Gilbert Walker, for example, developed ideas on natural variations in the climate system, sunspots, and relationships with Indian monsoon rainfall, and Edwin Quayle undertook work on climatic oscillation—but the nascent field of ENSO research was not fully developed until after the 1960s.[92]

While the study of climate science history provides many answers as to how contemporaries understood weather systems and their associated failures in properly predicting or adapting to local or regional atmospheric phenomena, a broad-brush historical study also provides a considerable methodological opportunity to illuminate and unpick the complex reasons underlying why a drought was so impactful. Indeed, here we see that weather was not the direct cause of the worst impacts, but certain failings of government to sufficiently prepare mitigation strategies and reserves, against a backdrop of severe social inequality. While it is certainly not new to study the climate in history, or the history of nature-induced disaster, this chapter has argued that a close reading of specific events through the lens of human-climatic-environmental interconnections and an understanding of the meteorology of each event, does enable a new avenue into the history of the IOW. First, the study of drought (or indeed flood or other disaster) allows for a different lens into the discussion of colonial science and the ways in which people conceptualised, understood, and responded to the world around them. As the discussion of Singapore suggests, such ideas fed directly into practical projects, such as new hydraulic schemes rather than scientific research, which allows insight into what the colonial government felt ought to be prioritised and why. Second, understanding the meteorology allows greater insight into the human role in creating disasters and, third, the study of a particular event enables insight into a far wider range of historical themes and how these respond directly or indirectly to climatic fluctuation. This includes urban planning, development and government responsiveness, socio-economic stress, resilience and social stratification, and intersections between climate

[92] George Adamson, 'Imperial Oscillations: Gilbert Walker and the Construction of the Southern Oscillation,' in *Weather, Climate and the Geographical Imagination: Placing Atmospheric Knowledges*, eds. Martin Mahony and Sam Randalls (Pittsburgh: University of Pittsburgh Press, 2020), 43–66; Ruth Morgan, 'Southern Skies: Australian Atmospheric Research and Global Climate Change,' *Disaster Prevention and Management*, 30, 1 (2021), 47–63.

and health, amongst others. Such studies are especially important in the IOW, given the macro-region's susceptibility to disaster and its intimate relationship with the vagaries of the monsoon. Arguably, useful lessons might be learned about social resilience and cascading disaster which go beyond climatic modelling or indeed social or political histories. Such an approach to the study of the IOW has the potential for shedding new light on the rich and multifaceted history of the region.

Bibliography

Archival Sources

Archive of the British Association for the Advancement of Science (ABAAS), Wiley Digital Archives.
National Archives of Singapore (NAS), Singapore.
National Archives, Kew, London, (TNA, UK) UK.
Wellcome Trust (WT), London, UK.

Newspapers

Singapore Daily Times.
Straits Times Overland Journal.
The Australasian.
The Singapore Free Press and Mercantile Advertiser.
The Straits Times.

Published and Unpublished Sources

Aceituno, Patricio, Maríadel del Rosario Prieto, María Eugenia Solari, Alejandra Martínez, Germán Poveda, and Mark Falvey. 'The 1877–1878 El Niño episode: Associated impacts in South America.' *Climatic Change*, 92 (2009): 389–416.

Adamson, George. 'Imperial Oscillations: Gilbert Walker and the Construction of the Southern Oscillation,' in *Weather, Climate and the Geographical Imagination,* eds. Mahony and Randalls: 43–66.

Aiken, S. Robert. 'Runaway Fires, Smoke-Haze Pollution, and Unnatural Disasters in Indonesia.' *Geographical Review*, 94, 1 (2004): 55–79.

Allan, Robert J. 'ENSO and Climatic Variability in the Past 150 Years,' in *ENSO: Multiscale Variability,* eds. Diaz and Markgraf: 3–55.

Allan, Robert J., Joëlle L. Gergis, and Rosanne D'Arrigo. 'Placing the AD 2014–2016 "Protracted" El Niño into a Long-term Term Context.' *The Holocene*, 30, 1 (2020): 90–105.

Alpers, Edward A. *The Indian Ocean in World History*. Oxford: Oxford University Press, 2014.

Amrith, Sunil. *Unruly Waters: How Mountain Rivers and Monsoons Have Shaped South Asia's History*. London: Penguin, 2018.

Bankoff, Greg, and Joseph Christensen, eds. *Natural Hazards and Peoples in the Indian Ocean World: Bordering on Danger*. New York: Palgrave Macmillan, 2016.

Blanford, Henry F. 'On the Barometric See-saw between Russia and India in the Sunspot Cycle.' *Nature*, 21 (1880): 477–82.

Boomgaard, Peter. *A World of Water: Rain, Rivers and Seas in Southeast Asian Histories*. Singapore: NUS Press, 2007.

Brázdil, Rudolf, Andrea Kiss, Jürg Luterbacher, David J. Nash, and Ladislava Rezníckovà, 'Documentary Data and the Study of Past Droughts: A Global State of the Art.' *Climate of the Past*, 14 (2018): 1915–1960.

Brooks, Nick, W. Neil Adger, and P. Mick Kelly. 'The Determinants of Vulnerability and Adaptive Capacity at the National Level and the Implications for Adaptation.' *Global Environmental Change*, 15, 2 (2005): 151–163.

Buckley, C.B. *An Anecdotal History of Old Times in Singapore*. 2 Vols. Singapore: Fraser and Neave, 1902.

Chakrabarty, Dipesh. 'The Climate of History: Four Theses.' *Critical Inquiry*, 35, 2 (2009): 197–222.

Chaudhuri, K.N. *Trade and Civilisation in the Indian Ocean: An Economic History from the Rise of Islam to 1750*. Cambridge: Cambridge University Press, 1985.

Chuah, C. Joon, Beatrice H. Ho, and Winston T.L. Chow. 'Trans-boundary Variations of Urban Drought Vulnerability and Its Impact on Water Resource Management in Singapore and Johor, Malaysia.' *Environmental Research Letters*, 13 (2018): 074011.

Cook, Edward R., Kevin J. Anchukaitis, Brendan M. Buckley, Rosanne D. D'Arrigo, Gordon C. Jacoby, and William E. Wright. 'Asian Monsoon Failure and Megadrought During the Last Millennium.' *Science*, 328, 5977 (2010): 486–9.

D'Arrigo, Rosanne, Edward R. Cook, Rob J. Wilson, Rob Allan, and Michael E. Mann. 'On the Variability of ENSO Over the Past six Centuries.' *Geophysical Research Letters*, 32, L03711 (2005): 1–4.

Diaz, H.E. and V. Markgraf, eds. *ENSO: Multiscale Variability and Global and Regional Impacts*. Cambridge: Cambridge University Press, 2000.

Erdkamp, P., J.G. Manning, and K. Verboven, eds. *Climate Change and Ancient Societies in Europe and the Near East: Diversity in Collapse and Resilience.* Cham, CH: Palgrave Macmillan, 2020.

Gergis, Joëlle L., and A. M. Fowler. 'A History of ENSO Events since AD 1525: Implications for Future Climate Change.' *Climatic Change*, 92 (2009): 343–87.

Grove, Richard, and George Adamson. *El Niño in World History.* Basingstoke: Palgrave, 2018.

Grove, Richard, and J. Chappell. *El Nino: History and Crisis: Studies from the Asia-Pacific Region.* Cambridge: White Horse Press, 2000.

Grove, Richard. *Ecology, Climate and Empire: Colonialism and Global Environmental History, 1400–1940.* Cambridge: White Horse Press, 1997.

Hall, Richard and John Stravinsky. *Empires of the Monsoon: A History of the Indian Ocean and Its Invaders.* London: HarperCollins, 1996.

Huang, Boyin, Michelle L'Heureux, Zeng-Zhen Hu, Xungang Yin, and Huai-Min Zhang. 'How Significant was the 1877/78 El Nino?' *Journal of Climate*, 33, 11 (2020): 4853–4869.

Hulme, Mike. *Weathered: Cultures of Climate.* London: Sage, 2017.

James, C.C. *Drainage Problems of the East.* Bombay: Times of India Office, 1906.

Janku, A.G.J. Schenk, and F. Mauelshagen, eds. *Historical Disasters in Context: Science, Religion and Politics.* New York and London: Routledge, 2012.

Kelman, I., J.C. Gaillard, J. Lewis, and J. Mercer. 'Learning from the History of Disaster Vulnerability and Resilience Research and Practice for Climate Change', *Natural Hazards,* 82 (2016): 129–143.

Lim, Tin Seng. 'Land from Sand: Singapore's Reclamation Story.' *BibioAsia*, 13, 1 (2017): 16–23.

Lough, J.M., K.D. Anderson, and T.P. Hughes. 'Increasing Thermal Stress for Tropical Coral Reefs: 1871–2017.' *Scientific Reports* 8, 6079 (2018): 1–8.

Lübken, U. and C. Mauch, 'Uncertain Environments: Natural Hazards, Risk and Insurance in Historical Perspective,' *Environment and History* 17 (2011): 1–12.

Mahony, Martin. 'For an Empire of "All Types of Climate": Meteorology as an Imperial Science.' *Journal of Historical Geography*, 51 (2016): 29–39.

Mahony, Martin, and Sam Randalls, eds. *Weather, Climate and the Geographical Imagination: Placing atmospheric knowledges.* Pittsburgh: University of Pittsburgh Press, 2020.

Mann, Michael E., R.S. Bradley, and M.K. Hughes. 'Long-term Variability in the ENSO and Associated Teleconnections,' in *ENSO: Multiscale variability*, eds. Diaz and Markgraf: 357–412.

Morgan, Ruth. 'Southern Skies: Australian Atmospheric Research and Global Climate Change.' *Disaster Prevention and Management*, 30, 1 (2021): 47–63.

Morgan, Ruth. 'Prophecy and Prediction: Forecasting Drought and Famine in British India and the Australian Colonies.' *Global Environment*, 13, 1 (2020): 95–132.

Quinn, W.H., D.O. Zoff, K.S. Short, and R.T.W. Kuo Yang. 'Historical Trends and Statistics of the Southern Oscillation, El Nino, and Indonesian Droughts.' *Fishery Bulletin*, 76, 2 (1978): 663–78.

Ropelewski, C.F. and P.D. Jones. 'An Extension of the Tahiti-Darwin Southern Oscillation Index.' *Monthly Weather Review*, 115 (1987): 2161–65.

Saw, Swee-Hock. 'Population Trends in Singapore, 1819–1967.' *Journal of Southeast Asian History*, 10, 1 (1969): 36–49.

Simpson, W.J. *Report on the Sanitary Condition of Singapore*. London: Waterlow, 1907.

Singh, Deepti, Richard Seager, Benjamin I. Cook, Mark Cane, Mingfang Ting, Edward Cook, and Mike Davis. 'Climate and the Global Famine of 1876–78.' *Journal of Climate*, 31, 23, (2018): 9445–67.

Skinner, A.M. 'Straits Meteorology.' *Journal of the Straits Branch of the Royal Asiatic Society*, 12 (1883): 245–55.

Subrahmanyam, Sanjay. 'Connected Histories: Notes Towards a Reconfiguration of Early Modern Eurasia.' *Modern Asian Studies*, 31, 3 (1997): 735–62.

Wade, Geoff and Li Tana, eds. *Anthony Reid and the Study of the Southeast Asian Past*. Singapore: Institute of Southeast Asian Studies, 2012.

Wang, Qianfeng, Jianjun Wu, Tianjie Lei, Bin He, Zhitao Wu, Ming Liu, Xinyu Mo, Guangpo Geng, Xiaohan Li, Hongkui Zhou, and Dachuan Liu. 'Temporal-spatial Characteristics of Severe Drought Events and their Impact on Agriculture on a Global Scale.' *Quaternary International*, 349 (2014): 10–21.

Warren, James F. 'Weather, History and Empire: The Typhoon Factor and the Manila Galleon Trade, 1565–1815,' in *Anthony Reid and the Study* eds. Wade and Tana: 183–220.

Warren, James F. *Rickshaw Coolie: A People's History of Singapore, 1880–1940*. Singapore: Singapore University Press, 2003.

Wasson, R., A. Saikia, P. Bansal, and J.C. Chong. 'Flood Mitigation, Climate Change Adaption, and Technological Lock-in in Assam,' *Ecology, Economy and Society—The INSEE Journal* 3, 2 (2020): 83–1–4.

White, Sam, Christian Pfister, and Francis Mauelshagan. 'Archives of Nature and Archives of Societies,' in *The Palgrave Handbook*, eds. White, Pfister, and Mauelshagen: 27–36.

White, Sam, Christian Pfister, and Francis Mauelshagan, eds. *The Palgrave Handbook of Climate History*. London: Palgrave Macmillan, 2018.

Williamson, Fiona. 'Responding to Extremes: Managing Urban Water Scarcity in the Late Nineteenth-century Straits Settlements.' *Water History*, 12, 3 (2020): 251–63.

Williamson, Fiona. 'Malaya's "Greatest Menace"? Slow Onset Disasters and the Politics of the Environment c. 1920–1950.' *International Review of Environmental History*, 4, 2 (2018): 45–68.

Williamson, Fiona. 'Weathering the British Empire: Meteorological Research in The Early Nineteenth-century Straits Settlements.' *The British Journal for the History of Science*, 48, 3 (2015): 475–92.

Yi Loo, Yen, Lawal Billa, and Ajit Singh. 'Effect of Climate Change on Seasonal Monsoon in Asia and Its Impact on the Variability of Monsoon Rainfall in Southeast Asia.' *Geoscience Frontiers*, 6, 6 (2015): 817–23.

Yeoh, Brenda. *Contesting Space in Colonial Singapore: Power Relations and the Urban Built Environment*. Singapore: NUS Press, 2003.

Yeoh, Brenda. 'Urban Sanitation, Health and Water Supply in Late Nineteenth and Early Twentieth century Colonial Singapore.' *Southeast Asia Research*, 1, 2 (1993): 143–72.

CHAPTER 9

ENSO, IOD, Drought, and Floods in Equatorial Eastern Africa, 1876–1878

Philip Gooding

Analysed extensively in the first section of Mike Davis' *Late Victorian Holocausts*, the positive El Niño Southern Oscillation (ENSO) event of 1877–1878 is probably the most well-known oscillation of sea surface temperatures (SSTs) in world history.[1] SSTs in the east-central Pacific

[1] Mike Davis, *Late Victorian Holocausts: El Niño Famines and the Making of the Third World* (London: Verso, 2002).

[2] Deepti Singh, Richard Seager, Benjamin I. Cook, Mark Cane, Mingfang Ting, Edward Cook, and Mike Davis, 'Climate and the Global Famine of 1876–78,' *Journal of Climate*, 31, 23, (2018), 9460.

The research for this article was funded by the Social Sciences and Humanities Research Council (SSHRC) of Canada.

P. Gooding (✉)
Indian Ocean World Centre, McGill University, Montreal, QC, Canada
e-mail: philip.gooding@mcgill.ca

P. Gooding (ed.), *Droughts, Floods, and Global Climatic Anomalies in the Indian Ocean World*, Palgrave Series in Indian Ocean World Studies, https://doi.org/10.1007/978-3-030-98198-3_9

Ocean reached El Niño levels (0.5 °C above normal) for sixteen consecutive months between February 1877 and August 1878.[2] At its peak in December 1877, the SST in the rough area in the tropics between the international dateline and the South American coast reached around 2.5 °C above normal.[3] This may represent the single largest positive ENSO anomaly to have occurred in the last 200 years; it could also be the longest that an El Niño anomaly of such or similar magnitude has endured during this period.[4] Additionally, in a possibly related phenomenon, the positive 1877–1878 ENSO anomaly occurred concurrently with what may have been, until 2019, the largest positive Indian Ocean Dipole (IOD) anomaly in known history. Following seasonal patterns consistent with other positive IODs, this positive IOD occurred in May–December 1877, and peaked in August–September of that year.[5] Taken in aggregate and coupled with concurrent anomalies of SSTs in the Atlantic, these oscillations in SSTs are associated with significant climatic anomalies across the globe.[6]

Traditionally, the 1877–1878 ENSO-IOD event has been associated with drought. By mid-1877, drought had set in across much of the Indian Ocean World (IOW), including in India, northern China, mainland and island southeast Asia, southern and northeastern Africa, and Australia, in addition to parts of South America and around the Mediterranean.[7]

[3] Ibid., 9461; Boyin Huang, Michelle L'Heureux, Zeng-Zhen Hu, Xungang Yin, and Huai-Min Zhang, 'How Significant Was the 1877/78 El Nino?' *Journal of Climate*, 33, 11 (2020), 4854.

[4] Huang et al., 'How Significant,' 4853–56; Patricio Aceituno, Maríadel del Rosario Prieto, María Eugenia Solari, Alejandra Martínez, Germán Poveda, and Mark Falvey, 'The 1877–1878 El Niño Episode: Associated Impacts in South America,' *Climatic Change*, 92 (2009), 411.

[5] Singh et al., 'Climate and Global Famine,' 9456.

[6] See also: Richard Grove and George Adamson, *El Niño in World History* (London: Palgrave Macmillan, 2018), 97–98.

[7] Singh et al., 'Climate and the Global Famine,' 9449; Aceituno et al., 'The 1877–1878 El Nino,' 389–416; ZhiXin Hao, JingYun Zheng, GuoFeng Wu, ZueZhen Zhang, and QuanSheng Ge, '1876–78 Severe Drought in North China: Facts, Impacts and Climatic Background,' *Chinese Science Bulletin*, 55 (2010), 3001–7; Vimal Mishra, Amar Deep Tiwari, Saran Aadhar, Reepal Shah, Mu Xiao, D.S. Pai, and Dennis Lettenmaier, 'Drought and Famine in India, 1870–2016,' *Geophysical Research Letters*, 46, 4 (2019), 2075–83; David J. Nash, Kathleen Pribyl Jørgen Klein, Raphael Neukom, Georgina H. Endfield, George C.D. Adamson, and Dominic Kniveton, 'Seasonal Rainfall Variability in Southeast Africa During the Nineteenth Century Reconstructed from Documentary Sources,'

This is in line with larger-scale models, which associate positive ENSO and IOD anomalies with below-average rainfall in these regions.[8] The significance of the 1877–1878 anomaly and its teleconnections with other concurrent global oscillations, however, contributed to the drought being anomalously severe in these years. Northern China only received 76% and 45% of its average annual rainfall in 1876 and 1877, respectively. The latter figure represents the lowest recording of annual rainfall in the region since at least the first half of eighteenth century.[9] India, meanwhile, received its third lowest ever recording of rainfall in October–December 1876, and its lowest ever in June–September 1877 (records representing 'all India' began in the early 1870s).[10] These droughts are associated with famine and demographic decline in the worst-hit regions. The population in northern China decreased by over 20 million in 1876–1878, caused by migration and deaths by starvation and disease.[11] Taken in aggregate, Deepti Singh and her colleagues referred to the effects of the 1877–1878 ENSO-IOD event as a 'global famine,' contributing to the deaths of around 50 million people worldwide.[12] Elsewhere, Richard Grove and George Adamson referred to global financial consequences—thus displaying the cascading effects that droughts have had on human societies and institutions.[13]

Equatorial Eastern Africa (EEA) has thus far been absent from 'global' discussions of the 1877–1878 ENSO-IOD event. One could discuss a general negligence of Africa (especially eastern Africa) in 'world' and

Climatic Change, 134 (2016), 610; Stefan Grab and Tizian Zumthurm, '"Everything Is Scorched by the Burning Sun": Missionary Perspectives and Experiences of 19th- and Early 20th-Century Droughts in Semi-Arid Namibia,' *Climate of the Past*, 16, 2 (2020), 686; David J. Nash, Kathleen Pribyl, Georgina H. Endfield, Jørgen Klein, and George C.D. Adamson, 'Rainfall Variability Over Malawi During the Late Nineteenth Century,' *International Journal of Climatology*, 38, S1 (2018), 629–42; Fiona Williamson, 'Responding to the Extremes: Managing Urban Water Scarcity in the Late Nineteenth-Century Straits Settlements,' *Water History* (2020), 1–10; Chapter by Williamson in this volume.

[8] See also: Singh et al., 'Climate and Global Famine,' 9450.

[9] Hao et al., '1876–78 Severe Drought in North China,' 3002.

[10] Singh et al., 'Climate and Global Famine,' 9451.

[11] Hao et al., '1876–1878 Severe Drought,' 3005.

[12] Singh et al., 'Climate and Global Famine,' 9446.

[13] Grove and Adamson, *El Niño in World History*, 97–98.

'global' histories before European colonial rule as reasoning here.[14] Yet, there are likely two further specific reasons why historians have not yet examined EEA in this context before.[15] The first relates to the nature of the rainfall anomaly associated with the 1877–1878 positive ENSO-IOD event. Although drought was a factor in 1876, the more striking anomaly in EEA was above-average rainfall and floods from October 1877, lasting in some places until May 1878. In this context, floods related to the 1877–1878 ENSO-IOD event have also been identified elsewhere, for example, in parts of South America (southern Ecuador, northern Peru, and central Chile), in central China, and in Oman.[16] However, these examples, and that of EEA, are exceptions to a broader global pattern towards drought in 1877–1878. Moreover, trends since the publication of Davis' *Late Victorian Holocausts* in the early 2000s have, to a certain degree, led to a 'drought myopia' when thinking about the effects of the 1877–1878 global climatic anomaly. Closer examination of the effects of rainfall anomalies in EEA (and Oman, Central China, and parts of South America) leads to a more diverse appreciation of the 1877–1878 ENSO-IOD's global effects. Floods, although less widespread than drought, were a part of this history.

The second likely reason for EEA's absence from discussions of the 1877–1878 ENSO-IOD event relates to the sources. There were only two rain gauges installed in EEA at the time of the global climatic anomaly—one at Zanzibar and one at Mombasa (present-day coastal Kenya).[17] None were stationed in any of EEA's interior regions—on

[14] Gwyn Campbell, *Africa and the Indian Ocean World from Early Times to circa 1900* (Cambridge: Cambridge University Press, 2019), 21.

[15] There is one exception here: Philip Gooding, 'Tsetse Flies, ENSO, and Murder: The Church Missionary Society's Failed Ox-Cart Experiment of 1876–78,' *Africa: Rivista semestrale di studi e ricerche*, 1, 2 (2019), 21–36. See also: *The Indian Ocean World Podcast*, 'Podcast Episode 4—Gooding, Tsetse Flies, ENSO, and Murder': https://www.appraisingrisk.com/2020/06/12/podcast-episode-4-gooding-tsetse-flies-enso-and-murder/ [accessed: 18 Sep. 2020].

[16] Aceituno et al., 'The 1877–1878 El Nino,' 400–2, 408–10; Thomas F. McDow, *Buying Time: Debt and Mobility in the Western Indian Ocean* (Athens, OH: Ohio University Press, 2018), 41; Grove and Adamson, *El Niño in World History*, 97.

[17] Sharon E. Nicholson, 'A Semi-Quantitative, Regional Precipitation Data Set for Studying African Climates of the Nineteenth Century, Part I. Overview of the Data Set,' *Climatic Change*, 50, 3 (2001), 317–53.

which most of this chapter focuses—until mid-1878. Additionally, Europeans, who were the first to write down observations of EEA's interior regions, peoples, and environments, were only just beginning to enter the region at the time of the anomaly. The key actors in this context were representatives of the Church Missionary Society (CMS) and the London Missionary Society (LMS), who both started entering EEA's interior in 1876. Both departed the coast and headed towards Mpwapwa, about 260 kms inland opposite Zanzibar. From there, the CMS headed mostly northwestwards via Usukuma, towards and around Lake Victoria's shores, and the LMS headed westwards through Unyamwezi towards the north-eastern shores of Lake Tanganyika. Neither were firmly established much beyond Mpwapwa, however, until after mid-1878—that is, after most of the 1877–1878 ENSO-IOD global climatic anomaly had passed. Even so, their writings of their experiences, even after mid-1878, are valuable sources for reconstructing regional effects of the anomaly. Read alongside recent limnological research and climatic models, they suggest a multitude of ways that drought and floods affected human societies in EEA during these years.

There are, of course, challenges that arise from using European missionary sources for writing African history, including its environmental history. Missionaries had a very limited and distorted understanding of the peoples, environments, and climatic phenomena they encountered as they entered the interior of EEA.[18] They were also writing for a specific audience. In later years, they had a tendency to overstate the hardship they experienced, as their reports acted partly as moral justification for their work and intervention in African people's lives—and therefore also in their environments.[19] However, an opposite trend is possibly observable in the archival materials under review in this chapter. In 1876–1878, missionaries were liable to understate the environmental challenges they faced. This was because they were only just establishing their stations in regions that were barely known to most Europeans, and so they were invested in reporting favourable conditions to emphasise the feasibility of their missions to their funders. Therefore, the occasional reports of 'abundance' and 'plenty' may be misplaced in certain

[18] Gooding, 'Tsetse Flies, ENSO, and Murder,' 22–23.

[19] Georgina H. Endfield and David J. Nash, 'Missionaries and Morals: Climatic Discourse in Nineteenth-Century Central Southern Africa,' *Annals of the Association of American Geographers*, 94, 2 (2002), 728–29.

circumstances. Thus, when they occur, care is taken to verify them against other evidence. Reports of successful rice or maize crops in the rain-fed conditions that characterised nineteenth-century eastern African agriculture may be a better indicator of abundant or regular rainfall, given these crops' lack of resistance to water stress caused by drought or floods. Additionally, the dates at which farmers started planting and harvesting, noted frequently by missionaries, are somewhat indicative of seasonal conditions. Late planting suggests late arrival of seasonal rainfall, and consequentially a rainfall deficit, and late harvesting suggests irregular rainfall during the preceding rainy seasons. Notwithstanding these methodological approaches, it is important to state that, given the nature of the materials analysed, some of the conclusions drawn in this chapter are probably more suggestive than definitive. As is explored in the following pages, further climatological investigations may confirm or challenge these interpretations of missionary evidence in the future (Fig. 9.1).

ENSO, IOD, and EEA's Climate in 1876–1878 and Beyond

Climatological research produced in the last 20–30 years has examined the relationship between ENSO, IOD, and rainfall patterns in EEA. Broadly speaking, both positive ENSO and positive IOD anomalies are associated with above-average rainfall in EEA. Teleconnections between these oscillations and rainfall in EEA are, however, stronger when they occur at the same time, such as in 1877–1878—though they are not always certain, and the degree of teleconnection appears to have changed over time.[20] Other comparable years include 1997–1998 and 2018–2019. It should be noted, however, that the largest positive rainfall anomaly to occur on record (that is, since the mid-1870s) in EEA was in 1961, when only the IOD was positive (ENSO was neutral).[21] The 1877–1878 anomaly is EEA's second-highest positive rainfall anomaly in this

[20] Sharon E. Nicholson, 'Climate and Climatic Variability of Rainfall Over Eastern Africa,' *Reviews of Geophysics*, 55, 3 (2017), 605; Sharon E. Nicholson, 'Long-Term Variability of the East African "Short Rains" and Its Links to Large-Scale Factors,' *International Journal of Climatology*, 35, 13 (2015), 3979–90.

[21] N.H. Saji, B.N. Goswami, P.N. Vinayachandran, and T. Yamagata, 'A Dipole Mode in the Tropical Indian Ocean,' *Nature*, 401 (1999), 361.

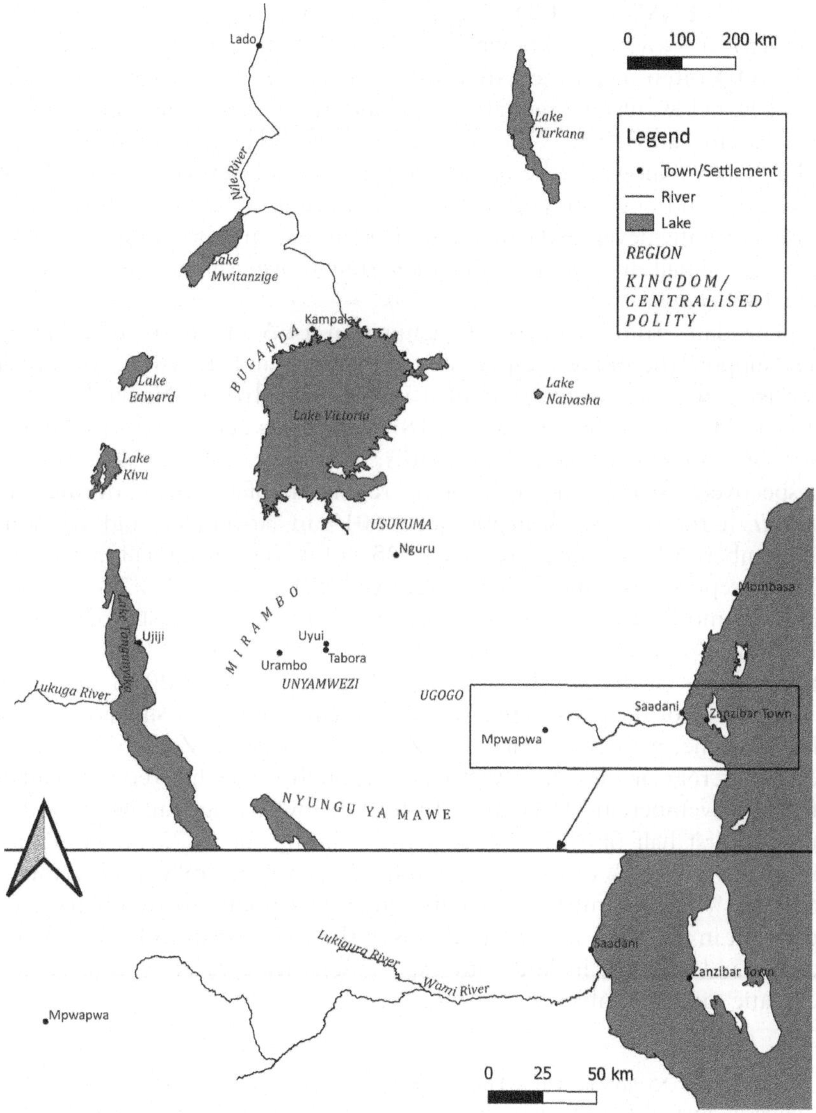

Fig. 9.1 Map of equatorial eastern Africa, with places and features mentioned in-text marked. Drawn by Philip Gooding

context.[22] ENSO and IOD affect rainfall during EEA's short rainy season in October–November, known as the *mvuli*. Positive anomalies of ENSO and IOD often mean the extension of the *mvuli* into December. Additionally, below-average rainfall is common in the *mvuli* season preceding the onset of an ENSO event, such as in 1876.[23] The long rainy season in March–May, known as the *masika*, is generally subject to less interannual variation, and does not appear to be modulated by ENSO or IOD.[24] In short, recently developed climate models suggest an expectation of below-average rainfall during the 1876 *mvuli* and above-average rainfall during the 1877 *mvuli*.

The data from the two rain gauges in EEA in 1876–1878 generally support the models, especially on the mainland. In 1876, Mombasa received 26% of its regular rainfall in October; 16% in November; and 26% in December. By contrast, in 1877, the rains began early. In August and September, normally dry months, of that year, Mombasa received, respectively, 361% and 415% of its regular rainfall. Then, during the *mvuli*, it received 331% in October, 601% in November, and 153% in December. Additionally, it received 725% of its regular rainfall in January 1878 (representing more than it received in December 1877)—January being a month in which rainfall is usually minimal in coastal regions of EEA. The pattern of above-average rainfall further endured into the first two months of the *masika*, with March 1878 receiving 201% and April 1878 159% of their respective regular rainfall. Somewhat similar, though less extreme, patterns occurred in Zanzibar. In 1876, Zanzibar received 3% of its regular rainfall in October, although somewhat regular rainfall fell in November and December. Instead, below-average rainfall prevailed for the first half of 1877. Subsequently, during the 1877 *mvuli*, Zanzibar received 164% of its regular rainfall in October, 169% in November, and 161% in December. This rain gauge data points to trends towards drought in the 1876 *mvuli* and floods in the 1877 *mvuli* in EEA's coastal and island regions, in line with expectations from recently constructed climatic models (Tables 9.1 and 9.2).

[22] Sharon E. Nicholson, Chris Funk, and Andreas H. Fink, 'Rainfall Over the African Continent from the 19th Through the Twenty-First Century,' *Global and Planetary Change*, 165 (2018), 116, 120.

[23] Sharon E. Nicholson and Jeeyoung Kim, 'The Relationship of the El-Niño Southern Oscillation to African Rainfall,' *International Journal of Climatology*, 17 (1997), 117–35.

[24] Nicholson, 'Climate and Climatic Variability,' 602–3.

Table 9.1 Monthly rainfall (mm) in Mombasa, 1876–1878 versus the average (avg.). Significant rainfall anomalies that are discussed in the text are shaded[25]

	Jan	*Feb*	*Mar*	*Apr*	*May*	*June*	*July*	*Aug*	*Sep*	*Oct*	*Nov*	*Dec*
Avg.	**27**	**16**	**59**	**200**	**305**	**117**	**85**	**69**	**66**	**95**	**92**	**58**
1876	10	5	91	152	411	79	117	69	69	25	15	15
1877	8	0	94	114	325	170	104	249	272	315	533	89
1878	196	13	122	318	114	206	48	109	76	76	53	8

Table 9.2 Monthly rainfall (mm) in Zanzibar, 1876–1878 versus the average (avg.). Significant rainfall anomalies that are discussed in the text are shaded[26]

	Jan	*Feb*	*Mar*	*Apr*	*May*	*June*	*July*	*Aug*	*Sep*	*Oct*	*Nov*	*Dec*
Avg.	**68**	**62**	**146**	**352**	**241**	**58**	**43**	**38**	**46**	**91**	**196**	**147**
1876	91	64	323	351	241	51	56	114	84	3	188	236
1877	48	6	107	269	127	79	53	99	43	150	333	300
1878	58	127	71	404	152	66	33	38	69	64	127	221

Climatologists, most notably Sharon Nicholson and her colleagues, have further constructed models that extrapolate the degree to which rain gauge readings such as those described are representative of rainfall in wider EEA. According to Nicholson, whose work was based on data from 1922 to 2012, Mombasa is in a region that correlates with a factor of 0.9 to the regional average.[27] This suggests that the rainfall anomalies described in Mombasa were broadly representative of patterns elsewhere in the region. Research into lake levels, of which Nicholson is also at the forefront, further supports this assessment for 1876–1878. Lakes Tanganyika, Rukwa, Victoria, Naivasha, and Turkana, which are all to varying degrees representative of rainfall in their catchment, reached sharp peaks in 1878, just after conclusion of the ENSO-IOD anomaly.[28]

[25] Nicholson, 'A Semi-Quantitative, Regional Precipitation Data Set,' 317–53.

[26] Ibid., 317–53.

[27] Nicholson, 'Climate and Climatic Variability,' 595.

[28] Sharon E. Nicholson and Xungang Yin, 'Rainfall Conditions in Equatorial East Africa During the Nineteenth Century as Inferred from the Record of Lake Victoria,' *Climatic Change*, 48 (2001), 388; Sharon E. Nicholson, 'Historical and Modern Fluctuations of Lakes Tanganyika and Rukwa and Their Relationship to Rainfall Variability,' *Climatic Change*, 41, 1 (1999), 53–71; Sharon E. Nicholson, 'Historical Fluctuations of Lake Victoria and Other Lakes in the Northern Rift Valley of East Africa,' in *Environmental Change and Response in East African Lakes*, ed. J.T. Lehman (Dordrecht: Kluwer

Somewhat notably, the levels of all of these lakes had been rising since c.1840, around the end of the Little Ice Age (LIA). Indeed, the last several decades of the LIA are broadly associated with drought and aridification in EEA.[29] The subsequent decades leading up to the 1877–1878 ENSO-IOD anomaly are associated with generally abundant rainfall, benign climatic conditions, and thus also rising lake levels.[30] To a certain degree, the rainfall anomaly during the *mvuli* of 1877 represented the culmination and extremity of this trend. After 1878, however, trends towards aridification returned to EEA, with notable droughts occurring, for example, in 1879–1880, 1883–1884, and 1887–1888.[31] This general pattern of benign followed by adverse climatic conditions pre- and post-1877–1878 ENSO-IOD anomaly is consistent with patterns in the wider IOW.[32] The 1877–1878 anomaly may represent a major change in the climate of EEA and the wider IOW.[33]

Returning to EEA during the 1877–1878 anomaly itself, missionary sources provide further precision on its extent and effects. The sources

Academic Publishers, 1998), 7–35; Stefan Hastenrath, 'Variations of East African Climate During the Past Two Centuries,' *Climatic Change*, 50 (2001), 209–17; Declan Conway, 'Extreme Rainfall Events and Lake Level Changes in East Africa: Recent Events and Historical Precedents,' in *The East African Great Lakes: Limnology, Palaeolimnology, and Biodiversity*, eds. Eric O. Odada and Daniel O. Olago (Boston, MA: Kluwer Academic Publishers, 2002), 63–92.

[29] Ilse Bessems, Dirk Verschuren, James M. Russell, Jozef Hus, Florias Mees, and Brian F. Cumming, 'Paleolimnological Evidence for Widespread Late Eighteenth Century Drought Across Equatorial East Africa,' *Paleogeography, Paleoclimatology, Paleoecology*, 259 (2008), 183–93; James M. Russell, Dirk Verschuren, and Hilde Eggermont, 'Spatial Complexity of "Little Ice Age" Climate in East Africa: Sedimentary Records from Two Crater Lake Basins in Western Uganda,' *The Holocene*, 17, 2 (2007), 183–93; Campbell, *Africa and the IOW*, 134–57.

[30] Campbell, *Africa and the IOW*, 184–85; Nicholson, 'Historical and Modern Fluctuations,' 53–71; Hastenrath, 'Variations of East African Climate,' 209–17.

[31] Campbell, *Africa and the IOW*, 245–54. For 1879–1880, see: Clive A. Spinage, *African Ecology: Benchmarks and Historical Perspectives* (New York: Springer, 2012), 139. In 1879 and 1880, Ujiji received 756 mm and 694 mm of annual rainfall respectively (Edward C. Hore, *Tanganyika: Eleven Years in Central Africa* [London: Edward Stanford, 1892], 145). This is significantly lower than its annual average of 952 mm, based on twentieth-century measurements. For 1883–1884, see: Chapter by Rockel in this volume.

[32] Campbell, *Africa and the IOW*, 176–254.

[33] Conway, 'Extreme Rainfall Events,' 85. A similar 'sea-change' in EEA climate may be observed in the aftermath of extreme positive rainfall anomalies in 1961, 1983, and 1997. See: Nicholson, 'Climate and Climatic Variability,' 611.

are most abundant for the region between the coast and Mpwapwa. In 1876, representatives of the LMS and CMS described parts of the region as 'swamps' that required wading through, which is what they expected due to reports made by European explorers in previous years.[34] Such descriptions in this context are also in line with similar ones made in the aftermath of other *mvuli* seasons of below-average rainfall, such as in 1879–1880.[35] Additionally, once they arrived in Mpwapwa, situated on the eastern edge of the arid region of Ugogo, they were struck by the unavailability of food, a phenomenon that they attributed to heightened demand, but which may also have been underpinned by the 1876 *mvuli* drought.[36] In 1877, however, rivers on the route to Mpwapwa were flooded, bridges and cornfields were entirely submerged, and swamps were covered in some places by knee-deep standing water.[37] The region between the Lukigura and Mkundi Rivers (both tributaries of the larger and more famous Wami) on the direct route between Saadani and Mpwapwa appears to have been especially affected. Beginning travel in January 1878, Edward Hore of the LMS described the region up to the Lukigura as comprised mostly of 'muddy path[s] and bogs.'[38] Once across the Lukigura, however, they were forced to wade through three feet of water.[39] Similarly, the Mkundi was 'too deep to ford.' It was only

[34] Church Missionary Society Archive (hereafter: CMS) C/A6/O/16 Mackay to Wright, 14 Oct. 1876; Roger Price and Joseph Mullens, 'A New Route and New Mode of Travelling into Central Africa,' *Proceedings of the Ryoal Geographical Society of London*, 2, 4 (1876–7), 234–35; Henry Morton Stanley, *How I Found Livingstone: Travels, Adventures, and Discoveries in Central Africa Including Four Months' Residence with Dr. Livingstone* (London: Sampson Low, Marston & Company, 1872), 114–20.

[35] CMS C/A6/O/20 Price to Wright, 25 Nov. 1879.

[36] CMS C/A6/O/7 Clark to Wright, 4 Nov. 1876; CMS C/A6/O/7 Clark to Wright, 6 Dec. 1876.

[37] Edward C. Hore, *Missionary to Tanganyika 1877-1888*, ed. James B. Wolf (London: F. Cass, 1971), 22–25.

[38] Ibid., 17.

[39] Ibid., 25. See also: CMS C/A6/O/16 Mackay to Wright, 2 Feb. 1878; CMS C/A6/O/9 Copplestone to Wright, 16 Feb. 1878; CMS C/A6/O/16 Mackay to Wright, 22 Mar. 1878; CMS C/A6/O/14 Last to Wright, 11 May 1878; Alexina Mackay Harrison, *The Story and Life of Mackay of Uganda: Pioneer Missionary* (London: Hodder & Stoughton, 1900), 103; Norman R. Bennett, 'Philippe Broyon: Pioneer Trader in East Africa,' *African Affairs*, 62, 247 (1963), 158; Norman R. Bennett, *From Zanzibar to Ujiji: The Journal of Arthur W. Dodgshun* (Boston: African Studies Center, 1969), 15–16.

when they took a significant detour and reached beyond its western side that the land became passable.[40] Additionally, in contrast to conditions in 1876, food was abundant in Mpwapwa in 1877–1878.[41] Between the coast and Mpwapwa, the above-average rainfall brought floods to swampy regions and fertility to more arid regions.

From Mpwapwa, representatives of the CMS mostly headed in a north-westerly direction towards Lake Victoria. This took them through a fertile tract in the region of Usukuma. In 1876, the region around Nguru appears to have been deeply affected by drought. Missionaries reported that farmers were only planting their millet and maize seeds in late December of that year, as the rains had only just come.[42] Given the dryness of January and February, especially in 1877 (according to Mombasa's rain gauge), this must have been disastrous for their agricultural output, especially of maize—a particularly thirsty crop.[43] Despite these challenges, the first CMS missionaries to the region characterised it as a region of abundance, at least compared to surrounding zones, suggesting high levels of resilience to seasonal drought.[44] In the ***mvuli*** of 1877, meanwhile, the level of rainfall appears to have been significantly above-average, though, unlike in Mombasa where the rains arrived early, significant levels do not appear to have fallen until October. Rivers feeding southeastern Lake Victoria were dry at the beginning of one missionary's October diary, for example. It was only after the 'commencement of the rains' in mid-October that these rivers dramatically rose to around 4 feet.[45] Between 5 November and 4 December, the same missionary recorded rain nearly every day, and, later, one of his peers stationed

[40] Hore, *Missionary to Tanganyika*, 26–27. See also: CMS C/A6/O/16 Mackay to Wright, 22 Mar. 1878.

[41] CMS C/A6/O/9 Copplestone to Wright, 16 May 1878.

[42] CMS C/A6/O/18 O'Neill to Wright, 29 Dec. 1876; CMS C/A6/O/21 Smith to Wright, 9 Feb. 1877.

[43] James C. McCann, *Maize and Grace: Africa's Encounter with a New World Crop, 1500-2000* (Cambridge, MA: Harvard University Press, 2009), 19; N. Mbava, M. Mutema, R. Zengeni H. Shimelis, and V. Chaplot, 'Factors Affecting Crop Water Use Efficiency: A Worldwide Meta-Analysis,' *Agricultural Water Management*, 228 (2020), 1–11.

[44] CMS C/A6/O/18 O'Neill to Wright, 29 Dec. 1876; CMS C/A6/O/16 Mackay to Wright, 9 Dec. 1877.

[45] CMS C/A6/O/22 Shergold Smith to Wright, 2 Nov. 1877.

on Lake Victoria's southern shores declared the season 'unusually' and 'excessively' wet.[46] This wetness flooded paths, which delayed travel, for example, from Lake Victoria to Mpwapwa and to Unyanyembe (Tabora), the latter place being in present-day west-central Tanzania.[47]

Moving into present-day west-central Tanzania, missionary and limnological data suggests rainfall patterns that were consistent with the wider region. In 1876, one missionary wrote in Unyanyembe that the rains only arrived on 10 December, suggesting failed rains during the *mvuli* season, as in Usukuma.[48] Lake-level research, which suggests a decline in Lake Tanganyika's level in the year-or-two immediately preceding the 1877–1878 ENSO-IOD event, supports this assessment.[49] In 1877–1878, meanwhile, there is evidence of floods. One CMS missionary, for example, was forced in May 1878 to rest in Uyui (northern Unyamwezi) to allow for the flooding to pass before onward travel.[50] Additionally, farmers were still growing rice and corn in July of that year, when normally the harvest would have occurred by the end of May.[51] This late growing season likely attests to the unusual wetness of the soil from deluges earlier in the year, and possibly to damaged crops planted at a similar time.[52] Further west, the excessive rains that caused the level of Lake Tanganyika to rise may have contributed to floods in Ujiji. In December 1878, a missionary reported that almost an entire district of Ujiji was now 'under' the lake.[53] This change was likely also linked to a gradual process of

[46] CMS C/A6/O/22 Journal of Lieut. Shergold Smith, 5 Nov.–4 Dec. 1877; CMS C/A6/O/25 Wilson to Wright, 15 Jan. 1878; CMS C/A6/O/25 Wilson to Wright, 19 Apr. 1878.

[47] CMS C/A6/O/25 Wilson to Wright, Mar. 1878; CMS C/A6/O/16 Mackay to Smith-MacKenzie & Co., 16 May 1878; CMS C/A6/O/25 Wilson to Wright, 20 May 1878.

[48] CMS C/A6/O/22 Shergold Smith to Wright, 1 Jan. 1877.

[49] Nicholson, 'Historical and Modern Fluctuations,' 57.

[50] CMS C/A6/O/16 Mackay to Smith-MacKenzie & Co., 16 May 1878.

[51] CMS C/A6/O/16 Mackay to Wright, 20 July 1878; Hore, *Missionary to Tanganyika*, 49.

[52] See also: CMS C/A6/M/M2 Thomson to Mackay, 31 July 1878.

[53] Council for World Missions/London Missionary Society (hereafter: CWM/LMS) 06/02/003 Hore, 'Kigoma Bay,' 9 Dec. 1878.

rising lake levels for most of the preceding 40 years.[54] Perhaps more significantly, though, is that this gradual rise and the 1877–1878 deluge contributed to the collapse of a naturally forming dam at the entrance of the Lukuga River, Lake Tanganyika's only outlet.[55] The increased current exiting the lake in following years exacerbated trends towards declining lake levels in the 1880s–90s.[56] This process exposed beaches, destroyed many ports, and linked several islands to the mainland.[57] Somewhat paradoxically, deluges associated with the 1877–1878 ENSO-IOD event contributed to the 'drying out' of Lake Tanganyika's shores over the long run.

The only other region in EEA for which missionary sources further illuminate the climatological record is the capital of Buganda (within present-day Kampala, Uganda), near Lake Victoria's northern shoreline. Here, though, is where the missionary data is at its thinnest—CMS missionaries only arrived there for the first time in July 1877. There is thus no primary account of possible drought conditions during the year before the onset of the 1877–1878 ENSO-IOD event. It is perhaps notable, however, that the first missionaries to arrive in Buganda commented on the abundance of produce available.[58] This suggests that drought conditions had not prevailed in the previous months, though it might also be

[54] See also: Henry Morton Stanley, *Through the Dark Continent or the Sources of the Nile Around the Great Lakes of Equatorial Africa and Down the Livingstone River to the Atlantic Ocean* (London: Sampson Low, Marston, Searle & Irvington, 1878), II, 11–12.

[55] Nicholson, 'Historical and Modern Fluctuations,' 62; Ruud C.M. Crul, 'Limnology and Hydrology of Lakes Tanganyika and Malawi,' *Comprehensive and Comparative Study of the Great Lakes* (Paris: UNESCO Publishing, 1997), 34; C. Gillman, 'The Hydrology of Lake Tanganyika,' *Tanganyika Territory Geological Survey Department*, 5 (1933), 6.

[56] Compare the accounts of the Lukuga outlet pre- and post-collapse of the dam: Stanley, *Dark Continent*, II, 45; Edward C. Hore, 'Lake Tanganyika,' *Proceedings of the Royal Geographical Society and Monthly Record of Geography*, 4, 1 (1882), 11–12; UK National Archives, Royal Geographical Society (hereafter: NA RGS) CB6/1167 Hore to RGS, 27 May 1879; Zanzibar National Archives (hereafter: ZNA) AA1/23 Hore to Kirk, 27 May 1879; ZNA BK1/12 Thomson to Kirk, 27 March 1880.

[57] See, for example: Beverly Bolser-Brown, 'Ujiji: The History of a Lakeside Town, c.1800-1914' (Unpublished PhD diss.: Boston University, 1973), 2; CWM/LMS/06/02/005 Griffith to Whitehouse, 28 Aug. 1880; CWM/LMS/06/02/006 Griffith to Thompson, 12 Aug. 1881; CWM/LMS/06/02/008 Hore to Whitehouse, 18–21 June 1883.

[58] CMS C/A6/O/22 Shergold Smith to Wright, 16 Aug. 1877; CMS C/A6/O/25 Wilson to Wright, 5 July 1877.

a function of this region receiving a higher average annual rainfall than regions of EEA south of Lake Victoria. Evidence from 1877 to 1878, meanwhile, also suggests that the positive rainfall anomaly felt elsewhere in EEA was not as strong there as it was further south. One missionary described Buganda in November 1877 as 'flourishing from an agricultural point of view,' suggesting that crops had not been adversely affected by floods.[59] The same missionary further noted the presence of 'swampy valleys,' which he believed would have been adequate for rice production.[60] The fact that rice was not yet grown in such valleys may indicate that these conditions were unusual, but this may also have been a question of taste. Buganda was known as banana-growing country, with plantains being crucial to its 'national' identity.[61] The missionary also attributed a rise in Lake Victoria's level in 1877–1878 to 'excessive' rain in Usukuma and its environs; not in Buganda, despite having visited the latter place.[62] Finally, the first missionary rain gauge established in the region recorded 100 mm of rain in April 1878 and 37 mm of rain in May 1878, representing, respectively, only 52% and 30% of normal, thus suggesting that excessive rainfall did not last into the *masika*, as it did further south and in Mombasa.[63] Missionary data suggests the effects of the 1877–1878 ENSO-IOD anomaly on rainfall were not as strong around Lake Victoria's northern shores as they were in other parts of EEA.

Taken in aggregate, missionary sources suggest that the 1877–1878 ENSO-IOD anomaly affected rainfall in EEA most significantly in coastal and hinterland regions, as well as across present-day Tanzania's central and northern latitudes. In these regions, the *mvuli* rains failed in 1876, leading to drought, but in 1877–1878, they started early and were abundant to the extent of causing floods. Additionally, in a possibly related phenomenon, the *masika* season in 1878 was more abundant and lasted longer than normal. Similar, though less extreme, patterns are visible from evidence further north in parts of EEA's interior. Evidence suggests

[59] CMS C/A6/O/25 Wilson to Wright, 21 Nov. 1877.

[60] Ibid.

[61] Richard J. Reid, *Political Power in Pre-Colonial Buganda: Economy, Society & Warfare in the Nineteenth Century* (Oxford: James Currey, 2002), 22–25.

[62] CMS C/A6/O/25 Wilson to Wright, 19 Apr. 1878.

[63] Nicholson, 'A Semi-Quantitative, Regional Precipitation Data Set,' 317–53. This is taken from the data at Rubaga, in present-day Kampala.

that regions around Lake Victoria's northern shores only experienced limited rainfall anomalies in 1876–1878. Other climatological investigations, however, suggest that these patterns located in close vicinity to Lake Victoria shores were exceptional. The level of the White Nile at Lado, a town in present-day southern South Sudan, caused floods and remained high until December 1878, suggesting an unusually high level for Lake Mwitanzige (Albert), which may at least be partly attributable to high levels of rainfall in its catchment, including in parts of present-day northern and central Uganda. Such patterns in the White Nile are corroborated by European reports from Gondokoro and by Nuer oral traditions, located mostly in northeastern regions of present-day South Sudan.[64] Further climatological research is necessary to ascertain the degree to which conditions around Lake Victoria's northern shores were really exceptional compared to the wider EEA in this context, as well as the possible drivers of such exceptions.

The Effects of ENSO-IOD-Related Drought and Floods in 1876–1878

The 1877–1878 ENSO-IOD event occurred during a particularly dynamic period in EEA's history. The second half of the nineteenth century was the height of the ivory trade, linked to the spread of capitalism from industrialing Europe and North America. Caravans, which facilitated this trade, often contained over 1000 people, and travelled between the EEA's Indian Ocean littoral and the eastern Congo rainforests. This trade brought several new phenomena to hinterland and interior regions of EEA, including crops, such as cassava, maize, and rice; diseases, such as smallpox; international commodities, such as industrially produced glass beads, cotton cloths, and firearms; and belief systems, especially Islam. Phenomena such as these contributed to the transformation of agricultural production, material cultures, and kinship connections across much of the region.[65] They also sparked heightened levels of competition for access to and control of trade, which manifested itself in new, militarised forms of state-building. The most prominent

[64] Nicholson, 'A Semi-Quantitative, Regional Precipitation Data Set,' 317–53.

[65] Philip Gooding, 'History, Politics, and Culture in Central Tanzania,' *Oxford Research Encyclopedia of African History* (2019), 5–9.

state-builders in this context were Nyungu ya Mawe (in present-day southwest-central Tanzania), Mirambo (west-central Tanzania), and the *kabakas* (kings) of Buganda (Uganda).[66] Given Mirambo's location in a part of EEA that appears to have been distinctly affected by rainfall anomalies associated with the 1877–1878 ENSO-IOD event, his state features prominently in the following discussion. This section also focuses on phenomena that are widely known to be affected by droughts and floods, namely, agriculture, the spread of disease, and political instability. The limits of the source material inhibit analysis beyond these themes.

Rainfall anomalies in 1876–1878 affected the seasonality of EEA's agricultural regimes. As described above, drought in 1876 caused the delay of the planting season, and floods in 1877–1878 likely contributed to the delay of the following harvest in much of present-day mainland Tanzania. Both these changes to the seasonal planting schedule likely caused significant hardship. The dryness of January and February (especially in 1877, according to the Mombasa's rain gauge data) almost certainly stunted the crops planted in December 1876, if it did not kill them outright. Moreover, reliance on a harvest in July likely caused a great deal of anxiety. Normally, June and July in EEA are too dry for most crops to mature into a bountiful crop. The fact that (again, according to Mombasa's rain gauge) June (although not May) 1878 appears to have been wetter than normal may have limited some of these effects. Nevertheless, drought and floods in these years likely affected farmers near caravan routes and commercial centres more than elsewhere. Many of these farmers had in previous years adopted maize and rice as staple crops.[67] These crops have high potential yields, which in good rainfall years helped to feed large populations in towns and caravans. However, they are also both significantly less resistant to water stress than the lower-potential, African staples they sometimes replaced, such as sorghum and millet.[68] By 1876–1878, the adoption of new crops and increased demand from urban populations

[66] Philip Gooding, 'The Ivory Trade and Political Power in Nineteenth-Century East Africa,' in *Animal Trade Histories in the Indian Ocean World*, eds. Martha Chaiklin, Philip Gooding, and Gwyn Campbell (Cham, CH: Palgrave, 2020), 251–60.

[67] Richard F. Burton, *The Lake Regions of Central Africa: A Picture of Exploration* (New York: Harper & Brothers, 1860), 354; Stanley, *How I Found*, 395; Stanley, *Dark Continent*, II, 4, 47.

[68] Mbava et al., 'Factors Affecting Crop Water Use Efficiency,' 1–11.

made the size of farmers' yields highly vulnerable to the effects of drought and floods.

Drought and floods in 1876–1878 also contributed to the spread of disease. In another publication, I argued that floods in 1877–1878 around the Lukigura and Mkundi Rivers contributed to a dramatic, temporary increase in the population of tsetse flies, leading to an epizootic of bovine trypanosomiasis (sleeping sickness), a disease fatal to cattle.[69] This was because tsetse flies (as with mosquitoes) reproduce faster in moist, shaded, humid conditions, such as around flooded rivers amongst dense vegetation.[70] The extent to which other regions in EEA experienced similarly dramatic trypanosomiasis epizootics, however, was probably limited. In the late nineteenth and early twentieth centuries, tsetse flies were sometimes so sparse in coastal and hinterland regions, such as around the Lukigura and Mkundi Rivers, that observers (mistakenly) claimed that there were none at all.[71] It was only during significant positive rainfall anomalies that they became abundant. Such patterns were probably not so apparent elsewhere, such as in west-central Tanzania, where tsetse fly populations grew during much of the second half of the nineteenth century as a result of human activities. Violence associated with the growth of militarised states caused people to live in larger, stockaded settlements, leaving the regions in between them to return to bush—the ideal environment for tsetse flies.[72] Thus, tsetse flies, and so also trypanosomiasis, were prevalent in this region well before the 1877–1878 ENSO-IOD event.[73] If the latter had any effect here at all, it

[69] Gooding, 'Tsetse Flies, ENSO, and Murder,' 21–36.

[70] William L. Krinsky, 'Tsetse Flies (Glossinidae),' in *Medical and Veterinary Entomology*, eds. Gary R. Mullen and Lance Durden (London: Academic Press, 2019), 369–82.

[71] Verney Lovett Cameron, *Across Africa* (New York: Harper & Brothers, 1877), 48; Price and Mullens, 'A New Route,' 241–43; Helge Kjekshus, *Ecology Control and Economic Development in East African History: The Case of Tanganyika 1850-1950*, 2nd ed. (London: James Currey, 1996), 164, map 8.1.

[72] Stephen J. Rockel, 'The Tutsi and the Nyamwezi: Cattle, Mobility, and the Transformation of Agro-Pastoralism in Nineteenth-Century Western Tanzania,' *History in Africa*, 46 (2019), 233; Richard J. Reid, *War in Pre-Colonial Eastern Africa: The Patterns and Meanings of State-Level Conflict in the Nineteenth Century* (Nairobi: The British Institute in Eastern Africa, 2007), 134; Stanley, *How I Found*, 178.

[73] Burton, *Lake Regions*, 386; Stanley, *How I Found*, 227, 254, 447; CWM/LMS/06/02/003 Thomson to LMS, 4 Aug. 1878.

was likely that the size of the tsetse fly population may have temporarily increased where they were already well-known.[74]

There is also some evidence for the ENSO-IOD's effects on human diseases. Later ENSO/IOD-related floods, such as in 1961 and 1997, have been associated with outbreaks of Rift Valley fever, as well as an increase in vectors, such as tsetse flies and mosquitoes.[75] Unfortunately, the archival material is not sufficient enough to be able to draw direct comparisons between these cases and that of 1877–1878. Perhaps more intriguing, though, is the possibility that drought in 1876 contributed to an epidemic of smallpox along EEA's caravan routes. Drought is known, for example, to have contributed to smallpox epidemics elsewhere in the IOW, including during the 1877–1878 ENSO-IOD event.[76] Outbreaks often turned into epidemics in these contexts as hungry, infected people migrated in search of food or wages. Similar patterns appear to have pervaded EEA in 1876. CMS missionaries and a British official in Zanzibar reported in this year that smallpox affected every caravan they encountered, whether heading to the coast or towards the interior.[77] Their reports on smallpox, however, ceased almost entirely from early 1877 onwards, suggesting it declined in prevalence as rainfall became more abundant and people returned to their farms during the *masika*. This is further supported by evidence from other drought years in west-central Tanzania. Ujiji, on Lake Tanganyika's northeastern shore, experienced outbreaks of smallpox in nearly every year of the 1880s, but it only appears to have reached epidemic levels in 1879–1880 and

[74] This is further suggested in: CMS C/A6/M/M2 Thomson to Mackay, 31 July 1878.

[75] Hussein Gadain, Nicolas Bidault, Linda Stephen, Ben Watkins, Maxx Dilley, and Nancy Mutunga, 'Reducing the Impacts of Floods Through Early Warning and Preparedness: A Pilot Study for Kenya,' in *Natural Disaster Hotspots: Case Studies*, eds. Margaret Arnold, Robert S. Chen, Uwe Deichmann, Maxx Dilley, Randolph E. Pullen, and Zoe Trohanis (Washington, DC: The World Bank, 2006), 178–79.

[76] For nineteenth-century IOW smallpox epidemics, see: Campbell, *Africa and the IOW*, 246–53. For examples related to drought during the 1877–1878 ENSO-IOD event, see: Anastácio Q. Sousa and Richard Pearson, 'Drought, Smallpox, and Emergence of *Leishmania braziliensis* in Northeastern Brazil,' *Emerging Infectious Diseases*, 15, 6 (2009), 917–19; Davis, *Late Victorian Holocausts*, 80–88; Williamson, 'Responding to the Extremes,' 4.

[77] CMS C/A6/M/M1 Holmwood to Hutchinson, 19 Aug. 1876; CMS C/A6/O/16 Mackay to Wright, 18 Sep. 1876; CMS C/A6/O/16 Mackay to Wright, 14 Oct. 1876; CMS C/A6/O/13 Kirk to Wright, 22 Nov. 1876.

1883–1884, two years of significant drought.[78] In those years, smallpox was so prevalent that piles of corpses were left to rot on the edge of town.[79] This, coupled with the evidence from caravans in 1876, suggests that food shortages in drought years increased levels of migrancy, which contributed to the spread of smallpox.

Moving onto the possible structural effects of drought and floods, much has been written in recent years on the relationship between global climate change and political instability. Environmental historians and political scientists alike have recently stressed the importance of adverse climatic and environmental factors for understanding some significant political changes in both the recent and deeper past.[80] Some such studies

[78] Beverly Bolser-Brown and Walter Brown, 'East African Trade Towns: A Shared Growth,' in *A Century of Change in Eastern Africa*, ed. W. Arens (The Hague: Morton Publishers, 1976), 190–93; Gerald W. Hartwig, 'Demographic Considerations in East Africa During the Nineteenth Century,' *International Journal of African Historical Studies*, 12, 4 (1979), 662–64; Kjekshus, *Ecology Control*, 132; CWM/LMS/06/02/005 Griffith to Whitehouse, 19 May 1880; CWM/LMS/06/02/005 Hore to Whitehouse, 20 July 1880; Walter Hutley, *The Central African Diaries of Walter Hutley*, ed. James B. Wolf (Boston: African Studies Center, 1976), 170, 173, 183, 193, 197, 213–14; CWM/LMS/06/02/008 Hore to Whitehouse, 18–21 June 1883, Jones to Whitehouse, 20 Aug. 1883; Royal Museum for Central Africa Emile Storms Archive (hereafter: RMCA ESA) HA.01.017-6 Storms to AIA, 1883; CWM/LMS/06/02/009 Jones to Whitehouse 24 June 1884. The European explorer, Henry Morton Stanley, reported an outbreak of smallpox in Ujiji in August 1876, but he left the town before the subsequent failed *mvuli* rains. Thus, it is unknown how or if the 1876 ENSO-IOD-related drought affected its spread there thereafter, although evidence from the coast and hinterland suggests it was prevalent along all caravan routes. See: Stanley, *Dark Continent*, II, 62; Royal Museum of Central Africa Henry Morton Stanley Archive (hereafter: RMCA HMSA) 33. Stanley to *Daily Telegraph* and *New York Herald*, 13 Aug. 1876. Smallpox was also especially prelavent in 1884 near the coast and in Buganda. See: Juhani Koponen, *People and Production in Late Precolonial Tanzania: History and Structures* (Helsinki: Finnish Society for Development Studies, 1988), 165–66, 173; CMS G/3/A/6/O Ashe to Lang, 25 Mar. 1884; CMS G/3/A/6/O O'Flaherty to CMS, 1 Apr. 1884; CMS G/3/A/6/O O'Flaherty to Wigram, July 1884; Archivio Generale dei Missionarie d'Africa (hereafter: A.G.M.Afr.) Vicariat apostolique du Nyanza, 11 July 1884, *Chronique Trimestrielles*, 23–24 (Oct 1884).

[79] CWM/LMS/06/02/005 Hore to Whitehouse, 20 July 1880; A.G.M.Afr. C.16-84. Guillet to White Fathers, 17 Apr. 1884.

[80] Any number of publications could be cited here, which have grown in prominence since the Arab Spring in 2011. For a recent example, see: Emrah Sofuoğlu and Ahmet Ay, 'The Relationship Between Climate Change and Political Instability: The Case of MENA Countries (1985:01-2016:12),' *Environmental Science and Pollution Research*, 27 (2020), 14033–43.

have even drawn links between significant ENSO anomalies and political instability in the IOW. These include, for example, the collapse of the Ming Dynasty in China and a decline in Mughal power in India as related to the effects of the strong 1629–1635 El Niño.[81] In these contexts, droughts and floods have been seen to contribute to food shortages, higher prices, and scarcity, and thus also to heightened levels of competition for resources and dissatisfaction with the state, leading to violence and political instability. It thus might not be surprising that the 1877–1878 ENSO-IOD-related droughts and floods occurred at the same time as heightened levels of violence in some parts of EEA. Specifically, in these years, Mirambo attacked and expanded into parts of Usukuma, possibly for the first time.[82] In this context, it is notable that Mirambo's political centre, at Urambo, had grown significantly in previous years, putting pressure on food production in his wider domain.[83] By the late 1870s, it had around 15,000 inhabitants, when two decades before, the largest settlement in the region had around 150 huts. Therefore, his movement into Usukuma may have been motivated to seek the resources of a region known for its abundance, despite drought conditions. Usukuma's apparent resilience to the effects of drought made it a prime target for expansion from militarised states centred on regions whose vulnerability may have grown in previous years.

The archival record supports this hypothesis up to a point. In attacking Usukuma (and in line with his tactics elsewhere), Mirambo sought its resources, including its people, cattle, and crops.[84] Many people were enslaved and forced to work on farms in Unyamwezi, thus boosting

[81] Grove and Adamson, *El Niño in World History*, 63–65; Campbell, *Africa and the IOW*, 146–47; Jingyun Zheng, Lingbo Xiao, Xiuqi Fang, Zhixin Hao, Quansheng Ge, and BeiBei Li, 'How Climate Impacted the Collapse of the Ming Dynasty,' *Climatic Change*, 127 (2014), 169–82.

[82] CMS C/A6/O/18 O'Neill to Wright, 29 Dec. 1876; CMS C/A6/M/M1 Morton to Smith MacKenzie & Co., 25 July 1877; CMS C/A6/O/18 O'Neill to Wright, Oct. 1877; CMS C/A6/O/22 Shergold Smith to Wright, 2 Nov. 1877.

[83] CWM/LMS/06/02/004 Southon to LMS, 8 Sept. 1879; Burton, *Lake Regions*, 270.

[84] Reid, *War in Pre-Colonial*, 120–21, 133–34; CMS C/A6/O/18 O'Neill to Wright, 29 Dec. 1876; CMS C/A6/O/22 Shergold Smith to Wright, 2 Nov. 1877; CMS C/A6/O/24 Stokes to Wright, 23 Oct. 1878.

agricultural output in Mirambo's domain.[85] Mirambo's armed following (known as *ruga ruga*) supplemented this process by also pillaging corn and other grains in newly invaded zones.[86] Cattle were used for their milk and as stores of wealth. It is possible that cattle were especially demanded in 1876–1878, owing to a depletion of stocks around this time caused by an increased population of tsetse flies in Unyamwezi. In any case, by October 1877, there was apparently 'no milk to be had [in Usukuma] because Mirambo had taken all the cattle.'[87] Nevertheless, Mirambo likely had broader strategic aims for invading Usukuma in these years. A core *raison d'être* of his state was the control and taxation of long-distance trade between EEA's interior and coast.[88] His state's core was located on the landward routes between Tabora and Buganda via the Tanganyika-Victoria corridor, and between Tabora and the eastern shore of Lake Tanganyika, including Ujiji. His expansion into Usukuma in 1876–1878, therefore, can partly be interpreted as a response to a contemporaneous development, that traders travelling between the coast and Buganda were increasingly going through Usukuma and across Lake Victoria by boat.[89] That this was key to his entrance into Usukuma is supported by his armed representatives violently extorting taxes from

[85] Stephen J. Rockel, *Carriers of Culture: Labor on the Road in Nineteenth-Century East Africa* (Portsmouth, NH: Heinemann, 2006), 6; Jan-Georg Deutsch, 'Notes on the Rise of Slavery & Social Change in Unyamwezi,' in *Slavery in the Great Lakes Region of East Africa*, eds. Henri Médard and Shane Doyle (Oxford: James Currey, 2007), 87; Reid, *War in Pre-Colonial*, 132.

[86] CMS C/A6/O/18 O'Neill to Wright, 29 Dec. 1876; CMS C/A6/O/21 Smith to Wright, 9 Feb. 1877.

[87] CMS C/A6/O/22 Shergold Smith to Wright, 2 Nov. 1877.

[88] Gooding, 'The Ivory Trade and Political Power,' 257–59.

[89] Reid, *War in Pre-Colonial*, 124; McDow, *Buying Time*, 190–214; Gerald W. Hartwig, 'The Victoria Nyanza as a Trade Route in the Nineteenth Century,' *Journal of African History*, 11, 4 (1970), 535–52; C.F. Holmes, 'Zanzibari Influence at the Southern End of Lake Victoria: The Lake Route,' *African Historical Studies*, 4, 3 (1971), 477–503; Richard J. Reid, 'The Ganda on Lake Victoria: A Nineteenth-Century East African Imperialism,' *Journal of African History*, 39, 3 (1998), 349–63; CMS C/A6/O/16 Mackay to Wright, 23 Aug. 1880.

passing caravans during these years.[90] Thus, if 1876–1878 ENSO-IOD-related rainfall anomalies in EEA contributed to violence in Usukuma, other political and commercial factors were likely at least just as important.

Conclusion

As recently developed climatic models suggested would be the case, this chapter shows that the 1877–1878 ENSO-IOD event significantly affected rainfall in EEA. 1876, the year before the onset of SST anomalies in the east-central Pacific and Indian Ocean, is generally associated with drought in the region; 1877–1878, as the anomalies took place, are associated with floods. According to evidence gleaned from the missionary archive and supported by limnological research, the regions in EEA most affected by these anomalies were in coastal, hinterland, and southern parts of EEA, as well as in present-day northern Uganda. These regions are those in which agricultural regimes were most disrupted. Additionally, these regions appear to have been affected by a drought-related smallpox epidemic in 1876, and parts of the coastal hinterland were affected by a flood-related, sudden epizootic of bovine trypanosomiasis in 1877–1878. The likely reason for the region roughly encapsulated by present-day west-central Tanzania not experiencing the latter epizootic in the same way owes itself to tsetse flies probably being already abundant there. Nevertheless, these adverse effects of drought and floods in southern regions of EEA had effects on more northern regions as well, including around Lake Victoria's southern shores. Although Usukuma appears to have been generally resilient to the effects of drought and floods, its wealth of grain, cattle, and people may have contributed to Mirambo's reasons for attacking it. Relative abundance in Usukuma contributed to it becoming a prime target for militarised state-builders based in more drought-and-flood-vulnerable regions, who sought additional resources.

The larger effects of the 1877–1878 ENSO-IOD event on EEA and the wider IOW may, however, have been longer term. In EEA, a significant long-term effect of excessive rainfall in 1877–1878 was the collapse of the reed and mud dam at the Lukuga outlet of Lake Tanganyika, which exacerbated trends towards declining lake-levels in the 1880s–90s. But

[90] CMS C/A6/O/18 O'Neill to Wright, 29 Dec. 1876; CMS C/A6/O/16 Mackay to Wright, 20 July 1878; CMS C/A6/O/9 Copplestone to Wright, 24 June 1879.

there are wider climatic teleconnections going on here. Leading up to the 1877–1878 ENSO-IOD event, climatic conditions in the IOW were generally benign, and rainfall in EEA was generally regular and abundant. The years afterwards, however, were characterised by volatile climate throughout the IOW, and trends towards aridification in EEA. The role, if any, that the 1877–1878 ENSO-IOD event had in contributing to this climatic shift is currently unknown. Stephen Rockel's chapter in this volume suggests that the eruption of Krakatau in 1883 may have further contributed to rainfall extremes.[91] Nevertheless, evidence from other, similarly large, and more recent ENSO and/or IOD-related anomalies, suggests that such events are capable of making subtle shifts in teleconnections between global climatic anomalies and rainfall in EEA. Further climatological research could examine if similar shifts in teleconnections are observable with rainfall in other IOW regions. Given the interconnected nature of the IOW and the centrality of the Indian Ocean monsoon system to conceiving of the macro-region, climatologists might expect to find such shifts, if there is adequate data.

Bibliography

Archival Sources

Archivio Generale dei Missionari d'Africa (A.G.M.Afr.): White Fathers' Archive, Rome, Italy.

Church Missionary Society (CMS): Birmingham, UK.

Council for World Missions/London Missionary Society (CWM/LMS): London, UK.

Royal Museum for Central Africa Emile Storms Archive (RMCA ESA): Tervuren, Belgium.

Royal Museum of Central Africa Henry Morton Stanley Archive (RMCA HMSA): Tervuren, Belgium.

UK National Archives, Royal Geographical Society (NA RGS): London, UK.

Zanzibar National Archives (ZNA): Zanzibar Town, Tanzania.

Published and Unpublished Sources

Aceituno, Patricio, Maríadel del Rosario Prieto, María Eugenia Solari, Alejandra Martínez, Germán Poveda, and Mark Falvey. 'The 1877–1878 El Niño

[91] Chapter by Rockel, this volume.

Episode: Associated Impacts in South America.' *Climatic Change*, 92 (2009): 389–416.

Arens, W., ed. *A Century of Change in Eastern Africa*. The Hague: Morton Publishers, 1976.

Arnold, Margaret, Robert S. Chen, Uwe Deichmann, Maxx Dilley, Randolph E. Pullen, and Zoe Trohanis, eds. *Natural Disaster Hotspots: Case Studies*. Washington, DC: The World Bank, 2006.

Bankoff, Greg, and Joseph Christensen, eds. *Natural Hazards and Peoples in the Indian Ocean World: Bordering on Danger*. New York: Palgrave, 2016.

Bennett, Norman R. 'Philippe Broyon: Pioneer Trader in East Africa.' *African Affairs*, 62, 247 (1963): 156–64.

Bennett, Norman R. *From Zanzibar to Ujiji: The Journal of Arthur W. Dodgshun*. Boston: African Studies Center, 1969.

Bessems, Ilse, Dirk Verschuren, James M. Russell, Jozef Hus, Florias Mees, and Brian F. Cumming. 'Paleolimnological Evidence for Widespread Late 18th Century Drought Across Equatorial East Africa.' *Paleogeography, Paleoclimatology, Paleoecology*, 259 (2008): 107–20.

Bolser-Brown, Beverly. 'Ujiji: The History of a Lakeside Town, c.1800-1914.' Unpublished PhD diss.: Boston University, 1973.

Bolser-Brown, Beverly, and Walter Brown. 'East African Trade Towns: A Shared Growth,' in *A Century of Change*, ed. Arens: 183–200.

Burton, Richard F. *The Lake Regions of Central Africa: A Picture of Exploration*. New York: Harper & Brothers, 1860.

Cameron, Verney Lovett. *Across Africa*. New York: Harper & Brothers, 1877.

Campbell, Gwyn. *Africa and the Indian Ocean World from Early Times to circa 1900*. Cambridge: Cambridge University Press, 2019.

Chaiklin, Martha, Philip Gooding, and Gwyn Campbell, eds. *Animal Trade Histories in the Indian Ocean World*. Cham, CH: Palgrave, 2020.

Conway, Declan. 'Extreme Rainfall Events and Lake Level Changes in East Africa: Recent Events and Historical Precedents,' in *The East African Great Lakes*, eds. Odada and Olago: 63–92.

Crul, Ruud C.M. 'Limnology and Hydrology of Lakes Tanganyika and Malawi,' in *Comprehensive and Comparative Study of the Great Lakes*. Paris: UNESCO Publishing, 1997: 1–111.

Davis, Mike. *Late Victorian Holocausts: El Niño Famines and the Making of the Third World*. London: Verso, 2002.

Deutsch, Jan-Georg. 'Notes on the Rise of Slavery & Social Change in Unyamwezi,' in *Slavery in the Great Lakes Region*, eds. Médard and Doyle: 76–110.

Endfield, Georgina H. and David J. Nash. 'Missionaries and Morals: Climatic Discourse in Nineteenth-Century Central Southern Africa.' *Annals of the Association of American Geographers*, 94, 2 (2002): 727–42.

Gadain, Hussein, Nicolas Bidault, Linda Stephen, Ben Watkins, Maxx Dilley, and Nancy Mutunga, 'Reducing the Impacts of Floods Through Early Warning and Preparedness: A Pilot Study for Kenya,' in *Natural Disaster Hotspots*, eds. Arnold et al.: 165–84.

Gillman, C. 'The Hydrology of Lake Tanganyika.' *Tanganyika Territory Geological Survey Department*, 5 (1933): 1–25.

Gooding, Philip. 'History, Politics, and Culture in Central Tanzania.' *Oxford Research Encyclopedia of African History* (2019)[Accessed: 15 Jan. 2021].

Gooding, Philip. 'Tsetse Flies, ENSO, and Murder: The Church Missionary Society's Failed Ox-Cart Experiment of 1876–78.' *Africa: Rivista semestrale di studi e ricerche*, 1, 2 (2019): 21–36.

Gooding, Philip. 'The Ivory Trade and Political Power in Nineteenth-Century East Africa,' in *Animal Trade Histories*, eds. Chaiklin, Gooding, and Campbell: 247–75.

Grab, Stefan, and Tizian Zumthurm. '"Everything Is Scorched by the Burning Sun": Missionary Perspectives and Experiences of 19th- and Early 20th-Century Droughts in Semi-Arid Namibia.' *Climate of the Past*, 16, 2 (2020): 679–97.

Grove, Richard, and George Adamson. *El Niño in World History*. London: Palgrave Macmillan, 2018.

Hao, ZhiXin, JingYun Zheng, GuoFeng Wu, XueZhen Zhang, and QuanSheng Ge. '1876-78 Severe Drought in North China: Facts, Impacts and Climatic Background.' *Chinese Science Bulletin*, 55 (2010): 3001–7.

Harrison, Alexina Mackay. *The Story and Life of Mackay of Uganda: Pioneer Missionary*. London: Hodder & Stoughton, 1900.

Hartwig, Gerald W. 'The Victoria Nyanza as a Trade Route in the Nineteenth Century.' *Journal of African History*, 11, 4 (1970): 535–52.

Hartwig, Gerald W. 'Demographic Considerations in East Africa During the Nineteenth Century.' *International Journal of African Historical Studies*, 12, 4 (1979): 653–72.

Hastenrath, Stefan. 'Variations of East African Climate During the Past Two Centuries.' *Climatic Change*, 50 (2001): 209–17.

Holmes, C.F. 'Zanzibari Influence at the Southern End of Lake Victoria: The Lake Route.' *African Historical Studies*, 4, 3 (1971): 477–503.

Hopper, Matthew S. 'Cyclones, Drought, and Slavery: Environment and Enslavement in the Western Indian Ocean, 1870s to 1920s,' in *Natural Hazards and Peoples*, eds. Bankoff and Christensen: 255–82.

Hore, Edward C. 'Lake Tanganyika.' *Proceedings of the Royal Geographical Society and Monthly Record of Geography*, 4, 1 (1882): 1–28.

Hore, Edward C. *Tanganyika: Eleven Years in Central Africa*. London: Edward Stanford, 1892.

Hore, Edward C. *Missionary to Tanganyika 1877–1888*, ed. James B. Wolf. London: F. Cass, 1971.

Huang, Boyin, Michelle L'Heureux, Zeng-Zhen Hu, Xungang Yin, and Huai-Min Zhang. 'How Significant Was the 1877/78 El Nino?' *Journal of Climate*, 33, 11 (2020): 4853–69.

Hutley, Walter. *The Central African Diaries of Walter Hutley*, ed. James B. Wolf. Boston: African Studies Center, 1976.

Kjekshus, Helge. *Ecology Control and Economic Development in East African History: The Case of Tanganyika 1850-1950*, 2nd ed. London: James Currey, 1996.

Koponen, Juhani. *People and Production in Late Precolonial Tanzania: History and Structures*. Helsinki: Finnish Society for Development Studies, 1988.

Krinsky, William L. 'Tsetse Flies (Glossinidae),' in *Medical and Veterinary Entomology*, eds. Mullen and Durden: 369–82.

Lehman, J.T., ed. *Environmental Change and Response in East African Lakes*. Dordrecht: Kluwer Academic Publishers, 1998.

Mbava, N., M. Mutema, R. Zengeni, H. Shimelis, and V. Chaplot. 'Factors Affecting Crop Water Use Efficiency: A Worldwide Meta-Analysis.' *Agricultural Water Management*, 228 (2020): 1–11.

McCann, James C. *Maize and Grace: Africa's Encounter with a New World Crop, 1500-2000*. Cambridge, MA: Harvard University Press, 2009.

McDow, Thomas F. *Buying Time: Debt and Mobility in the Western Indian Ocean*. Athens, OH: Ohio University Press, 2018.

Médard, Henri, and Shane Doyle, eds. *Slavery in the Great Lakes Region of East Africa*. Oxford: James Currey, 2007.

Mishra, Vimal, Amar Deep Tiwari, Saran Aadhar, Reepal Shah, Mu Xiao, D.S. Pai, and Dennis Lettenmaier. 'Drought and Famine in India, 1870–2016.' *Geophysical Research Letters*, 46, 4 (2019): 2075–83.

Mullen, Gary R., and Lance Durden, eds. *Medical and Veterinary Entomology*. London: Academic Press, 2019.

Nash, David J., Kathleen Pribyl, Jørgen Klein, Raphael Neukom, Georgina H. Endfield, George C.D. Adamson, and Dominic Kniveton. 'Seasonal Rainfall Variability in Southeast Africa During the Nineteenth Century Reconstructed from Documentary Sources.' *Climatic Change*, 134 (2016): 605–19.

Nash, David J., Kathleen Pribyl, Georgina H. Endfield, Jørgen Klein, and George C.D. Adamson. 'Rainfall Variability Over Malawi During the Late 19th Century.' *International Journal of Climatology*, 38, S1 (2018): 629–42.

Nicholson, Sharon E., and Jeeyoung Kim. 'The Relationship of the El-Niño Southern Oscillation to African Rainfall.' *International Journal of Climatology*, 17 (1997): 117–35.

Nicholson, Sharon E. 'Historical Fluctuations of Lake Victoria and Other Lakes in the Northern Rift Valley of East Africa,' in *Environmental Change and Response*, ed. Lehman: 7–35.

Nicholson, Sharon E. 'Historical and Modern Fluctuations of Lakes Tanganyika and Rukwa and Their Relationship to Rainfall Variability.' *Climatic Change*, 41, 1 (1999): 53–71.

Nicholson, Sharon E. 'A Semi-Quantitative, Regional Precipitation Data Set for Studying African Climates of the Nineteenth Century, Part I. Overview of the Data Set.' *Climatic Change*, 50, 3 (2001): 317–53.

Nicholson, Sharon E. and Xungang Yin. 'Rainfall Conditions in Equatorial East Africa During the Nineteenth Century as Inferred from the Record of Lake Victoria.' *Climatic Change*, 48 (2001): 387–98.

Nicholson, Sharon E. 'Long-Term Variability of the East African "Short Rains" and Its Links to Large-Scale Factors.' *International Journal of Climatology*, 35, 13 (2015): 3979–90.

Nicholson, Sharon E. 'Climate and Climatic Variability of Rainfall Over Eastern Africa.' *Reviews of Geophysics*, 55, 3 (2017): 590–635.

Nicholson, Sharon E., Chris Funk, and Andreas H. Fink. 'Rainfall Over the African Continent from the 19th Through the 21st Century.' *Global and Planetary Change*, 165 (2018): 114–27.

Odada, Eric O. and Daniel O. Olago, eds. *The East African Great Lakes: Limnology, Palaeolimnology, and Biodiversity*. Boston, MA: Kluwer Academic Publishers, 2002.

Price, Roger and Joseph Mullens. 'A New Route and New Mode of Travelling into Central Africa.' *Proceedings of the Royal Geographical Society of London*, 2, 4 (1876–7): 233–48.

Reid, Richard J. 'The Ganda on Lake Victoria: A Nineteenth-Century East African Imperialism.' *Journal of African History*, 39, 3 (1998): 349–63.

Reid, Richard J. *Political Power in Pre-Colonial Buganda: Economy, Society & Warfare in the Nineteenth Century*. Oxford: James Currey, 2002.

Reid, Richard J. *War in Pre-Colonial Eastern Africa: The Patterns and Meanings of State-Level Conflict in the Nineteenth Century*. Nairobi: The British Institute in Eastern Africa, 2007.

Rockel, Stephen J. *Carriers of Culture: Labor on the Road in Nineteenth-Century East Africa*. Portsmouth, NH: Heinemann, 2006.

Rockel, Stephen J. 'The Tutsi and the Nyamwezi: Cattle, Mobility, and the Transformation of Agro-Pastoralism in Nineteenth-Century Western Tanzania.' *History in Africa*, 46 (2019): 231–61.

Russell, James M., Dirk Verschuren, and Hilde Eggermont. 'Spatial Complexity of "Little Ice Age" Climate in East Africa: Sedimentary Records from Two Crater Lake Basins in Western Uganda.' *The Holocene*, 17, 2 (2007): 183–93.

Saji, N.H., B.N. Goswami, P.N. Vinayachandran, and T. Yamagata. 'A Dipole Mode in the Tropical Indian Ocean.' *Nature*, 401 (1999): 360–63.

Singh, Deepti, Richard Seager, Benjamin I. Cook, Mark Cane, Mingfang Ting, Edward Cook, and Mike Davis. 'Climate and the Global Famine of 1876-78.' *Journal of Climate*, 31, 23, (2018): 9445–67.

Sofuoğlu, Emrah, and Ahmet Ay. 'The Relationship Between Climate Change and Political Instability: The Case of MENA Countries (1985:01–2016:12).' *Environmental Science and Pollution Research*, 27 (2020): 14033–43.

Sousa, Anastácio Q., and Richard Pearson. 'Drought, Smallpox, and Emergence of *Leishmania braziliensis* in Northeastern Brazil.' *Emerging Infectious Diseases*, 15, 6 (2009): 916–21.

Spinage, Clive A. *African Ecology: Benchmarks and Historical Perspectives*. New York: Springer, 2012.

Stanley, Henry Morton. *How I Found Livingstone: Travels, Adventures, and Discoveries in Central Africa Including Four Months' Residence with Dr. Livingstone*. London: Sampson Low, Marston & Company, 1872.

Stanley, Henry Morton. *Through the Dark Continent or the Sources of the Nile Around the Great Lakes of Equatorial Africa and Down the Livingstone River to the Atlantic Ocean*. 2 Vols. London: Sampson Low, Marston, Searle & Irvington, 1878.

The Indian Ocean World Podcast. 'Podcast Episode 4—Gooding, Tsetse Flies, ENSO, and Murder': https://www.appraisingrisk.com/2020/06/12/podcast-episode-4-gooding-tsetse-flies-enso-and-murder/ [Accessed: 18 Sep. 2020].

White, Sam, Christian Pfister, and Franz Mauelshagen, eds. *The Palgrave Handbook of Climate History*. London: Palgrave Macmillan, 2018.

Williamson, Fiona. 'Responding to the Extremes: Managing Urban Water Scarcity in the Late Nineteenth-Century Straits Settlements.' *Water History* (2020): 1–10.

Zheng, Jingyun, Lingbo Xiao, Xiuqi Fang, Zhixin Hao, Quansheng Ge, and BeiBei Li. 'How Climate Impacted the Collapse of the Ming Dynasty.' *Climatic Change*, 127 (2014): 169–82.

CHAPTER 10

A Forgotten Drought and Famine in East Africa, 1883–1885

Stephen J. Rockel

The widespread drought and famine of 1883–1885 that hit much of East Africa has never been examined in depth. Historians usually treat it as a mere *entrée* to the wave of environmental catastrophes that afflicted the region over the subsequent two decades, coinciding with the beginnings of colonial domination.[1] Most of the now extensive literature on climate

[1] For pioneering work in the Tanzanian context see: Juhani Koponen, *People and Production in Late Precolonial Tanzania: History and Structures* (Jyväskylä: Finnish Society for Development Studies, Finnish Historical Society, Finnish Anthropological Society, 1988); Juhani Koponen, 'War, Famine, and Pestilence in Late Precolonial Tanzania: A Case for a Heightened Mortality,' *International Journal of African Historical Studies*, 21, 4 (1988), 637–76. Koponen gives greater attention to combinations of human and animal diseases, drought, and warfare as causes of famine than the direct effects of climate change, and has surprisingly little to say about the 1883–1885 drought and famine.

S. J. Rockel (✉)
Department of Historical and Cultural Studies,
University of Toronto, Toronto, ON, Canada
e-mail: Stephen.rockel@utoronto.ca

P. Gooding (ed.), *Droughts, Floods, and Global Climatic Anomalies in the Indian Ocean World*, Palgrave Series in Indian Ocean World Studies, https://doi.org/10.1007/978-3-030-98198-3_10

change in the Indian Ocean and East Africa ignores it and its causes.[2] Yet the drought and resulting famine has its own history, causes, and consequences. As a whole, the naturally occurring upheavals of the mid-1880s–1890s were interlinked with and overlapped man-made processes and events, particularly those associated with increasing commercialization and colonial invasions. Epidemics affecting both people and livestock and episodes of drought and famine coincided with migrations and mobility on a greater scale than before, as well as exposure to the economic and political forces of violent western intrusion, and the longer-term impact of industrialization and commercial expansion. The East African region became more closely tied to Europe, North America, and the wider western Indian Ocean World.

Some of these events have been well documented, others much less so. We now know quite a lot about the passage of the cattle diseases, rinderpest and east coast fever, and their consequences on pastoral peoples and other livestock herders.[3] The terrible sleeping sickness pandemics in Uganda, the Congo, and elsewhere, and their treatments have been

[2] As in, for example: Stefan Hastenrath, 'Variations of East African Climate During the Past Two Centuries,' *Climatic Change*, 50 (2001), 209–17.

[3] Helge Kjekshus, *Ecology Control and Economic Development in East African History* (London: Heinemann, 1977), 126–32; Koponen, *People and Production*, 168–70; Richard Waller, 'Emutai: Crisis and Response in Maasailand 1883–1902,' in *The Ecology of Survival: Case Studies from Northeast African History*, eds. Douglas H. Johnson and David M. Anderson (London: Lester Crook, 1988), 73–112; Charles H. Ambler, *Kenyan Communities in the Age of Imperialism: The Central Region in the Late Nineteenth Century* (New Haven and London: Yale University Press, 1988), 96–100; Jan Bender Shetler, 'Interpreting Rupture in Oral Memory: The Regional Context for Changes in Western Serengeti Age Organization (1850–1895),' *Journal of African History*, 44, 3 (2003), 385–412; Thaddeus Sunseri, "The Entangled History of *sadoka* (rinderpest) and Veterinary Science in Tanzania and the Wider World, 1891–1901,' *Bulletin of the History of Medicine*, 89, 1 (2015), 92–121; James L. Giblin, 'East Coast Fever in Socio-Historical Context: A Case Study from Tanzania,' *International Journal of African Historical Studies*, 23, 3 (1990), 401–21.

analyzed.[4] Cholera and smallpox have been studied in several contexts.[5] Some attention has been given to the impact of new diseases introduced by colonial invaders, such as yaws and jiggers. Even where environmental crisis occurred somewhat independently of human activity, such as in the El Niño and Indian Ocean Dipole (IOD) events of 1876–1878, the rapid socio-economic transformations associated with British, German, and Italian expansion made the consequences of drought, famine, and disease more difficult to escape as environmental conditions and security deteriorated.

In this chapter, rather than making a general statement about declining ecological and environmental conditions in conjunction with the violence and disruptions of the colonial conquest of East-Central Africas by British and German colonizers from the mid-1880s, the specific causes, extent, and impact of the 1883–1885 drought and famine are examined. Its geographical extent and impact on African peoples are then described across three broad East African regions. The complex adaptations and coping strategies of various African cultural groups along with the consequences of drought and famine in the context of colonial conquests will be analyzed in a separate publication. After a general overview of some of the historical literature on late precolonial drought, I discuss aspects

[4] Among others: Koponen, *People and Production*, 156–58; Jonathan Musere, *African Sleeping Sickness: Political Ecology, Colonialism and Control in Uganda* (Lewiston, NY: Edwin Mellen Press, 1990); James L. Giblin, 'Trypanosomiasis Control in African History: An Evaded Issue?' *Journal of African History*, 31, 1 (1990), 59–80; James L. Giblin, *The Politics of Environmental Control in Northeastern Tanzania, 1840–1940* (Philadelphia: University of Pennsylvania Press, 1992); Richard Waller, 'Tsetse Fly in Western Narok, Kenya,' *Journal of African History*, 31, 1 (1990), 81–101; Shetler, 'Interpreting Rupture'; Kirk Arden Hoppe, *Lords of the Fly: Sleeping Sickness Control in British East Africa, 1900–1960* (Westport, CN, Praeger, 2003); Mari K. Webel, 'Ziba Politics and the German Sleeping Sickness Camp at Kigarama, Tanzania, 1907–14,' *International Journal of African Historical Studies*, 47, 3 (2014), 399–423; Mari K. Webel, *The Politics of Disease Control: Sleeping Sickness in Eastern Africa, 1890–1920* (Athens, OH: Ohio University Press, 2019).

[5] Important studies include: Gerald W. Hartwig, 'Social Consequences of Epidemic Diseases: The Nineteenth Century in Eastern Africa,' in *Disease in African History: An Introductory Survey and Case Studies*, eds. Gerald W. Hartwig and K. David Patterson (Durham, NC: Duke University Press, 1978); Koponen, *People and Production*, 162–68, 173–77; Myron Echenberg, *Africa in the Time of Cholera: A History of Pandemics from 1817 to the Present* (Cambridge: Cambridge University Press, 2012). Echenberg rightly highlights Dr. James Christie's pioneering, *Cholera Epidemics in East Africa, from 1821 till 1872* (London: Macmillan & Co., 1876).

of the relevant literature on climate change in the Indian Ocean and the East African mainland. It will be shown that for this drought and famine, the direct cause was almost certainly the series of eruptions at Krakatau in Indonesia, culminating on 27 August 1883, which due to the effect of volcanic aerosols and ash in the atmosphere and stratosphere, reduced sea surface temperatures in the Indian Ocean and drastically disrupted the usual monsoon cycle. The chronology and other evidence support this argument because of the timing of the eruption, which was one to two months before the usual onset of the small rains along the East African coast in October, when the southwest monsoon cycle moves into its transitional stage before the winds reverse in January.[6]

The study of droughts and famines in precolonial East Africa was a key theme in the wider growth of environmental history.[7] However, currently active historians have published relatively little on droughts occurring before the 1890s, and have not yet incorporated a scientific understanding of climate change into their work.[8] In the twenty-first century, in the face

[6] See: Gooding's introduction, this volume.

[7] In addition to the sources in fn. 1 see: Clarke Brooke, 'The Heritage of Famine in Central Tanzania,' *Tanzania Notes and Records*, 67 (1967), 15–22; Eric Ten Raa, 'Bush Foraging and Agricultural Development: A History of Sandawe Famines,' *Tanzania Notes and Records*, 69 (1968), 33–40; Peter Rigby, *Cattle and Kinship Among the Gogo: A Semi-Pastoral Society of Central Tanzania* (Ithaca and London: Cornell University Press, 1969), 20–22; Kjekshus, *Ecology Control*, 137–42; J.B. Webster (ed.), *Chronology, Migration and Drought in Interlacustrine Africa* (Halifax, Nova Scotia: Dalhousie University Press, 1979); Johnson and Anderson (eds.), *The Ecology of Survival*, several chapters; Ambler, *Kenyan Communities in the Age of Imperialism*, esp. Ch. 6; Gregory Maddox, 'Mtunya: Famine in Central Tanzania, 1917–20,' *Journal of African History*, 31, 2 (1990), 181–97; Thaddeus Sunseri, 'Famine and Wild Pigs: Gender Struggles and the Outbreak of the Majimaji War in Uzaramo (Tanzania),' *Journal of African History*, 38, 2 (1997), 235–59. James C. McCann provides an early discussion of how environmental historians in African might think about climate change and history in: 'Climate and Causation in African History,' *International Journal of African Historical Studies*, 32, 2/3 (1999), 261–79.

[8] See: Marek Pawelczak, *The State and the Stateless. The Sultanate of Zanzibar and the East African Mainland: Politics, Economy and Society, 1837–1888* (Warszawa: Instytut Historyczny Uniwersytetu Warsawskiego, 2010), 170–82; David Anderson, 'The Beginning of Time? Evidence for Catastrophic Drought in Baringo in the Early Nineteenth Century,' *Journal of Eastern African Studies*, 10, 1 (2016). Not surprisingly for northern Kenya, drought is a central theme in: George L. Simpson and Peter Waweru, 'Becoming Samburu: The Ethnogenesis of a Pastoral People in Nineteenth-Century Northern Kenya,' *Journal of the Middle East and Africa*, 3 (2012), 175–97.

of rapid climate change, the assumption of ever-rising agricultural production and more secure food supplies seems to be untenable. Drought is a great risk, although modern storage and distribution systems mean that famine can be avoided. Even so, in 2020–2021 there were massive locust invasions in the Horn and in northern parts of eastern Africa, a blight that most countries in the region had largely forgotten about and were not prepared for.[9]

Primary sources for this study include early colonial records, missionary archives, oral histories recorded by other researchers, and the journals and diaries of explorers and other travellers. It also relies on key findings of climatologists, palaeoclimatologists, limnologists, and geophysicists. The pioneering study by the German climatologist Eduard Kremer is important, as is the early collection of materials collected by the British Royal Society related to the global impact of the series of volcanic eruptions of Krakatau in Indonesia in the middle of 1883.[10] Mission sources are particularly valuable. Members of the London Missionary Society (LMS) and the Church Missionary Society (CMS) were frequent travellers along the main caravan routes, and they sometimes ventured well away from them into remote regions. It should not be assumed that European travellers' accounts give only a limited snapshot. Many missionaries were long-term residents and spoke African languages. Well informed and experienced European travellers made sure to collect details from local people and other travellers along the caravan tracks about geographic information, the conditions of travel, deviations from the norm in water and food supplies, as well as ethnographic, botanical, and natural history.[11] Unfortunately, the limited number of late nineteenth-century Swahili texts are silent on the drought and famine of 1883–1885, although Tippu Tip's autobiography frequently refers to famine in the eastern Congo and other regions.[12]

[9] See also: Gooding's introduction, this volume.

[10] Eduard Kremer, *Die unperiodischen Schwankungen der Niederschläge und die Hungersnöte in Deutsch-Ost-Afrika* (Hamburg: Archiv der Deutschen Seewarte, 1910); G.J. Symons (ed.), *The Eruption of Krakatoa and Subsequent Phenomena: Report of the Krakatoa Committee of the Royal Society* (London: Harrisons & Sons, 1888).

[11] A point made by Kremer in 1910.

[12] François Bontinck (ed.), *L'autobiographie de Hamed ben Mohammed el-Murjebi Tippo Tip* (Bruxelles: Académie royale des Sciences d'Outre Mer, 1974).

The specific historiography of the drought and subsequent famine beginning in late 1883 is limited. Marek Pawelczak, in his detailed study of the Zanzibar Sultanate and its mainland hinterland between 1837 and 1888, provides a useful summary of the impact of the drought and famine among people immediately inland from the coast of southern Kenya and northern Tanzania, especially the Mijikenda. Establishing the regional context, he argues that the rapidly changing social and economic conditions of the period—with the incipient development of market relations and the major impact of the long-distance caravan trade—drastically changed living conditions for East Africans, especially as the regions within 300 kms from the coast and most exposed to the Indian Ocean world were those with fewest natural resources to sell, gum copal and wild rubber excepted. The development of plantation agriculture producing food crops with the labour of enslaved Africans at many points along the southern coast of Kenya and the northern and central coast of Tanzania created greater risks for local populations, as much of the produce was commercially sold, including for export.[13] Many of these districts were among those badly affected from late 1883, although there was no substantial plantation system in central regions such as Usagara and Ugogo, where the drought was even more severe. The increasing cultivation of maize with the aim of maximizing output in many areas perhaps contributed to a reduction of crop diversity, creating risk for maize farmers. Maize also had disadvantages over the cultivation of indigenous crops, including sorghum and millet, as it was more difficult to store and less drought resistant. Pawelczak notes, however, that maize farmers sometimes brought entirely new areas into cultivation, such as the foothills of Usambara and Upare in northern Tanzania.[14] This development may have been stimulated by the demands of the expanding caravan trade in the lowlands. Whatever the case, the range of available cereals was relatively narrow, although cassava, a drought resistant introduction from South America was becoming better known to East Africans, and sweet potatoes were also commonly grown in suitable areas.

[13] Pawelczak, *The State and the Stateless*, 170–71, 181. Detailed studies include: Frederick Cooper, *Plantation Slavery on the East Coast of Africa* (New Haven: Yale University Press, 1977); Jonathon Glassman, *Feasts and Riot: Revelry, Rebellion, and Popular Consciousness on the Swahili Coast, 1856–1888* (Portsmouth, NH: Heinemann, 1995).

[14] Pawelczak, *The State and the Stateless*, 174–76.

When taking a broader view of the events of 1883–1885, it is clear that the failure of the *mvuli* 'short rains' in late 1883 and then the *masika* 'long rains' (March–May) of 1884 was catastrophic for the majority agriculturalists and mixed farmers across a broad swathe of East-Central Africa. The failure of even one of the two rainy seasons could be disastrous given the limited storage and transport infrastructure in most parts of late precolonial East Africa, and its impact extended into the following year even if normal precipitation resumed. Mountain or lakeside regions with a wide range of ecosystems, crop options, and more reliable rainfall and irrigation options, such as in Kilimanjaro, Taveta, and in the Uluguru mountains south of Morogoro, tended to do better, although this was not always the case in some of the more exposed highland regions.[15] The people of the Taita mountains (sometimes Teita), in southeastern Kenya, and the Pare and Nguu mountains in northern Tanzania, were particularly badly affected from late 1883, as we will see. But the inhabitants of Ukambani, many of the Mijikenda localities, Uzigua, Uzaramo, and neighbouring areas in northeastern Tanzania—in other words the entire *nyika* lowlands between the coast and the major mountain chains and massifs of the interior—were also hard hit by hunger and starvation.[16] However, this was not the extent of the impact, as the drought and famine were devastating in central Tanzania (Ugogo) and eastern Tanzania as a whole, notably in areas adjacent to the central caravan routes that linked the coast to the Great Lakes, the eastern Congo and other parts of central Africa.

In Indian Ocean world studies and also in some recent work on famine, much has been said about *longue durée* factors.[17] Even though East Africa anchored one side of the Indian Ocean monsoon system, its

[15] Ibid., 171.

[16] E. Hollis Merrit, 'A History of the Taita of Kenya to 1900' (Unpublished Ph.D. diss.: Indiana University, 1975); Bill Bravman, *Making Ethnic Ways: Communities and Their Transformations in Taita, Kenya, 1800–1950* (Portsmouth, NH: Heinemann, 1998); Isaria N. Kimambo, 'The Political History of the Pare People to 1900' (Unpublished Ph.D. diss.: Northwestern University, 1967); Isaria N. Kimambo, 'Environmental Control and Hunger in the Mountains and Plains of Northeastern Tanzania,' in *Custodians of the Land: Ecology and Culture in the History of Tanzania*, eds. Gregory Maddox, James L. Giblin, and Isaria N. Kimambo (London: James Currey, 1996), 71–95; Giblin, *The Politics of Environmental Control*.

[17] See: Gooding's introduction, this volume; David Arnold, *Famine: Social Crisis and Historical Change* (Oxford: Basil Blackwell, 1988), 5–28.

geography, physical and biological environments, ecology, and patterns of human settlement and socio-political organization were and remain vastly different than those in South Asia or further east. Conditions have always been difficult for humans, even though *homo sapiens* have lived in East Africa longer than in any other region on earth. For one thing, the land is generally drier than other Indian Ocean regions, except in the vicinity of the Great Lakes, which lie beyond the region discussed here. Parts of East Africa, especially northern Kenya and into southern Somalia, are very marginal for agriculture, and left to pastoral peoples. The 50 cm rainfall line is a good marker. However, given the generally low population density and the vastness of the land mass, in late precolonial times intrepid migrants with their families could easily find unoccupied land in lowland farming regions, and in some of the foothills of the mountain ranges, or they could attach themselves to established communities. New settlements were constantly being founded in frontier districts, particularly as long-distance trade expanded.[18] Lack of access to land, even if the most productive districts were permanently occupied, was a barrier only in times of crisis. Nevertheless, even though land was generally available, its quality was often poor, and, except in the most fertile areas, it had to be cultivated using various systems of shifting cultivation, intercropping, and crop rotation, and, where possible, a combination of highland and lowland farming for different crop regimes. Most cultivation was rain fed. Only in well-watered mountainous regions such as Kilimanjaro and Upare was irrigation possible. This reliance on rain-fed cereal and vegetable production made any drought particularly devastating.

To add to the relatively dry climate (except on east facing slopes) there are virtually no navigable rivers in East Africa. Between the Juba in southern Somalia and the Ruvuma, the modern border between Tanzania and Mozambique, there are only four major rivers reaching the Indian Ocean: the Tana, the Pangani, the Ruvu (Kingani), and the Rufiji. Virtually all rivers in the interior are reduced to muddy pools or dry up completely during the dry season (which is extended during periods of drought). The lack of permanent navigable rivers has made the transport of any marketable surpluses by boat into the interior impossible, except for short stretches. Along the coast, *dhows* and smaller boats were the main form of transportation, and so coastal agricultural production could

[18] Igor Kopytoff (ed.), *The African Frontier: The Reproduction of Traditional African Societies* (Bloomington: Indiana University Press, 1987).

be moved north and south as well as to offshore islands, including Zanzibar and Pemba. Food imports were possible for coastal regions, but the worst affected regions were generally a little inland and further into the interior where food had to be carried by porters. On the coast, some food was indeed imported to places like Mombasa where the famine had rather less of an impact anyway.

The transportation of heavy goods, such as grain, has historically been severely limited by an additional environmental factor. The use of pack and draft animals was generally impossible due to the prevalence of tsetse fly (*glossina*) which carries the parasitic animal trypanosome. The effect of the parasite is to kill domestic animals including horses, oxen, cattle, and most donkeys as they lack immunity. The only other option was the engagement of human porters who could carry some relief supplies, but with quickly diminishing returns. One only needs to remember the song of the conscripted porters supplying the British forces during the arduous East African campaign of the First World War: 'We are the porters who carry the food of the porters who carry the food of the porters who carry the food.'[19] Before the introduction of mechanized transport, including railways and motor vehicles, the transportation of food supplies in large quantities from well-supplied districts to others, where there was an absolute shortage of or a lack entitlement to food, was moot. In such circumstances, drought across regions could lead to absolute shortages and starvation.

Even in the generally difficult environments of East Africa and taking into account the rarity of large-scale states compared with South Asia, African peoples had worked out balancing mechanisms, networks of exchange, local and regional markets, and structures of social and political management of productive land and food resources that, since the development of agriculture and pastoralism, had generally succeeded in keeping the third horseman of the apocalypse at bay. Social collapse and famine across vast distances were quite rare. That is what makes the events of the 1880s and 1890s particularly significant, particularly in combination with cattle diseases that wiped out assets. Even if people had prior resources to buy food, east coast fever and then rinderpest wiped out cattle herds, which were money in the bank for pastoralists in parts

[19] Geoffrey Hodges, 'Military Labour in East Africa and Its Impact on Kenya,' in *Africa and the First World War*, ed. Melvin E. Page (New York: Palgrave Macmillan, 1987), 145.

of central and especially northern regions. There was also no way of transporting large-scale grain imports into places hundreds of kilometres inland from the coast due to tsetse and reliance on porterage. By the middle of 1884 people were starving fifty kilometres inland from coastal towns: There was not enough food except for the most powerful within the most powerful chiefdoms. The response was generally migration for those who survived. In many cases whole communities moved, even if temporarily. People were mobile, food from other regions much less so.

Cormac Ó. Gráda, a leading historian of famine, states that 'famine refers to a shortage of food or purchasing power that leads directly to excess mortality from starvation or hunger-induced diseases.'[20] This was certainly the case in late nineteenth-century East Africa. The concept of entitlement, developed by Amartya Sen and much discussed in many recent analyses of the causes of famine, is best applied to stratified state societies with relatively good transport systems, such as mid-nineteenth-century Ireland or 1940s Bengal. In much of Africa, it was only after the First World War, when the structure of the typical colonial state became more firmly founded and mechanical transport systems were established, that efficient endowment transfers during periods of disaster could be made on a large scale. By that time the numerous colonial famines could be put down to a multiplicity of causes.[21] In precolonial Africa, failures of production were typically more important than obstacles to satisfying demand, although very marginalized groups such as enslaved Africans certainly suffered from entitlement deficits. In East-Central Africa in 1883–1884, the reliance on rain-fed agriculture, the limited development

[20] Cormac Ó. Gráda, *Famine: A Short History* (Princeton, NJ: Princeton University Press, 2009), 4.

[21] Amartya Sen, *Poverty and Famines: An Essay on Entitlement and Deprivation* (Oxford: Oxford University Press, 1983). For criticism and debate, see among many others: Arnold, *Famine*, 42–46; Megan Vaughan, *The Story of an African Famine* (Cambridge: Cambridge University Press, 1987); Alex De Waal, 'A Re-Assessment of Entitlement Theory in the Light of the Recent Famines in Africa,' *Development and Change*, 21 (1990), 469–90; S.R. Osmani, 'Comments on Alex de Waal's Re-Assessment of Entitlement Theory in the Light of Recent Famines in Africa,' *Development and Change*, 22, 3 (1991), 587–96; Mike Davis, *Late Victorian Holocausts* (London and New York: Verso, 2002), 19–21; Getnet Alemu, 'Revisiting the Entitlement Approach to Famine: Taking a Closer Look at the Supply Factor—A Critical Survey of the Literature,' *Eastern Africa Social Science Research Review*, 23, 2 (2007), 95–129; Olivier Rubin, 'The Entitlement Approach: A Case for Framework Development Rather Than Demolition,' *Journal of Development Studies*, 45, 4 (2009), 621–40.

of centralized states except in certain highland regions, and the crippling transport problem that necessitated carriage of all goods by human porterage meant that the collapse of food production combined with a general inability to access alternative supplies was catastrophic across most communities, except in a few highland chiefdoms and coastal urban centres. Widespread food shortages could be managed for some months, but not over consecutive failed rainy seasons, especially when the destruction of cattle herds—the means for some to acquire food—coincided with the crisis.

Climate Change and Drought in Nineteenth-Century East Africa

The climatological study of the patterns and impact of El Niño and La Nina events, the Indian Ocean Dipole (IOD), and other factors related to long-term climate change and short-term climatic fluctuations in East Africa has made great strides in recent decades. Science has demonstrated that in equatorial East Africa, rainfall variations are to some extent correlated with El Niño Southern Oscillation (ENSO) events, but more closely linked to positive or negative sea surface temperature (SST) oscillations in the Indian Ocean Dipole as well as zonal atmospheric conditions, with impacts varying between the long and short rainy seasons and trending over years.[22] In general terms, high rainfall and flooding occurs in El Niño years, and the opposite—lower rainfall and droughts—occurs in La Niña years, as colder SSTs result in moisture laden air currents moving away from the region and towards zones with warmer surface temperatures. In East Africa, during La Niña years precipitation sometimes diminishes with negative effects on water sources and food production and security.[23] However, climatologists have established a more direct connection between drought in East Africa and negative IODs, especially during the transitional months of the short (*mvuli*) rains in October and November, as the Eurasian landmass starts to go into winter. In negative IOD years, when the regular transition in the monsoon winds from the south west to

[22] Sharon E. Nicholson, 'Climate and Climatic Variability of Rainfall Over Eastern Africa,' *Reviews of Geophysics*, 55 (2017), 605–11.

[23] See: Gooding's introduction and chapter, this volume; Abdul Sheriff, 'Globalization with a Difference,' in *The Indian Ocean: Oceanic Connections and the Creation of New Societies*, eds. Abdul Sheriff and Engseng Ho (London: Hurst and Co., 2014), 15 (map).

the north east is disrupted by cooling SSTs in the western Indian Ocean, reducing the humidification of the dry winter winds blowing from the Eurasian land mass, the result can be a failure of the *mvuli* rains along the Swahili coast and interior regions.[24] The consequent droughts can affect a large region up to 500 kms from the coast and perhaps 2000 kms from north to south. The opposite pattern for both IOD and ENSO is associated with excessive rains and heavy flooding, a frequent event in recent years. There are many further complicating factors including oscillations in and migrations of the Intertropical Convergence Zone (ITCZ), the low-pressure belt roughly straddling the equator,[25] and changes in atmospheric conditions, including air pressure and the circulation of dust or ash, a factor that was particularly significant in 1883.

Recent studies of the impacts of the accumulation of vast quantities of charged volcanic ash in the atmosphere after large volcanic eruptions show that the disturbance to normal atmospheric electrical potential suppresses rain cloud formation. This combined with a lowering of air temperature leads to a decrease in rainfall precipitation. This seems to be what happened after the colossal explosions at Krakatau in late August 1883. Air temperature records collected at 47 stations in Europe, North America, Russia, and Australia show a decrease from the norms of the previous eight years over the period from 22 August to 12 September 1883, by which time the temperature decrease had reached an average of 4 degrees.[26] The changes in the atmosphere, including, unusual colouration at sunrise and sunset, and twilight afterglows caused by the circulation of ash around the globe, were observed and recorded in many parts of the world, and as late as November 1883 in Surrey, England.[27] On the side of caution geophysicist, Matthew Genge, notes that: 'The

[24] For a brief overview, see: Peter Beaumont and Graham Readfearn, 'Global Heating Supercharging Indian Ocean Climate System,' *The Guardian* (19 Nov. 2019). The unreliability of the short rains is the focus of: Stefan Hastenrath, Dierk Polzin, and Pierre Camberlin, 'Exploring the Predictability of the 'short rains' at the Coast of East Africa,' *International Journal of Climatology*, 24 (2004), 1333–43.

[25] For the intertropical convergence zone (ITCZ), see: Gooding's introduction, this volume.

[26] Matthew J. Genge, 'Electrostatic Levitation of Volcanic Ash into the Ionosphere and Its Abrupt Effect on Climate,' *Geology*, 46, 10 (2018), 837.

[27] Genge, 'Electrostatic Levitation of Volcanic Ash,' 837. See also: Symons (ed.), *The Eruption of Krakatoa*; Tom Simpkin and Richard S. Fiske (eds.), *Krakatoa 1883: The Volcanic Eruption and Its Effects* (Washington, DC: Smithsonian Institution Press, 1983);

coincidence of a low average temperature and precipitation with the eruption is consistent with the predicted effect of levitated ash but not conclusive evidence, considering the inherent variability of weather data on such short time scales.'[28]

Furthermore, the Krakatau eruption probably affected SSTs in the Indian Ocean, contributing to or exacerbating a negative IOD anomaly in 1883, and perhaps in subsequent years as well. Significantly, a team of scientists led by Peter Gleckler, most of whom are based at the Program for Climate Model Diagnosis and Intercomparison at the Lawrence Livermore National Laboratory in California, have argued that the impact of the reduction of SSTs caused by the Krakatau eruptions was not only short term, but that it lasted for decades into the twentieth century.[29] Their conclusions are as follows:

> We have analysed a suite of 12 state-of-the-art climate models and show that the ocean warming and sea level rise in the twentieth century were substantially reduced by the colossal eruption in 1883 of the volcano Krakatoa in the Sunda strait, Indonesia. Volcanically induced cooling of the ocean surface penetrated into deeper layers where it persisted for decades after the event. This remarkable effect on oceanic thermal structure is longer lasting than has been previously suspected and is sufficient to offset a large fraction of ocean warming and sea level rise caused by anthropogenic influences.[30]

Wilfried Schröder, 'The Krakatoa Event and Associated Phenomena: A Historical Review,' *Earth Sciences History*, 21, 2 (2002), 199–206.

[28] Genge, 'Electrostatic Levitation of Volcanic Ash,' 837.

[29] P.J. Gleckler, K. AchutaRao, J.M. Gregory, B.D. Santer, K.E. Taylor, and T.M.L. Wigley, 'Krakatoa Lives: The Effect of Volcanic Eruptions on Ocean Heat Content and Thermal Expansion,' *Geophysical Research Letters*, 33, 17 (2006); P.J. Gleckler, T.M.L. Wigley, B.D. Santer, J.M. Gregory, K. AchutaRao, and K.E. Taylor, 'Krakatoa's Signature Persists in the Ocean,' *Nature*, 439 (9 Feb. 2006).

[30] Gleckler, et al., 'Krakatoa's Signature Persists.' Surprisingly, neither Genge's work nor that of Glecker and his colleagues is referred to in: Shayne McGregor, Miriam Khodri, Nicola Maher, Masamichi Ohba, Francesco S.R. Pausata, and Samantha Stevenson, 'The Effect of Strong Volcanic Eruptions on ENSO,' in *El Niño Southern Oscillation in a Changing Climate*, eds. Michael J. McPhaden, Agus Santoso, and Wenju Cai (Hoboken, NJ: John Wiley Sons, 2021), 267–87. Nevertheless, their conclusions (p. 266) concerning the longer-term impact of volcanic eruptions on SSTs are broadly similar to those of Gleckler, et al., except that they do not consider eruptions in relation to IODs, only ENSOs: 'Global average surface temperatures have been shown to reach the maximum cooling 6–18 months after the eruptions peak optical depth and return to normal values

For historians of East Africa, there is the implication here that perhaps the subsequent droughts and famines in eastern and central regions of Tanzania and Kenya of 1888–1890, 1892–1894, and 1898–1900 were also linked to the cooling impact of the Krakatau eruptions in both the atmosphere and the Indian Ocean itself. This is a possibility that has not yet been explored.

If we turn to the contemporary observations related to the eruptions collected in the report of the Royal Society Committee, we find that despite its length and thoroughness, no direct evidence from East Africa is included, which is surprising given the importance of Zanzibar as the key commercial centre with a significant and active British consulate.[31] However, numerous observations were collected from the Seychelles, Mauritius, and other locations in the Indian Ocean, indicating the effect of the eruptions in the atmosphere in the western IOW, such as the following: 'At Mauritius, on the 27th [of August 1883]...the sunset was smoky in the west; and on the 28th the sunset was gorgeous...Captain Loyseau of the *Salazic*, lat. 9° 15' S., long. 93° E, talks of encountering blinding showers of sand, while the sun was reddish and the sky white. On August 28th the haze continued at Diego Garcia, Rodriquez, the Seychelles, and Mauritius, and was observed by the *Simla*, 6° 12' S., 88° 17' E., "At 2 p.m., sky very hazy, a fine white powder falling in a constant shower like snow." At 8 p.m., sky still very hazy and dust falling. On August 29th the *Simla*, 6° 26' S., 87° 52' E., reports:– "A very large quantity of dust fell in the past night. Very hazy still, and dust falling. At 5 p.m., sun completely obscured 15° above the horizon, owing to haze."'[32] Further reports indicate the 'general spread [of the dusty haze] over the tropical part of the Indian Ocean,' and a 'red glare,' 'crimson dawn' in Rodriguez, Diego Garcia, the Seychelles, and Mauritius. Similar unusual atmospheric phenomena were observed more than a year later in October 1884, in various places.[33] These reports are suggestive of the atmospheric conditions disrupting the normal monsoon cycle in both the eastern and western Indian Ocean, contributing to below-average rainfall

approximately 5–6 years after the eruption... The strong surface temperature cooling influence of these events has also seen them at least partly implicated in many past hiatuses of global surface warming.'

[31] There is, however, good rainfall evidence from Zanzibar that will be discussed below.

[32] *Report of the Krakatoa Committee*, 220, 315, 319, 322, 335.

[33] Ibid., 228.

on the east African mainland.[34] This represents the core global climatic context for drought and subsequent famine in East Africa in 1883–1885.

The next step is to consider scientific evidence for drought in east African drylands from the evidence of variations in lake levels. One set of methodologies available for measuring climate change and variations in annual rainfall over the longer term is the scientific examination of lake sediments.[35] Dirk Verschuren and his colleagues in the fields of palaeoclimatology and palaeoecology have developed an approximate chronology of decadal changes in fluctuations of water levels in East African lakes based on examinations of 'sedimentary archives' that allow reasonably accurate measurements. In broad terms, this evidence is particularly valuable if it correlates well with archaeological, historical, and cultural sources.[36] This data is collected at individual sites, aggregated to identify longer-term trends, and compared with similar research at other lakes. The incompleteness of this type of research across the entire intertropical

[34] For an explanation of the unstable transition from the summer monsoon in the Indian Ocean to the winter monsoon that in 1883 carried volcanic aerosols and ash towards East Africa see: Hastenrath, Polzin, and Camberlin, 'Exploring the Predictability of the "short rains",' 1336.

[35] Dirk Verschuren, 'Lake-Based Climate Reconstruction in Africa: Progress and Challenges,' *Hydrobiologia*, 500, 1 (2003), 315–30.

[36] Dirk Verschuren, John Tibby, Peter R. Leavitt, and C. Neil Roberts, 'The Environmental History of a Climate-Sensitive Lake in the Former "White Highlands" of Central Kenya,' *Ambio*, 28, 6 (1999), 494–501; Dirk Verschuren, Kathleen R. Laird, and Brian F. Cumming, 'Rainfall and Drought in Equatorial East Africa During the Past 1,100 Years,' *Nature*, 403, 6768 (2000); Vanessa Gelorini and Dirk Verschuren, 'Historical Climate-Human-Ecosystem Interaction in East Africa: A Review,' *African Journal of Ecology*, 51 (2012), 409–21. For Lake Baringo region in northern Kenya, see: Ilse Bessems, Dirk Verschuren, James M. Russell, Jozef Hus, Florias Mees, and Brian F. Cumming, 'Palaeolimnological Evidence for Widespread Late Eighteenth Century Drought Across Equatorial East Africa,' *Palaeogeography, Palaeoclimatology, Palaeoecology*, 259 (2008), 107–20; and for historical interpretations: Anderson, 'The Beginning of Time?' 47–49. For recent work see Gijs De Cort, Ilse Bessems, Edward Keppens, Florias Mees, Brian Cumming, and Dirk Verschuren, 'Late-Holocene and Recent Hydroclimatic Variability in the Central Kenya Rift Valley: The Sediment Record of Hypersaline Lakes Bogoria, Nakuru and Elementeita,' *Palaeogeograply, Palaeoclimatology, Palaeoecology*, 388 (2013), 69–80; Christian Wolff, Iris Kristen-Jenny, Georg Schletter, Birgit Plessen, Hano Meyer, Peter Dulski, Rudolph Naumann, Achim Brauer, Dirk Verschuren, and Gerald H. Haug, 'Modern Seasonality in Lake Challa (Kenya/Tanzania) and Its Sedimentary Documentation in Recent Lake Sediments,' *Limnology and Oceanography*, 59, 5 (2014), 1621–36. Lake Challa is the only lake studied for sediment records located in the 1883–1885 drought affected regions.

region means the extent of the measured variables in precipitation levels over the long term cannot be extrapolated for all East African regions. There are vast areas, particular in the eastern half of East Africa, where there are very few lakes. Conclusions from palaeoclimatology and palaeoecology are very suggestive for longer trends although they do not fully account for shorter periods of drought or sharp fluctuations within longer periods of above-average or below-average precipitation, such as can be measured through rain gauge measurements and historical observations. Thus, there was little indication from paleoclimatology or paleoecology that in October 1883 much of East-Central Africa was about to enter a crisis period that was to last until at least 1885 or 1886.

If evidence from lake sediments is lacking, what climate-related evidence can we find 'on the ground' to better understand the drought and famine beginning in late 1883? As numerous researchers have noted, there is very little in the way of continuous meteorological recordings for specific places from the late precolonial period in East Africa. Yet there are rainfall measurements from 1850 taken intermittently at the British Consulate in Zanzibar, including a series from 1880 to 1884 recorded by Consul Sir John Kirk. His average over these years was 1241 mm, which although including several normal years, also includes 1883 and 1884, which were drought years. The figure for 1883 was about 300 mm below the average, although the drought on the mainland started only at the end of the year with the failure of the short rains.[37] Annual fluctuations shown in the rainfall records from Zanzibar, even though the island rarely suffers from extreme drought, do follow quite closely those from the mainland stations in what later became German East Africa. Here, we can turn to the work of German meteorologist, Dr. Eduard Kremer.

One of the most important aspects of Kremer's comprehensive study was his demarcation of four rainfall regions in the then colony of German East Africa (his study concentrates on German territory). Although this was a high level division with very little data available for the vast southern (region III) and western regions (IV) stretching respectively to the Ruvuma river in the south (the border with Mozambique) and Lake Tanganyika in the west, regions I and II, the central and north coast running about 400 kms into the interior, largely coincide with eastern

[37] Clive A. Spinage, *African Ecology: Benchmarks and Historical Perspectives* (Berlin: Springer, 2012), 122–23; Kremer, *Die unperiodischen Schwankungen*, 23, 13 (graph).

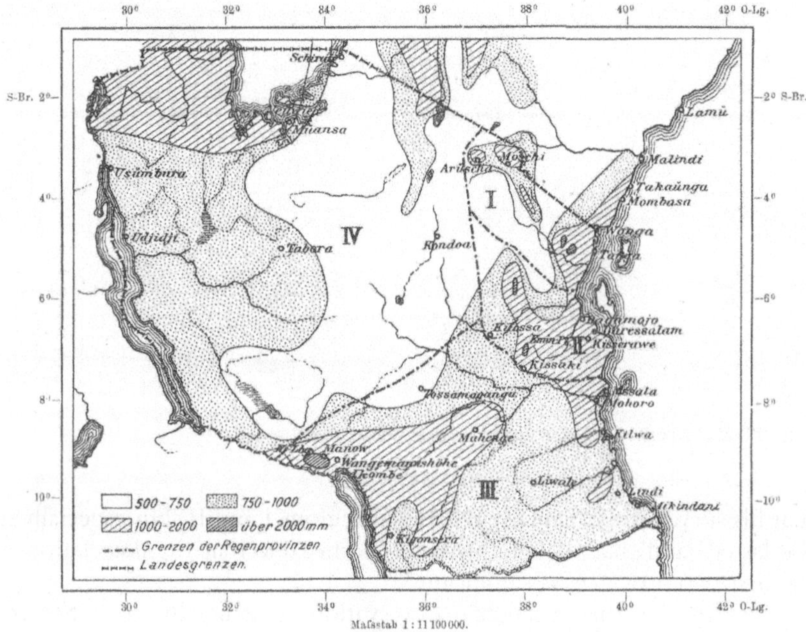

Fig. 10.1 Eduard Kremer, *Die unperiodischen Schwankungen*, 4

parts of the territory most affected by the drought and famine of 1883–1885.[38] The main exceptions are Ugogo in central Tanzania, a very badly affected territory, which is the eastern part of Kremer's region IV, some areas further to the north, in Kenya, and in the south of German East Africa, beyond Kilwa. In addition, it is regions I and II near the coast that provide by far the bulk of Kremer's data given that German colonialism in East Africa had barely began as the drought was taking hold, and even by the early 1890s there were few rainfall records from the far interior (Fig. 10.1). The German records in fact begin only in 1892 and then only for a small number of stations, but Kremer was able to obtain comparative British rainfall data for the 1880s from Zanzibar.[39] He notes

[38] See the map: Kremer, *Die unperiodischen Schwankungen*, 4.

[39] Ibid., 23.

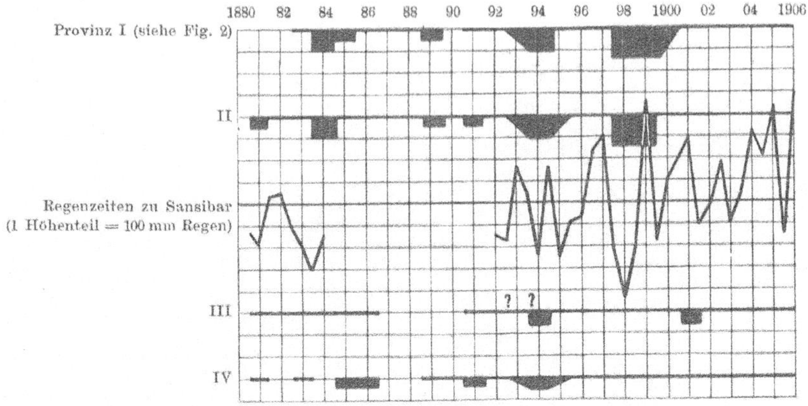

Fig. 10.2 Kremer, *Die unperiodischen Schwankungen*, 13

that his survey is 'particularly reliable in regions I and II, but especially in the latter,' and that 'It is immediately noticeable that the correlation of the first two provinces is significant.'[40]

We will see that Kremer's observation is generally true for the 1883–1885 drought and famine. Further, the rain gauge data showing severe declines in regions I and II during major droughts over the period 1880–1901 (1883–1885, 1888–1889, 1892–1894, 1897–1901) is closely matched by significant drops in the corresponding years in the records from Zanzibar.[41] Similar analysis comparing rainfall in Zanzibar with that in Mombasa over the period 1892–1900 shows a very close correlation (Fig. 10.2). Using rainfall records from various stations along the coast of British East Africa, from Kismayu south to Wanga near the border with German East Africa for the period 1890–1895, Kremer shows a further correlation with the record from Zanzibar, although this breaks down at the northern extremity (Lamu and Kismayu).[42]

Extensive data on nineteenth-century rainfall for the African continent collected and analyzed over several decades by meteorologist Sharon E. Nicholson supports Kremer's conclusions, and shows a series of extremely

[40] Ibid., 13.

[41] Ibid., 13 (graph).

[42] Ibid., 25–26.

dry years from 1883 until the early 1890s over virtually the whole of East Africa (excluding the Horn).[43] This is highly aggregated data that goes beyond the region and years covered in this chapter, but it also points to the widespread effect of the Krakatau eruption, perhaps in conjunction with other factors, and the severity of the 1883–1885 drought and famine.

The Extent of the Drought and Famine

The following sections attempt to establish at least in approximate form the extent of the drought and subsequent famine both in terms of geography and in terms of human impact up until the return of the monsoon rains, thus corroborating and sometimes adding to the scientific data. It is of course impossible to ascertain the number of deaths and the degree of hunger. The impact within particular regions was variable because of huge differences in ecological, meteorological, and political conditions over relatively short distances, and between and within highland and lowland areas. As is often the case, the impact on agricultural production, incomes from other sources, the ability of hard-hit households and families to retain assets, effects on infant mortality, and other measures, is impossible to accurately assess. Nevertheless, food prices, which are known for coastal towns, patterns of enslavement and self-enslavement, and migration can be used as proxies for scarcity and hardship. Most of what follows is descriptive, and there is a bias in the evidence towards localities near the coast, along caravan routes, and relatively accessible to mission stations owing to the relative abundance of information for these zones compared to others.

In large parts of East Africa, the 1883–1885 drought and famine was a key historical event.[44] It had a devastating impact, coinciding with

[43] Sharon E. Nicholson, 'A Semi-Quantitative, Regional, Precipitation Dataset for Studying African Climates of the Nineteenth-Century. Part I Overview of the Data Set,' *Climatic Change*, 50 (2001), 317–53; Sharon E. Nicholson, Douglas Klotter, and Amin K. Dezfuli, 'Spatial Reconstruction of Semi-Quantitative Precipitation Fields Over Africa During the Nineteenth Century from Documentary Evidence and Gauge Data,' *Quaternary Research*, 78, (2012), 21, Fig. 8.

[44] Although it is missing from Rigby's list of historic famines in Ugogo, a region that nevertheless suffered greatly. See: Rigby, *Cattle and Kinship*, 21. It is not mentioned, either, by Kjekshus in his discussion of early colonial famine: Kjekshus, *Ecology Control*, 137–42.

the beginnings of colonial rule, and foreshadowing the environmental collapse of the 1890s. Yet drought and famine had been relatively rare in the preceding 50 years—a phenomenon that may have undermined people's and communities' preparedness. The following three sections consider different regions of East Africa, and they are ordered north to south. The first is southeastern Kenya and northeastern Tanzania; the second, eastern and central Tanzania along the central caravan route, and the third, southeastern Tanzania to the Ruvuma River. More specifically this includes, first, Mombasa up to Lamu, Rabai to Taita, Taveta, Ukambani, and the nearer parts of Maasailand, along with the Pare mountains, Nguu, Uzigua, Bonde; second, the Indian Ocean caravan towns of Saadani and Bagamoyo through Uzaramo and Ukwere to Morogoro, Usagara, Ukaguru, Mpwapwa, Mamboya and Ugogo; and third, southeastern Tanzania, including Kilwa and as far south as Makua and Makonde territory near Masasi and Newala, just north of the Ruvuma river, as well as Lindi and its neighbourhood. There is little or no information available regarding the vast interior regions of the Kilombero and Ruaha valleys, Mahenge, the southern highlands, or Matumbi, Mwera, and Ngindo territory.

Southeastern Kenya/Northeastern Tanzania

In mid-March 1884 a Scottish explorer, Joseph Thomson, with his African caravan, was on his return journey to the coast, having undertaken the first crossing of Maasailand to the shores of Lake Victoria by a European. Although aiming south southeast for the important caravan stop and market centre at Taveta, just east of Mount Kilimanjaro, Thomson was very ill with severe dysentery. He was forced to remain at Mianzini—at a height of well over 2500 m and far to the north north-west of Taveta, west of the Aberdare range and the Mau escarpment—for several weeks while recovering his strength and no doubt that of the expedition's coast-based porters. The cattle of the local Maasai, normally inhabiting the plains below, were dying of disease and there was no rain. At this time, the drought seems to have extended at least two thirds of the distance from the Indian Ocean to Lake Victoria, although we lack further evidence. Thomson later wrote:

> The Masai of the surrounding district were in despair through the almost utter loss of their cattle, and from the absence of rains in the low-lying

> district causing them to remain up in the cold bleak highlands. They were greatly disposed to ascribe their misfortunes to our presence. "What do you want here?" they would ask. "You have no goods left; you can't give our young warriors their customary presents. The rain is not coming, and the grass has not sprung up. Our cattle are dying off. You must be the cause of all this."[45]

Although Thomson published his account two years later, this is perhaps the first report of the drought from the East African interior. In his detailed history of the environmental collapse and existential crisis that gripped northern Maasailand from 1883 to 1902, Richard Waller plays down drought as a precipitating factor, and instead concentrates on the impact of bovine pleuropneumonia, rinderpest, and smallpox.[46] This may be the right approach when considering pastoral societies like the Maasai, to whom cattle were far more significant than cultivation, although it is important to remember that the Maasai economy also relied on exchanges with non-pastoral peoples who were in many cases hard hit by the drought.[47] Most East Africans were not pastoralists (Fig. 10.3).

A more detailed early description comes from British Vice-Consul C.E. Gissing, based at Mombasa, who in May 1884 with his staff and 23 porters undertook a journey of about 150 kms to the northwest from Mombasa via Rabai to Mounts Ndara and Kasigao in the Taita hills.[48] He explained, 'It was necessary to have a large number of porters, as all the food for the party had to be carried, there being none obtainable on the way.'[49] This should have been the middle of the *masika*, and indeed, the rain continued from the landing place on the mainland until the party had crossed the Rabai Hills. From that point, the caravan route had to cross over 100 km of dry, relatively lightly inhabited lowland country known as the *nyika*, which lay between the Swahili coastal fringe

[45] Joseph Thomson, *Through Masai Land* (London: Sampson Low, Marston, Searle and Rivington, new and revised edition, 1887), 337.

[46] Waller, 'Emutai,' 73.

[47] A question well studied by Waller himself in: 'Economic and Social Relations in the Central Rift Valley: The Maa Speakers and Their Neighbours in the Nineteenth Century,' in *Kenya in the Nineteenth Century*, ed. B.A. Ogot (Nairobi, 1985), 83–151.

[48] C.E. Gissing, 'A Journey from Mombasa to Mounts Ndara and Kasigao,' *Proceedings of the Royal Geographical Society*, 6, 10 (1884), 551–66.

[49] Ibid., 551.

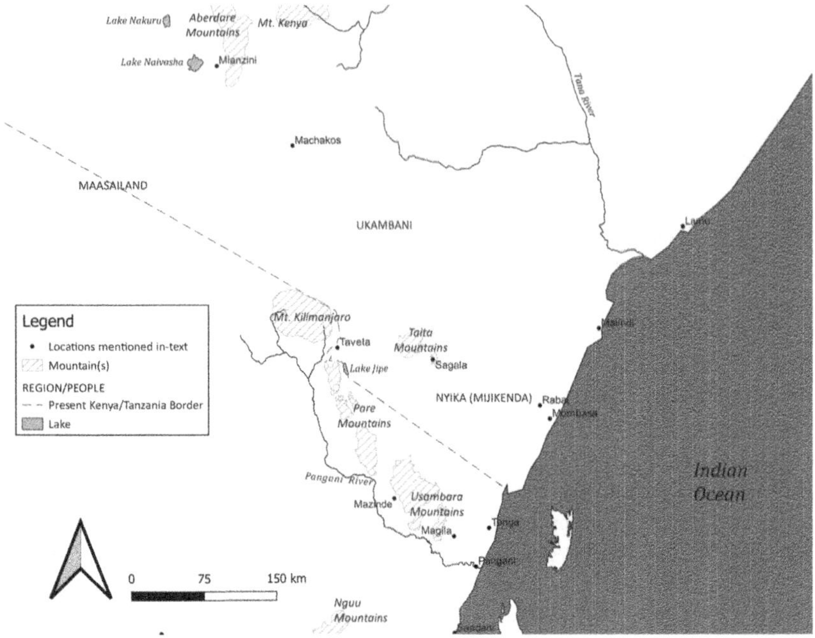

Fig. 10.3 Map of the region encapsulated by present-day southeastern Kenya/northeastern Tanzania in the late nineteenth century. Drawn by Philip Gooding

and the first mountain ranges of the interior of southern Kenya and much of Tanzania. The *nyika* was always a challenge, especially in the dry season, for heavily laden porters.[50] Gissing reported on two simultaneous disasters that had befallen the inhabitants of the *nyika*, beginning with the Duruma people, one of the Mijikenda group. Maasai raiders, no doubt aiming to compensate for cattle and other losses caused by the outbreak of bovine pleuropneumonia and the drought had attacked numerous villages:

[50] For details see: Stephen J. Rockel, 'Caravan Porters of the *Nyika*: Labour, Culture and Society in Nineteenth Century Tanzania' (Unpublished PhD diss.: University of Toronto, 1997), Ch. 2.

> All along the route were deserted villages and *shambas*, or cultivated patches, the inhabitants having been killed off by these marauders [the Maasai]. Mwache [the first Duruma village beyond Rabai] is quite deserted, nothing marks the spot but a few coco-nut trees and two or three mango trees that have escaped the general destruction. After passing through the thorny part of the road we came on many ruined shambas, and occasionally met a M-duruma, who had always the same tale to tell—he was very hungry, had nothing to eat, the Masai had destroyed everything, &c.[51]

At Gorah, the next camping place, the impact of the drought became clearer:

> We expected to find plenty of water, as we had been assured we should, but there was not a drop...The wells at Gorah...are...dug out of the clayey soil at the foot of the hills to a depth of about 15 feet; when rain falls the water from the hills fills them and remains in them for about six weeks; when I arrived there they were quite dry. Here the immense difference between the climate of the coast and the interior first becomes apparent...when you leave the hills and descend to the plain you leave the fertility behind you; here everything was dried up, evidently no rain had fallen for a long time, for the grass was withered, the trees leafless.[52]

Beyond Gorah, normal agricultural activities had become impossible:

> After leaving the forest the road passes through many miles of plantations of the people of Gorah: they were quite dry when I saw them; a few people in the hollows were growing a little mahindi, or Indian corn, and millet seed; also some beans were struggling to come up, but on the whole they looked very poor, and the people we met looked thin, and complained much of hunger. I obtained a guide, who...[took] us to a place where there was water...the place is called Mfufuni...the water was certainly unfit to drink, [but] we all drank largely of it. The place was formed by the rocks...cropping up above the soil some six feet, for about two acres, and inclining downwards to a central place ...into a deep hole about 16 feet deep...These places are most important on this route, as there are no rivers, no wells, and no other places whatever for water. It is very curious to see the beds of sandstone worn into...perfectly circular holes, about

[51] Gissing, 'A Journey from Mombasa,' 551–52.

[52] Ibid., 553.

two to six feet diameter, and varying in depth from two to 12 feet; these holes fill during rain, and water remains for three or four months. They are the natural wells of the country, and without them it would be quite impassable.[53]

The bone-dry conditions remained through the remaining days of the march to Taita. Gissing noted the high status of the ritual rain-making experts among the Taita people at Mount Ndara, who though living high on the slopes of the Taita mountains, were surrounded by dry drought prone lowlands.

> The rain-maker is a great personage amongst them. While I was there it was a time of great drought; no rain had fallen, as it should have done, early in June, and consequently the shambas [cultivated fields] were bare; the rain-maker had said that the presence of Mr. Wray [a CMS missionary] was the reason why there was no rain, and many were the hints that gentleman got that if he moved off the rain would come.[54]

At Mount Kasigao, a long march across the dry savannah lands south of Ndara, Gissing found conditions as difficult as those at Gorah and Ndara:

> Crossing the plain everything was dried up, animals scarce on account of want of water...On the morning of the 7th of June we arrived at the foot of the mountain...a huge mass, rising almost perpendicularly out the plain....On mounting to the village, which is about 1500 feet up....most of the houses are empty, the people having gone elsewhere in search of food."[55] After a few days at Kasigau, the caravan turned north northeast across the desiccated plain to the Maungu mountain, where the usual water source at the summit for both the local Taita people and passing caravans had shrunk to a small muddy pool of polluted water.[56]

[53] Ibid., 553–54.

[54] Ibid., 556. Rainmaking specialists became powerful political leaders among many of the peoples of the eastern interior during the long period of drought and famines. See among others: Isaria N. Kimambo, 'The Political History of the Pare People to 1900' (Unpublished Ph.D. diss.: Northwestern University, 1967).

[55] Gissing, 'A Journey from Mombasa,' 559.

[56] Ibid., 562.

Other reports from May 1884 point to the same level of distress. Thomson's caravan had made progress on its return to the coast. Crossing Kikumbulyu, a semi-arid district in Kamba territory, west of what is now the Tsavo East National Park, they found the people 'dying of famine' and 'no food was to be got there.' By the time the caravan had reached Ndi in Taita, the porters had completely run out of food. Thomson writes:

> At Ndi we found the famine also devastating the land. No food was to be got. Ndara was reached on the 21st of May....We stayed at Ndara only one day. Famine was the cry everywhere, and my men at Ndara could get nothing but sugar-cane, not a very nutritious article of food taken by itself.[57]

Not every locality in Taita was suffering to this extent, however, in June 1884. The same month, the explorer H.H. Johnston with a caravan about 120 strong also passed Ndara mountain, *en route* to Kilimanjaro. Leaving Ndara, he wrote:

> We started on a day's journey to Mwatate, passing through dense thorny bush, rocky defiles, and dried-up stream courses, and finally reaching, to my joy, the first channel of running water we had met with since leaving the vicinity of the coast. The ground along its banks had been cleared and cultivated; we crossed the stream on a rickety wooden bridge, and passed up through fields of maize and brakes of sugar-cane to our camping place.... The next day we were travelling through a Swiss-like country of mountain passes and richly fertile valleys, and at length came to Bura... Here we were nearly having a skirmish with the natives, who demanded an exorbitant payment for the right of drinking from their river. However... we were soon on excellent terms with the Wa-taita, buying fowls, maize, and honey at a great rate.

Bura in June 1884 was an oasis, however, and the next stages of the march to Taveta were waterless.[58]

The extremely dry conditions encountered on Gissing's return journey via Gorah to Mombasa were no different than the experience of the

[57] Thomson, *Through Masai Land*, 339–40.

[58] H.H. Johnston, *The Kilima-Njaro Expedition* (London: Kegan Paul, Trench & Company, 1886), 64–65.

outward stages. Miles of *shambas* (farms) of the Duruma were lying uncultivated, with no crops growing:

> It was pitiable to see where the Indian corn and other things had come up above the ground and then withered up; but there was absolutely no food growing for the people. We passed some Duruma villages, but saw no one, and from what I heard the people have gone to the coast to look for food...the wells at Gorah were quite dry, no rain having fallen since I passed there some weeks before.[59]

Further evidence on the progress of the drought and famine in this region comes from the reports and letters of the CMS missionaries based at Rabai as well as at Ndara and at Frere Town, just across the harbour from Mombasa Island. Rev. J.W. Handford notes in his annual report for the year 1884 an increase in church attendance at Frere Town, resulting from the famine: 'The people of the settlement have improved wonderfully in their attendance upon the means of grace; but the cause of the great and sudden increase, swelling the congregation from an average of 250 to 400 has been the influx of strangers from famine.'[60] He continues:

> The famine, besides teaching our people here to value more highly their privileges, has been the means in God's hands of bringing hundreds within the sound of the gospel. Wa-Digo, Wa-Nyika, Wa-Giriama, Wa-Taita, have come into the place by droves for no other purposes than to obtain food, and to pick up the fallen mangoes, which I allowed to become common property for the time; but we felt that God had directed them here for another purpose, and we gathered them together in the church, and made very possible arrangement for their instruction. Husbands had lost their wives from starvation, and wives their husbands; others had fled their country to escape capture from the prowling Swahilis, who, taking advantage of their present distress, have reopened their nefarious trade with impunity. It would take too long, and require greater powers of description that I possess, to picture the scenes that we daily witnessed, and at the same time we knew that what came before our immediate notice represented only a drop in the ocean as it were of the universal distress.[61]

[59] Gissing, 'A Journey from Mombasa,' 563.

[60] *Church Missionary Intelligencer* (hereafter: CMI), Mar. 1885, 160: 'From the Rev. J.W. Handford's Report,' Frere Town, 1884.

[61] *CMI*, Mar. 1885, 162: 'From the Rev. J.W. Handford's Report,' Frere Town, 1884.

If most of the evidence for drought and famine conditions in southern Kenya concerns the homelands of inland peoples including the Taita and Kamba, as well as districts near the coast inhabited by Mijikenda groups including the Duruma and Rabai, it is clear that localities further up the coast were also suffering. The British Consul at Zanzibar, Sir John Kirk, whose interest was more particularly the suppression of the revival of the slave trade, reported that:

> The drought, which over the whole of east tropical Africa has this year been severe, has caused famine and death in some of the inland districts, and everywhere raised the prices of all articles of food. The coast towns have been saved from actual want only through the mail service, which fortunately has enabled the merchants to import grain from abroad, a service which I found highly appreciated and spoken of by all classes. At the time of our visit [August 1884] the new crop was just ripening near the coast, and already this is relieving the famine in the interior, where the harvest has proved a total failure.[62]

Further official reports show that the drought and consequent famine was felt through 1884, at least as far north as the Lamu archipelago and probably well into the northern interior. At Takaunga, the crops had failed, and 'a famine has followed with all its horrors.' Missionaries reported that when travelling, 'the stench is sometimes dreadful from the dead bodies lying in the bush by the wayside.' Indian merchants at Takaungu and other coastal towns, forbidden as British subjects from owning slaves, requested instructions from consular officials on what to do when starving Wanyika [Mijikenda] mothers begged them 'to take their children in exchange for food, and so save the lives of both.'[63] Gissing reported on the conditions along the northern coast, as well as some of the interior districts: 'This year is perhaps the worst that has been the experience here for fifty years; in the upcountry districts there has been a total failure of the usual rains; in Giriama the people have a harvest that will barely suffice till the next crop; round the district of Kwilili the crops arc spoilt; the

[62] House of Commons Parliamentary Papers (hereafter: HCPP) 1884–85. Correspondence with British representatives and agents abroad and reports from naval officers ...Relative to the slave trade. Sir J. Kirk to Earl Granville (Extract.), 2–3 Sep. 1884.

[63] HCPP 1884–85. Correspondence with British representatives and agents abroad and reports from naval officers ...Relative to the slave trade: Vice-Consul Haggard to Sir J. Kirk, 8 Sep. 1884.

Malindi and Mambrui districts have had their 'mtama' [sorghum] much destroyed by the insects.'[64] Gissing also pointed to the desperation that led to parents selling their children for food, or perhaps pawning them:

> This year has been a trying one, terribly hard for a starving people to see their crops daily drying up in the fields. To show how severely it has been felt at this time, there should be hundreds of thousands of cocoa-nuts [sic] for sale or export, now it is hard to buy one for money, the people have eaten them all before they had time to ripen. The coast tribes have been selling their children for food; this is deplorable, as it increases the number of slaves, but after all it is better than the poor things dying of hunger, as they otherwise would; and I believe the custom is for the parents to redeem them in good years.[65]

Mombasa and Malindi districts had in fact received enough rain to produce a harvest by October 1884 but, as Kirk reported, few in the interior districts benefitted, as their inhabitants had already lost or sold any assets: 'The difficulty is that the people of the interior have no means of buying, while the high prices that prevail elsewhere encourage shippers to send the grain away.'[66]

In the interior, the famine persisted, and conditions were still extremely harsh through 1885. Taita had been particularly hard hit, and death accompanied almost all families. Handford wrote in November 1884 that, 'Taita die in their houses and are left to rot there.' The source of this information was probably his colleague, Wray, based at Sagalla on Mount Ndara in Taita, who years later wrote: 'They died in their houses, on the roadside, in their gardens, and they were left unburied, no one having the strength to dig a grave.' 'The residents of Ndara, at least, had been reduced by the famine itself and migration to only about thirty families by

64 HCPP 1884–85. Correspondence with British representatives and agents abroad and reports from naval officers ...Relative to the slave trade: Vice-Consul Gissing to Sir J. Kirk, 16 Sep. 1884.

65 Vice-Consul Gissing to Sir J. Kirk, Mombasa, 16 Sep. 1884; Gissing, 'A Journey from Mombasa,' 565.

66 HCPP 1884–85. Correspondence with British representatives and agents abroad and reports from naval officers ...Relative to the slave trade: Sir J. Kirk to Earl Granville, 24 Oct. 1884.

early 1885.'[67] In March the newly appointed Anglican Bishop of Eastern Equatorial Africa, James Hannington, accompanied Handford on a visit to Taita and Taveta to assess local conditions. At Mount Taro one of the porters encountered two Swahilis from the coast, with an enslaved woman and seven children. They were 'half-starved.' Their Swahili masters having fled, the missionaries sent the captives down to the coast with some of their own porters. They were formally freed by the British Vice-Consul in Mombasa and handed over to the mission. It was too late, however, and all but one of the freed slaves soon died.[68] At Mount Ndara, the crisis continued, and many of the remaining Taita people around the recently established mission station blamed the resident missionary, J.A. Wray, for their plight. Hannington recorded in his journal:

> The people were dying of starvation, and inclined to curse him as the evil author of their troubles; but he had, nevertheless, won the confidence and affection of those with whom he was able to come into contact. ... Yet the famine presses hard. In spite of our supplies of food, many have died, many have left, and many have been killed or captured and sold for slaves. Thus, all the villages, except those immediately under Mr. Wray's wing, are utterly abandoned. The people around him number less than a hundred. These I assembled to hear their opinion upon the situation. They are most decided. 'We do not wish to desert Bwana, but we cannot stop here. Sometimes you feed us, sometimes you do not, and then we have to return to eating grass and insects. Not one neighbour have we left. Even if you gave us seed to-morrow, it would be four months before we could get any food.' It seemed then to all of us that, in the face of this, the station could not be continued....I therefore arranged that they should be received at Rabai.[69]

Hannington no doubt privileged the role of the mission in the maintenance of the remaining residents of Ndara, yet this was the worst drought and famine in the region for many decades and desperation

[67] Handford to Lang, 26 Nov. 1884, in Merrit, 'A History of the Taita,' 102; J. Alfred Wray, *Kenya: Our Newest Colony* (London: Marshall Brothers, 1928), 56, in Merrit, 'A History of the Taita,' 102–3; Henry Morris, 'The Switzerland of Africa,' *CMI*, July 1885, 512.

[68] E.C. Dawson, *James Hannington, First Bishop of Eastern Equatorial Africa: A History of His Life and Work, 1847–1885* (London: Selley & Co., 1891), 289–90.

[69] Ibid., 293.

was all around. The station at Rabai was in a much better position and about thirty of the 'half-starved' Ndara people did make the journey to find some kind of relief nearer the coast. On March 15, Hannington's caravan reached Taveta after trekking across the intervening waterless plain. Taveta was usually safe from drought and famine due to its location on a slight depression near the south eastern slopes of Kilimanjaro and adjacent to Lake Jipe. In the 1880s it was forested and well-watered by the underground passage of the Lumi river through subterranean channels.[70] However, even here, conditions had worsened, as Hannington noted: 'Usually this would be a land of plenty, but this terrible famine has driven a large number of starving neighbours within their bounds, and they too are feeling the pinch.'[71]

In July 1885, almost two years after the Krakatau eruptions, travellers and the communities they passed through found conditions in south-eastern and central Kenya just as difficult. In that month, Hannington was on his last journey, accompanied by the recently ordained, but vastly experienced missionary and talented linguist, Rev. William Henry Jones, a freed slave of Yao origins, and a caravan of 200 porters. Once again in Taita, Hannington wrote to his wife, 'Food seems the great difficulty, as the country has not recovered from the famine.'[72] Further on, there were still shortages in Ukambani. On 9 August, after passing through Kikumbulyu, food was still 'exceedingly short' in the south eastern localities of Ukambani, although the next day provisions were available for the porters. On 19 August, there was a new tone in the journal entries. At Machakos, Hannington (or was it Jones?) wrote of the district of Ulu, through which they had just passed: 'It is difficult to say how beautiful the Ulu country is. It is a country full of water and cattle. All provisions are plentiful and cheap.'[73] At Machakos the caravan had 'arrived in a densely populated district....Heaps of provisions were poured into the camp– butter, milk, and Indian corn. We found that we must buy for the men, so we opened a market. I found that two yards of cloth would

[70] For a discussion of the geology and great fertility and productivity of Taveta, see: Thomson, *Through Masai Land*, 58–62.

[71] Dawson, *James Hannington*, 297.

[72] Ibid., 334.

[73] Ibid., 336. Dawson had access to Jones' journal as well as Hannington's.

buy food for one man for ten days.'[74] The famine, at least in the north western parts of Ukambani, in Ulu and in Machakos, appeared to be over. If we consider the evidence of lake levels, which act somewhat as a proxy for rainfall, there seems to be little evidence that the crisis in precipitation and food production in 1883–1885 reached as far north-west as Lakes Nakuru, Bogoria, and Naivasha in the Kenyan section of the great rift valley. In fact, water levels in Lake Naivasha and nearby Lake Oloidien were rapidly rising during the early and mid-1880s to record levels by the end of the 1890s and the beginning of white settlement in this region in 1901.[75]

In reality, the impact was to be long lasting as far as agricultural production was concerned. In March 1887, Bishop Parker visiting Wray at Sagalla found that 'The long broad stretch of shambas, with various kinds of corn, &c., which I hear used to clothe the plain…beneath Sagala…is now given up to the rank growth of the coarsest grasses, thorns and tangle.'[76] There was, however, some recovery of the mixed farming economy based on the evidence of a report of an extensive tour of Taita and Tsavo by a colonial official, from September to December 1892.[77] Its author, C.W. Hobley, reported that as a consequence of famine deaths, the migration of many of the Taita, and the loss of more to enslavement, much formerly utilized farm land had been lost:

> Large areas of ground at the base of various mountains – Ndi, Mbololo, Ndara, Kisigau, etc. – were under complete cultivation; but the people have been so reduced in numbers that only about one-fifth to one-sixth of the former areas is at present under cultivation, the remainder having lapsed into tracks [tracts?] covered with grass and here and there dotted with light scrub, the most extensive of these tracks being at Ndara.[78]

[74] E.C. Dawson (ed.), *The Last Journals of Bishop Hannington* (London: Seeley and Co., 1888), 156: Refers to 20 Aug. 1885.

[75] Dirk Verschuren, John Tibby, Peter R. Leavitt, and C. Neil Roberts, 'The Environmental History of a Climate-Sensitive Lake in the Former "White Highlands" of Central Kenya,' *Ambio*, 28, 6 (1999), 494–501; Dirk Verschuren, 'Lake-Based Climate Reconstruction in Africa: Progress and Challenges,' *Hydrobiologia*, 500, 1 (2003), 315–30.

[76] Bishop H.P. Parker to CMS, 3 Mar. 1887 in *CMI* (July 1887), 425.

[77] C.W. Hobley, 'Upon a Visit to Tsavo and the Taita Highlands,' *Geographical Journal*, 5, 6 (1895), 545–61.

[78] Hobley, 'Upon a Visit to Tsavo,' 553. For further details relating to Taita, see: Merrit, 'A History of the Taita.'

Further south, in what is now northeast Tanzania, the drought also bit hard, including in the highland regions of Usambara and especially the Pare mountains, to the south east of Kilimanjaro. Isaria Kimambo, a pioneer of historical research in Tanzania, points out that the people of Taita and Pare had a shared history through their responses to drought, and the words for both hunger and famine *njaa* (in Chagga), *dhaa* (Pare), and *saa* (Shambaa) were almost interchangeable.[79] Referring to the famine of about ten generations before the 1960s (the time of his original research) known as *Mbofu* (ants, eaten for survival), he argues that disparate groups from Taita found refuge in different parts of the Pare mountains because they had greater prestige as rainmakers than the Pare, and famine was less severe there.[80] In fact, all the peoples of what is now southeast Kenya and northeast Tanzania east of the high mountain ranges were interlinked through migration over several centuries as a result of drought and famine, as well as complementary economies of production and exchange. By the middle decades of the nineteenth century the local and regional economies of the highlands and *nyika* of northeast Tanzania were well connected to the long-distance caravan system and the ivory trade that underlay it, and food production was in part geared towards feeding thousands of caravan porters trekking along the lowland routes connecting the coastal towns of Tanga, Pangani, and Saadani with the far interior. Thus, greater food production and commercialization of markets did not protect local populations in the 1880s.[81] In Upare the

[79] Kimambo, 'The Political History of the Pare People,' 50–51; Kimambo, 'Environmental Control and Hunger,' 71. Taita refugees from famine also migrated to the Usambara mountains several times in their history: Merrit, 'A History of the Taita,' 75–76.

[80] Kimambo, 'The Political History of the Pare People,' 51, 85. See pp. 7, 119–27 for details of rain-making rituals and use of symbols, medicines, sacrifices, rain-making groves of the 'ritual experts' and rain-making ruling lineages in Upare. Rain makers collected tribute which buttressed their political power and the emergence of a ruling group.

[81] Kimambo, 'Environmental Control and Hunger.' See: Stephen J. Rockel, *Carriers of Culture: Labor on the Road in Nineteenth-Century East Africa* (Portsmouth, NH: Heinemann, 2006), Ch. 5, for caravan provisioning; Stephen J. Rockel, 'Forgotten Caravan Towns in Nineteenth Century Tanzania: Mbwamaji and Mpwapwa,' *Azania*, 41 (2006), 1–25, for the breakdown of relations between caravan personnel and peasant communities during the late 1880s and 1890s.

consequences of the drought and famine were devastating, with constant raiding for slaves and food in an 'atmosphere of plunder.'[82]

As we have seen in the case of southeast Kenya, in the nineteenth century the most basic ecological and cultural distinction was between the plains and the mountains, or the *nyika* and—in the case of the Shambaa people of the Usambara mountains—*Shambaai*, with its emphasis on banana cultivation.[83] However, the dualism of Steven Feierman's categories can be overstated.[84] Many highlands peoples also had significant investments in cattle raising, agriculture, and trading relations in the lowlands, and during the commercializing era of the second half of the nineteenth century, power in Shambaai shifted to the lowlands, as Sembodja, a son of the *Simbamwene*, the Shambaai king, moved his base to control the caravan town of Mazinde, and hence much of the trade along the Pangani valley route.[85] Nevertheless, the Shambaai people in the highlands were not as badly affected by the drought as most other populations in the region, although it is likely that the poorest and those without patrons suffered great hunger. This contrasts with the devasting famine of 1897–1899.[86]

A little to the south southeast of the Usambara range and north of the Pangani River lies the country of the Bonde people, for which we have more direct evidence of the impact of the drought and famine. At the beginning of April 1884, the people in a village neighbouring Magila, the local centre of the Universities Mission to Central Africa (UMCA), about three days' march from the port of Pangani, were already in fear

[82] Kimambo, 'The Political History of the Pare People,' 276–83.

[83] See: Steven Feierman, *The Shambaa Kingdom: A History* (Madison: University of Wisconsin Press, 1974), 17–31, for the overlap between ecological and cultural zones in this part of East Africa, and the meanings and changing subsistence strategies with the spread of new crops for each. An updated view of late nineteenth-century change is provided in: Frans D. Huijzendveld, 'Changes in Political Economy and Ecology in West-Usambara, Tanzania: ca. 1850–1950,' *International Journal of African Historical Studies*, 41, 3 (2008), 383–409.

[84] Huijzendveld, 'Changes in Political Economy,' 384–85.

[85] Feierman, *The Shambaa Kingdom*; Kimambo, 'Environmental Control and Hunger,' 90–92; Huijzendveld, 'Changes in Political Economy,' 392–94.

[86] Huijzendveld, 'Changes in Political Economy and Ecology,' 395.

of imminent famine, and proposed consulting the most powerful rain-maker, the chief Kibanga.[87] On 19 May the Rev. Woodward, based at Magila, wrote of 'famine everywhere' and the impossibility of purchasing food supplies while travelling beyond the station.[88] This was confirmed in detail by Archdeacon Farler in a report to London of 17 June:

> I have sad news to tell you of Magila – there is a great famine there, and we are doing all we can to meet it...since the beginning of this month we have sent to Magila in rice and money more than $1000 [Maria Theresa *thalers*], and there is yet two months to the harvest. We are sending up rice from Bombay as fast as we can, and...distributing it to the famished people. Already many are dead. ... The people have no food and they cook weeds and eat them together with a poisonous root called Mdiga, which has to be well soaked in water before it can be safely eaten. The Bondeis are coming from all parts of the country to Magila to beg for food. Every morning there are some two hundred men besieging our store room. Some bring money, others bring bills, i.e. promises to pay back when they gather their harvest, some have nothing and plead *in forma pauperis*. The harvest, through delay of the rains, is very late, but it promises to be a good one.

Farler then emphasizes the usefulness of the crisis to the mission. 'If therefore our friends will help us for Christ's sake to feed these hungry ones for the next few weeks the result will be a great gathering of souls to God. Already the natives are speaking of our goodness and love.'[89] Many local Bonde people found themselves bound to the mission either through labour or debt, but lives were no doubt saved. Two months later the situation was still grim, although there were 'fair' crops of maize and millet by early September. Farmers carried some of this down to the coast to take advantage of high prices. Woodward reported, 'Very few people are well off, but some are beginning to repay money lent during the famine.' Many of the local Bondei people were sick with 'stomach complaints' and

[87] Charles Alan, 'The Bishop's Visit to Magila,' *Central Africa: A Monthly Record of the Work of the Universities Mission*, 18 (June 1884), 101.

[88] 'Visit to the Wadigo,' *Central Africa*, 20 (Aug. 1884), 134.

[89] 'Famine around Magila,' *Central Africa*, 20 (Aug. 1884), 134–35. The 'poisonous root,' *mdiga*, that Farler refers to, was probably a variety of bitter cassava eaten as famine food.

other illnesses as a result of the famine.[90] By the end of May 1885, the situation in Bondei had improved markedly after an 'unusually protracted' wet season. At Umba, one of the mission outstations not far from Magila, the maize crop—'a very large one'—was being harvested. The resident missionary wrote: 'There will certainly be no famine this year.'[91]

For the people of Uzigua (now largely in Handeni district)—the lowlands south of the Pangani river in northeastern Tanzania down to the Wami river, and between the coastal strip and the Nguu mountains—the drought and famine was known as *lugala*, or star. Like Mijikenda, Taita, and Ukambani to the north, Uzigua was well connected to the networks of the long-distance caravan trade that linked the coast with the far interior, including the well-travelled Pangani river valley routes that led to Kilimanjaro, Maasailand, and the shores of Lake Victoria.[92] The historian of the region, Jim Giblin, dates the famine from 1884 to 1886, indicating the relatively long duration of the distress, although its impact was not as severe as the subsequent famines of the German colonial period that led to high death rates, mass flight, the loss of productive agricultural land, epizootics, and depopulation.[93] As elsewhere, famine followed the drought of late 1883 and failure of the long rains in 1884. However, in Uzigua the real impact came with the failure of the 'second line of defence,' the pumpkins that were intercropped with sorghum and maize as an insurance crop. From 1884 it was the poor and those without good connections to powerful families or mission stations who suffered the most. For their descendants, memory of the famine lived on at least to the 1980s, the time of Giblin's oral history research, and possibility longer. Giblin writes, 'for the destitute, the cost of survival was servility or dependence upon the chieftains whose powers and commercial involvement were at their height.'[94] This was also the case in parts of the Nguu mountains, where famine conditions prevailed in the periphery of the chiefdoms, beyond the more secure core areas, and less well-connected

[90] 'News from Magila,' *Central Africa*, 22 (Oct. 1884), 167; 'Magila,' *Central Africa*, 24 (Dec. 1884), 199.

[91] 'A Notable Conversion,' *Central Africa*, 32, (Aug. 1885), 114.

[92] Giblin, *The Politics of Environmental Control*, 22–28.

[93] Ibid., 121.

[94] Ibid., 122.

highland farmers were not able to secure or purchase food after the highland crop of May 1884. Many northern Nguu families were forced to relocate themselves to more favoured lower slopes and river valleys in northern Uzigua. But the price was submission to Zigua chiefs or the Spiritan missionaries, and 'the status of client, pawn, or slave.'[95]

Famine on the Caravan Routes

Over the nineteenth century, East Africa underwent a commercial revolution, as the forces of financial and industrial capitalism stimulated a massive increase in the demand for tropical products, including ivory, slaves, and cloves, and the labourers that produced and transported them. As elephant populations near the coast were destroyed for their valuable tusks, East African elephant hunters and trading caravans trekked further and further in the vast interior of the continent. Going up country they carried imported trade goods including an astonishing variety of manufactured cloths, textiles, and beads, as well as guns and other metal goods. Free wage labourers from a variety of ethnic groups, especially the Nyamwezi and related peoples from the western Tanzanian interior, coastal Swahili, and enslaved peoples of a variety of origins collectively known as the *Waungwana* ('Gentlemen') provided the caravan workforce of tens of thousands. They were professionals, frequently on the road, who undertook journeys of many months or even years on foot, due to the impossibility in nearly all of East Africa south of the Horn of using draft or pack animals. This was due to the trypanosome parasite that caused sleeping sickness and death, and was carried by the tsetse fly that attacked domestic animals such as oxen, donkeys, and horses, which had no immunity to the disease.[96] This and the lack of navigable waterways meant that human porterage dominated the transport system. Huge caravans could consist of 2000–3000 personnel, although most included two or three hundred porters, typically travelling with womenfolk. Professional caravan porters, or *wapagazi*, were highly skilled, multilingual, and adapted to a lifestyle of mobility and travel in tough conditions. They developed a unique labour culture with its own codes and customs. They knew how to survive far from their homelands and built up networks

[95] Ibid., 67, 123.

[96] See also: chapter by Gooding, this volume.

of mutual support with peasant and pastoral societies they traded with and often resided among.[97] Responsible caravan leaders had to plan their itinerary and possible alternative routes with care and make preparations accordingly. As the colonial meteorologist Kremer wrote with reference to his own research:

> If one nevertheless passed through areas of famine, the provisioning of the caravan was exposed to the greatest difficulties, and that is why the diaries of such expeditions, even if they contain so little scientific information, are ...used for this investigation.[98]

Yet in 1884, thousands of caravan porters starved to death. Because they were outsiders in the eastern and central lands of what is now Kenya and Tanzania where the people were hungry and often starving, caravan porters often found it impossible to buy staple provisions and, even if they could, prices were vastly inflated. For a caravan on a journey of several months, drought and famine could mean disaster.[99] An obvious option was to not travel at such times. When the famine was widespread, caravan traffic was considerably reduced. But travelling was often necessary if famine conditions prevailed over a large region. Travelling could not always be avoided when a caravan might be two or more years away from home. And the dry season was the normal season for travel. This was the Achilles' heel of the long-distance caravan system (Fig. 10.4).

Even in normal times, a porter's diet was monotonous, although there were occasional treats. Despite preferences according to region of origin, religion, and customary taboos, survival, and health depended on adaptability. Preferences and taboos were ignored when necessary. When marching, porters ate only one or two meals per day, but on rest days they ate as frequently and as much as possible. Porters arriving at a well-stocked village after a march through the desert or famine-stricken country were known to consume enormous quantities of grain, vegetables, meat, honey,

[97] Rockel, *Carriers of Culture*.

[98] Kremer, *Die unperiodischen Schwankungen*, 27.

[99] The rest of this section is largely based on: Rockel, *Carriers of Culture*, 148–50, 153–59.

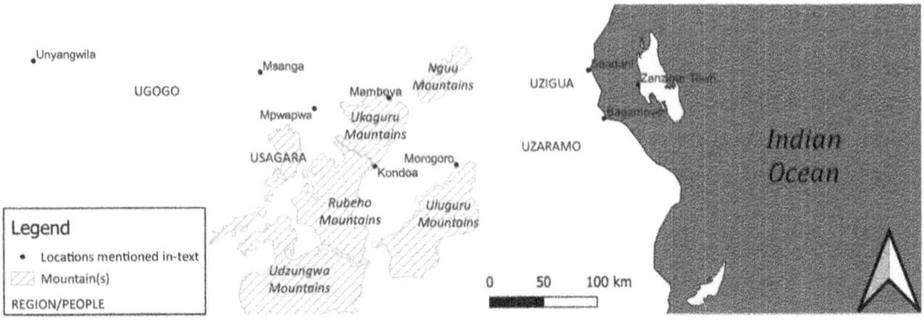

Fig. 10.4 Map of regions on the eastern portion of the central caravan route. Drawn by Philip Gooding

and other foodstuffs.[100] Porters added whatever vegetables, sweet potatoes, nuts, and other foods they could obtain to the high-bulk staple. Rations (*posho*) went further with the addition of edible herbs and roots collected in the vicinity of camps. During the rainy season, travellers could collect mushrooms in the *pori*, or bush country. When locust swarms appeared, Nyamwezi and Sukuma porters ate the insects. Hungry porters ate wild fruits and termites as a famine food.[101] But famines or even shortages could be much more devastating for caravan porters, who, as short-term visitors and marginal outsiders with little claim on extremely scarce local resources, and often having only their *posho* disbursements to pay for food selling at prices beyond their means, were the first to starve.[102]

[100] Richard F. Burton, *The Lake Regions of Central Africa* (Michigan: Michigan Scholarly Press, 1971 [1960]), 246; Joseph Thomson, *To the Central African Lakes and Back* (London: S. Low, Marston, Searle, & Rivington, 1881), I, 208–9; Rachel Stuart Watt, *In the Heart of Savagedom* (London, n.d. [1912]), 157; J.A. Moloney, *With Captain Stairs to Katanga* (London: Sampson Low, Marston & Co, 1893), 33; R. Stanley and A. Neame (eds.), *The Exploration Diaries of H. M. Stanley* (London: W. Kimber, 1961), 55.

[101] Burton, *Lake Regions*, 495; Church Missionary Society Archive (hereafter: CMS) G3A6/01 Copplestone to Lang, 4 Mar. 1882; Mtoro bin Mwinyi Bakari, *The Customs of the Swahili People*, trans. & ed. J.W.T. Allen (Berkeley, Los Angeles, London, 1981), 140; Speke, 15–20 Dec. 1860, in: John Hanning Speke, *Journal of the Discovery of the Source of the Nile* (New York: Harper, 1864), 88.

[102] Carol Jane Sissons, 'Economic Prosperity in Ugogo, East Africa, 1860–1890' (Unpublished Ph.D. diss., University of Toronto, 1984), 8.

By 1884, the CMS mission stations at Mpwapwa and Mamboya in Ukaguru were well established, and in that year several Europeans were travelling to other parts of the interior, passing through the affected regions. The documentary record allows, therefore, a discussion of the impact of the famine on caravan porters. By February 1884, the shortage of rain in the Mpwapwa area portended the approach of famine. By July and early August, famine conditions were apparent along central and eastern sections of the caravan routes. Ugogo, a drought prone region, was struck early.[103] In the east, at Mpwapwa and in Usagara, there was still a little food, and many Gogo left their homes for these places. This is consistent with the Gogo name for the famine, *Chilemu*, meaning migration. But by November, most people in Mamboya were on the verge of starvation, and some were dying. A little to the north in the Nguru mountains, and in Uzaramo, as we have seen for parts of Uzigua, many poor farmers were forced to turn as supplicants to big men and chiefs, and lost their independence. Others had no choice but to pawn or enslave themselves or family members. Everywhere Maasai, Baraguyu, and Gogo herders had great trouble finding pasture and water for their cattle.[104] In the west, Unyamwezi was less affected, and when famine conditions continued into 1885, some Gogo people went to Unyanyembe, the most powerful Nyamwezi chiefdom, for relief. The harvest there was good, and small groups of Nyamwezi carried surplus grain through the waterless Mgunda Mkali region to sell in famine-stricken Ugogo and to caravans.

At the coast, porters from Unyamwezi and Usukuma arrived on their last legs. In July 1884, Charles Stokes, an Irish trader who integrated into Sukuma society through his marriage into a chiefly family, was in Saadani and then Bagamoyo, organizing his caravan to Uyui, near Tabora. He had to feed his upcountry porters for 15 days before they were fit to carry

[103] In addition to the work of Maddox, cited above, for an overview of Gogo management of food shortages caused by drought or famine, see: Doris Schmied, 'Managing Food Shortages in Central Tanzania,' *GeoJournal*, 30, 2 (1993), 153–58.

[104] Lettre de M. Ch. Ledoulx, consul de France à Zanzibar, 18 Dec. 1884, *Compte rendu des séances de la Société de géographie de Paris* (Feb. 1885), 105; CMS G3A6/01 Price to Lang, 5 Feb. 1884; CMS G3A6/01 Price to Lang, 5 Aug. 1884; CMS G3A6/02 Stokes to Lang, 6 Oct. 1884; Sissons, 'Economic prosperity,' 136, 143–44; Brooke, 'The Heritage of Famine,' 20; CMS G3A6/01 Roscoe to Wigram, 1 Nov. 1884; Giblin, *Politics of Environmental Control*, 121–24; Cooper, *Plantation Slavery*, 126–28; Annie Hore, *To Lake Tanganyika in a Bath Chair* (London: Sampson Low, Marston, Searle & Rivington, 1886), 120, 125, 127.

loads, given the famine conditions. 'A more miserable set of skeletons you could not see,' he wrote. At Bagamoyo, the main coastal terminus of the central routes, the death rate was 'enormous'; porters who arrived from the interior in bad condition were left to starve. Missionaries of the Holy Ghost Fathers did what they could, 'having something like 200 at a time in a dying state.' A few porters recovered, but most were too far gone.[105]

The scarcity of food earlier in the year and the impossibility of obtaining adequate provisions affected the labour supply at the coast. Few porters were available compared with normal years. Thousands of Nyamwezi and Sukuma on their way to Bagamoyo and other towns turned back. Stokes believed thousands died along the roads. The Indian touts who normally found men for travellers 'had not a man to sell.' The most powerful merchants of the interior, such as Chief Mirambo, had to leave large quantities of goods at the coast because there was no one to carry them.[106] In late July, Stokes' caravan left for Uyui, a journey of at least 850 kms. Only one of the Ugogo routes was passable and had food for sale. The journey was marked by extreme hunger and many cases of dysentery among his porters. Forty to fifty sick porters had to be left at various points along the road. Many others died, and others again deserted. A week after the caravan's arrival at Uyui at the end of September, his half-starved porters were beginning to 'pick up' given that there was 'plenty of food' there.[107]

In early October 1884, an LMS caravan set off from Saadani, opposite Zanzibar, for Ujiji, 1200 kms away. At Kondoa, near the eastern entrance to the Mukondokwa valley, a pass through the Rubeho mountains, the porters were able to buy food. Before descending from the mountains to the plains of Ugogo they heard of hard times ahead, and of caravans broken up or forced to return due to the grim conditions.[108] Marching on, the caravan passed through the arid region on the east side of Mpwapwa. Lake Gombo was a dry basin. Annie Hore, on her second attempt to reach Ujiji with her husband, wrote:

[105] CMS G3A6/01 Stokes to Lang, 26 July 1884.

[106] CMS G3A6/01 Stokes to Lang, 26 July 1884; CMS G3A6/01 Baxter to Lang, 4 Aug. 1884; CMS G3A6/01 ? to Wigram, 17 Jan. 1885; Sissons, 'Economic Prosperity,' 143.

[107] CMS G3A6/02 Stokes to Lang, 6 Oct. 1884; CMS G3A6/02 Stokes to Stock, n.d.

[108] Hore, *To Lake Tanganyika*, 76, 77.

> Every now and then I saw curious dark objects lying on, or beside the path, and shortly afterwards became aware that they were the dead bodies of helpless laggards from the various hungry caravans that had passed that way. The heat and drought had been so great, that these bodies were perfectly hardened and preserved. It was a terrible sight.[109]

In Ugogo the famine was at its height. At Msanga and elsewhere the people were eating husks and, in a reversal of normal conditions, came to the camp to try to purchase food from the caravan. By the time the caravan reached Mpara, many of the porters had finished their rations and were in despair on being told that there was no water. Some were obviously starving and sick. Others, perhaps more careful, or with private resources, were able to manage with less hardship. A little muddy water would be bought with tobacco and cloth.[110] The condition of the porters deteriorated the rest of the way through Ugogo, despite them having been issued with an extra cloth to buy provisions. The only food procurable was sheep and goat meat, which caused dysentery for the weakest. Some had to be left behind with a little cloth for maintenance.[111] During this journey, the Hores lost 80 porters out of 200, 50 by desertion, and there were several deaths.[112]

Between Bagamoyo on the coast, and Kwamamba, 250 kms inland, near Morogoro, there was virtually nothing for travellers to eat. In Mamboya, where there was a CMS station, between December 1884 and March 1885, hunger killed many of the resident poor and elderly, who could not survive on a diet of grasses and wild fruits. In January 1885, a European traveller reported that the people between the coast and Mpwapwa existed on 'poisonous roots,' probably toxic varieties of cassava, 'for which they scour the jungles.' The porters of one missionary searched the countryside for food for up to two days at a time, sometimes returning to camp empty handed. They survived by buying plantain trees, 'which they cut down and ate for lack of anything better.'[113] In this

[109] Ibid., 80–81.

[110] Ibid., 97–98, 100–2.

[111] Ibid., 102–6.

[112] Sissons, 'Economic Prosperity,' 143.

[113] A. Bloyet, 'De Zanzibar à la station de Kondoa,' *Bulletin de la Société de Géographie*, 7, 11 (1890), 361; CMS G3A6/02 Roscoe to Lang, 7 Sep. 1885; CMS G3A6/02? to Wigram, 17 Jan. 1885.

region, famine strengthened slavery; a reversal seen by a missionary 'as though Satan was making another struggle against the kingdom of God.' People were 'carried off by night,' women disappeared while working their fields, and clients' children were sold by their chiefly patrons.[114]

The drought was also devastating for pastoralists and mixed farmers in central Tanzania. 'Many of the Wagogo are leaving their homes, some have come here [Mpwapwa, on the eastern edge of Ugogo], and many others have gone to Usagara,' Price of the CMS reported. 'The Masai & Wahumba, too, find great difficulty in getting either water or pasture for their cattle; and the Wagogo who have cattle of course suffer in the same way.' In parts of Ugogo, famine conditions prevailed through to late 1885, not so much from continuing lack of rain, but because of the earlier consumption of seed, and debility. Distressed by what he had seen, Price wrote of conditions in February: 'The famine, especially in some parts of Ugogo proper, is very very distressing. The prospects for the coming harvest are anything but very bright. May the good Lord have mercy upon these people.'[115] Later in the year, he travelled across the region to its western extent. At Unyangwila, in October 1885, he noted:

> Now there is great scarcity of food. We meet and pass whole families with their goods and chattels hung about all over their persons, so that the individual is scarcely visible amidst pots, calabashes, skins, bedding, &c., going to Muhalala [western Ugogo] and elsewhere, where food is more plentiful.

A little later, an LMS missionary travelling from Unyamwezi through Ugogo in November, confirmed the difficult situation, 'All through Ugogo there was a great scarcity of food though it did not quite amount to famine.'[116] As elsewhere, survival for many Gogo people was only possible through migration to more favourable areas. Even with these reports, it must not be forgotten that most of the suffering was away from

[114] CMS G3A6/02 Roscoe to Wigram, 1 June 1885.

[115] CMS G3A6/01 Price to Lang, 15 Aug. 1884; CMS G3A6/02 Price to Lang, 13 Feb. 1885.

[116] J.C. Price, Journal, 22 Oct. 1885, in *CMI*, 31 Oct. 1886; Sissons, 'Economic Prosperity,' 144. Price also notes that on his return (eastward) journey to Mpwapwa, Gogo women at Njinje, west of Dodoma, were selling maize to passing caravans. Clearly some parts of Ugogo were recovering at a faster rate than others.

the coast and main caravan routes and was out of the sight of officials of the Zanzibar Sultanate or foreign observers. People in more remote districts had fewer options or means to buy grain, and there was little chance of food supplies being brought in from outside.

The South

Written sources are fewer for southern Tanzania, a vast region much of which was very lightly populated, although Swahili and occasionally Yao trading caravans crossed between Kilwa, Lindi, Mikindani, and other coastal towns to the Lake Malawi region and the southern highlands. The rubber boom of the 1880s had also begun, and regular caravans from Ngindo territory in the interior brought wild rubber down to Kilwa.[117] In the early to mid-1880s, European travellers rarely crossed the wild expanses of the southern interior except for the localities closer to what is now Mtwara region north of the Ruvuma River, where once again, we rely largely on reports from the Universities Mission. Evidence shows that the drought extended at least as far south as Masasi and Mtua, although its impact seems to have been less severe than in regions further north (Fig. 10.5).

In mid-March 1884 at Mtua, about 60 kms southwest of Lindi on the Lukuledi River, Joseph Williams of the UMCA reported that 'dry-weather' had already damaged the maturing rice and millet crops. The result was that basic foods stuffs were 'scarce and dear,' although there was still a young crop of maize for the local people to fall back on. In the mixed Makonde, Yao, and Makua communities around Newala, 'all seems to be flourishing.' Six months later, in October, at Machemba's and Chionda's near Lindi, there was 'great scarcity.' In early March 1885, the people of Mtua were still suffering from food shortages. Using almost the same words as a year earlier, Williams wrote: 'The rains are very late this year, and all the early crops of maize and rice have been dried up by the drought.' There was a positive addition, however: 'At present we

[117] For an excellent overview of the Ngindo and their country see: Lorne Larson, 'The Ngindo: Exploring the Center of the Maji Maji Rebellion,' in *Maji Maji: Lifting the Fog of War*, eds. James Giblin and Jamie Monson (Leiden: Brill, 2010), 71–114.

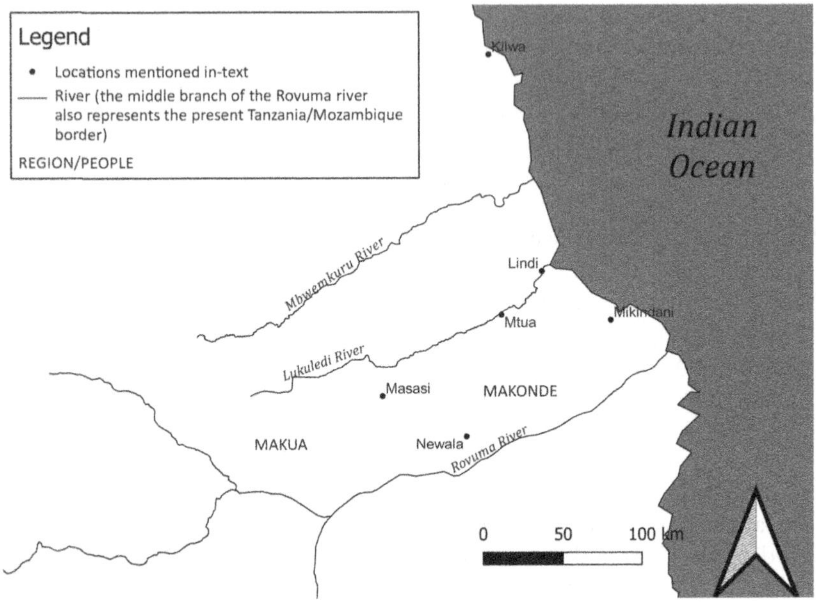

Fig. 10.5 Map of the region encapsulated by present-day southeastern Tanzania/northeastern Mozambique in the late nineteenth century. Drawn by Philip Gooding

are having good refreshing showers, and the Mtama crops [sorghum], though small, are promising.'[118]

More detail comes from the reports of journeys made by British Vice-Consul C.S. Smith. Starting in May from Kilwa Kivinje, once the major export port of the East African slave trade, Smith made several journeys exploring the southern coastal regions. Early in that month, normally the middle of the *masika* rains, Smith reported to Consul Kirk at Zanzibar that Kilwa and its agricultural hinterland was recovering from the worst effects of the drought and famine:

[118] 'Our African Postbag,' *Central Africa*, 18 (June 1884), 110; Charles Alan, 'A Letter from the Bishop,' *Central Africa*, 27 (Mar. 1885), 34; *Central Africa*, 29 (May 1885), 78.

> Two months ago I anticipated scarcity on this part of the coast, owing to the extreme lateness of the rains, which gave rise to fears that they might be altogether withheld. I am happy to be able to inform you that during the last month, and especially during the first week of my stay here, rain fell in abundance, and the crops seem now to be doing very well; even the rice, which was especially despaired of, is in some parts reviving. It is the general opinion that [the] price of food will this year be somewhat high, but I believe there is now no further reason to expect famine.[119]

The important wild rubber trade had been disrupted, but traders were optimistic that caravans would arrive soon at the coast, particularly from Donde, where the best rubber was found.[120]

Smith's first journey, to Mpuemu, a few days march to the west from Kilwa Kivinje, was marked by 'the extreme dryness of the season.'[121] His second, much longer journey through September and October 1884, was to districts along the Ruvuma River, about 400 kms south west of Kilwa Kivinje. Well aware of the famine conditions in the northern regions of what he called 'Zanzibar dominions,' Smith reported:

> Travelling was occasionally difficult on account of the scarcity of water, and we had once or twice to march more than 26 miles from one watering-place to another, only to exchange s slimy pool for a muddy pit. The season had been extraordinarily dry, and besides leaving no water in many of the rivers and wells which in ordinary years are not dried, the drought had in parts caused distress by the failure of the food supply. But there was no such suffering as was felt further north, and the scarcity seemed not to have extended more than 30 to 40 miles from the coast – indeed, to the southward of Newala the harvest was extraordinarily good. The Mtua district, where the people were selling their children for food, seemed to be in the worst state. At Lindi I was surprised to find the price of millet 50 per cent greater than at Kilwa, and double that at Zanzibar. Until the smallness of the resources of the people and the difficulties of transport

[119] HCPP 1884–85, Correspondence with British representatives and agents abroad and reports from naval officers ...Relative to the slave trade: Vice-Consul Smith to Sir J. Kirk, 5 May 1884.

[120] Smith to Sir J. Kirk, 5 May 1884.

[121] C.S. Smith, 'Explorations in Zanzibar Dominions,' Royal Geographical Society, *Supplementary Papers* (1889), 102–3.

> have been considered, the sharpness of the boundary between want and plenty is amazing.[122]

He also notes 'a considerable movement of slaves to the coast in the months of July, August, and September in 1884' which had been reported to him by UMCA missionaries and Chief Matola in Newala district. How much of this movement was considered unusual and perhaps linked to the drought is not stated. On 25 September 1884, at the Yao village of Kungwanga, on an island in the Ruvuma, he wrote:

> Food is easily obtained, there having been no drought, but it is as dear as at Kilwa, probably because of the large number of caravans that pass through. Much labour seems spent on the cultivation of the islands, which yield a good return in millet, Indian corn, rice, and tobacco.[123]

Clearly, the island security and the waters and fertility of the Ruvuma floodplains assisted the local people to maintain production in the face of a heavy demand.

Conclusion

One conclusion from this study is that it is vital that environmental historians of East Africa and other vulnerable world regions engage with the scientific literature on climate change. Looked at from another angle, climate change is not a theoretical issue, and climate scientists and others ought to look closely at the real historical experience of human societies under extreme climate stresses, as well as the physical changes in the lands that they inhabit. It is a truism to state that climate change has massive implications for human societies along with all other living creatures in terms of ecological balance, healthy habitats, landscape use, and water resources. I write as Lytton, a small town in the 'temperate' lands of central British Columbia, Canada, has just recorded a record-breaking air temperature of more than 49 degrees Celsius. The next day (1 July

[122] Smith, 'Explorations,' 103. He gives the price of millet at Lindi, in late October 1884 as 60 rupees 'a quarter.'

[123] Ibid., 114.

2021) the town was burnt to the ground in a 'volcanic' forest fire.[124] Second, even before the various branches of climate change science built a consensus that there is a close interaction between human activities and historical climate change over the last one or two centuries, East African peoples had to learn to live in a difficult and challenging physical environment, in part because of the irregularities of the Indian Ocean monsoon cycle. This chapter does not, of course, suggest that we should turn away from human responsibility including issues of colonialism, conflict, and entitlement, when we consider natural factors contributing to droughts and famines, such as volcanic eruptions. On the contrary, the impact of the Krakatau eruption on East Africans and peoples in other parts of the world and the resulting great drought and famine of 1883–1885 shows that if we in effect try to surpass the biggest volcanic eruption in modern times by continuing to pump greenhouse gases, smoke, ash, and particles from the burning of fossil fuels into the atmosphere, we will certainly cause a much larger catastrophe that threatens all life.

On the plane of documented history, it is difficult to reach secure conclusions concerning the human impact of the 1883–1885 drought and famine. Many Africans blamed western explorers, missionaries, and colonists for bringing drought with them. As we saw, in March 1884, the Maasai at Mianzini blamed Thomson. Wray wrote, when he returned to Sagalla in 1887 after being carried out in 1885, 'They accused us of having come to bewitch them.' Fearing a return to famine they urged us to 'move out of this place.'[125] Such beliefs were no doubt common. This chapter shows that the impact was widespread, and far beyond what the rather haphazard and localized historical research so far undertaken has suggested. The eastern half of east central Africa, from Lamu almost down to the Ruvuma river, was very badly affected. Research in southern Somalia and Borana territory in southern Ethiopia might show a more northerly impact. South eastern Kenya is perhaps the best documented of the most devastated regions. Merrit summarizes several estimates from the early colonial period of the demographic cost for the Taita people. In March 1887, the population of Sagalla was thought by a missionary to

[124] A term used by a forest fire expert when describing the new kinds of fires stimulated by climate change on CBC (Canadian Broadcasting Corporation) national radio, 2 July 2021.

[125] Wray to Lang, 26 Nov. 1887, in *CMI*, 13, New Series (1888), 84–85; Merrit, 'A History of the Taita,' 110. For details, see: Bravman, *Making Ethnic Ways*, 73–79.

be fewer than 1000, in contrast to a pre-famine population of 10,000. Wray suggested a loss of two thirds of the population, while a big game hunter visiting Sagalla in late December 1886 thought that only 1500 people were currently resident. On a wider scale, Sir Arthur H. Hardinge, Commissioner of the British East Africa Protectorate, in 1897 believed that the coastal province as a whole, which included Taita, had lost 'about half' of its population during the 'great famine' of 1884.[126] From these estimates, which were no more than educated guesses, and with the addition of corresponding or somewhat lower figures for the other badly affected regions further south in Tanzania, as well as in other lowland districts in central Kenya and Tanzania, we can imagine that the total loss of life through starvation and other causes must have been in the several hundreds of thousands.

There were other impacts that also had long-term consequences. Enslavement of the most vulnerable expanded after a decline from the mid-1870s, and sections of the Mijikenda, the Swahili, the Somali, and other groups were the beneficiaries. Population distribution changed as farming people abandoned their homes and trekked away from drought-stricken areas to find food and protection in less affected highlands, coastal towns, and political capitals of powerful chiefs, where they ended up as low level clients or slaves, and sometimes at mission stations. Agricultural production collapsed in some places, and fields in the dryer lowlands that had been cultivated for grain production were abandoned as there was insufficient rain and few hands to work them. High grass, bush, and tsetse replaced *shambas* previously used to diversify crop options, as we saw in regions such as Taita and Shambaai. Commerce along the main caravan routes was severely disrupted. Peasant farmers who had for decades entered the market by supplying caravans with food could no longer do so. In the wake of drought and famine came violence and raiding by the Maasai, by the Somali, in places by the Kamba, by the Mijikenda, by the Zigua, in the south by the Wagwangara (the Mbunga). These events along with European conquest set the stage for the environmental disasters of the later 1880s and 1890s.

[126] Bishop H.P. Parker to CMS, 3 Mar. 1887 in *CMI* (July, 1887), 425; John Willoughby, *East Africa and Its Big Game: Narrative of a Sporting Trip from Zanzibar to Kilimanjaro and the Borders of the Masai* (London: Longmans, Green & Co., 1889), 58, 60; Merrit, 'A History of the Taita,' 110–1.

Bibliography

Archival Sources

Church Missionary Society (CMS): Birmingham, UK.

House of Commons Parliamentary Papers 1884–85 (HCPP).

London Missionary Society (LMS): School of Oriental and African Studies, London, UK.

Published and Unpublished Sources

Alpers, Edward A. *Ivory and Slaves in East Central Africa*. Berkeley, CA: University of California Press, 1975.

Ambler, Charles H. *Kenyan Communities in the Age of Imperialism: The Central Region in the Late Nineteenth Century*. New Haven and London: Yale University Press, 1988.

Anderson, David M. 'Cultivation Pastoralists: Ecology and Economy among the Il Chamus of Baringo, 1840–1980,' in *The Ecology of Survival*, eds. Johnson and Anderson: 241–60.

Anderson, David M. 'The Beginning of Time? Evidence for Catastrophic Drought in Baringo in the Early Nineteenth Century.' *Journal of Eastern African Studies*, 10, 1 (2016): 45–66.

Arnold, David. *Famine: Social Crisis and Historical Change*. Oxford: Basil Blackwell, 1988.

Beaumont, Peter and Graham Readfearn. 'Global Heating Supercharging Indian Ocean Climate System.' *The Guardian* (19 Nov. 2019).

Bessems, Ilse, Dirk Verschuren, James M. Russell, Jozef Hus, Florias Mees, and Brian F. Cumming. 'Palaeolimnological Evidence for Widespread Late 18th Century Drought Across Equatorial East Africa.' *Palaeogeography, Palaeoclimatology, Palaeoecology*, 259 (2008): 107–20.

Bloyet, A. 'De Zanzibar à la station de Kondoa.' *Bulletin de la Société de Géographie*, 7, 11 (1890): 350–64.

Bontinck, François, ed. *L'autobiographie de Hamed ben Mohammed el-Murjebi Tippo Tip*. Bruxelles: Académie royale des Sciences d'Outre Mer, 1974.

Bravman, Bill. *Making Ethnic Ways: Communities and Their Transformation in Taita, Kenya, 1800–1950*. Portsmouth, NH: Heinemann, 1998.

Brooke, Clarke. 'The Heritage of Famine in Central Tanzania.' *Tanzania Notes and Records*, 67 (1967): 15–22.

Burton, Richard F. *The Lake Regions of Central Africa*. Michigan: Michigan Scholarly Press, 1971 [1860].

Central Africa: A Monthly Record of the Work of the Universities Mission. 1884–5.

Church Missionary Intelligencer. 1883–7.

Cooper, Frederick. *Plantation Slavery on the East Coast of Africa*. New Haven: Yale University Press, 1977.

Davis, Mike. *Late Victorian Holocausts*. London and New York: Verso, 2002.

Dawson, E.C. ed. *The Last Journals of Bishop Hannington*. London, Seeley and Co. 1888.

Dawson, E.C. *James Hannington, First Bishop of Eastern Equatorial Africa: A History of His Life and Work, 1847–1885*. London: Selley & Co., 1891.

De Cort, Gijs, Ilse Bessems, Edward Keppens, Florias Mees, Brian Cumming, and Dirk Verschuren. 'Late-Holocene and Recent Hydroclimatic Variability in the Central Kenya Rift Valley: The Sediment Record of Hypersaline Lakes Bogoria, Nakuru and Elementeita.' *Palaeogeograply, Palaeoclimatology, Palaeoecology*, 388 (2013): 69–80.

De Waal, Alex. 'A Re-Assessment of Entitlement Theory in the Light of the Recent Famines in Africa.' *Development and Change*, 21 (1990): 469–90.

Dingley, Zebulon. 'Rumor and History Revisited: 'Mumiani' in Coastal Kenya, 1945,' *Journal of African History*, 59, 3 (2018): 391–98.

Echenberg, Myron. *Africa in the Time of Cholera: A History of Pandemics from 1817 to the Present*. Cambridge: Cambridge University Press, 2012.

Feierman, Steven. *The Shambaa Kingdom: A History*. Madison: University of Wisconsin Press, 1974.

Gelorini, Vanessa and Dirk Verschuren. 'Historical Climate-Human-Ecosystem Interaction in East Africa: A Review.' *African Journal of Ecology*, 51 (2012): 409–21.

Genge, Matthew J. 'Electrostatic Levitation of Volcanic Ash into the Ionosphere and Its Abrupt Effect on Climate.' *Geology*, 46, 10 (2018): 835–38.

Getnet, Alemu. 'Revisiting the Entitlement Approach to Famine: Taking a Closer Look at the Supply Factor—A Critical Survey of the Literature.' *Eastern Africa Social Science Research Review*, 23, 2 (2007): 95–129.

Giblin, James L. 'East Coast Fever in Socio-Historical Context: A Case Study from Tanzania.' *International Journal of African Historical Studies*, 23, 3 (1990): 401–21.

Giblin, James L. 'Trypanosomiasis Control in African History: An Evaded Issue?' *Journal of African History*, 31, 1 (1990): 59–80.

Giblin, James L. *The Politics of Environmental Control in Northeastern Tanzania, 1840–1940*. Philadelphia: University of Pennsylvania Press, 1992.

Giblin, James L. and Jamie Monson, eds. *Maji Maji: Lifting the Fog of War*. Leiden: Brill, 2010.

Gissing, C.E. 'A Journey from Mombasa to Mounts Ndara and Kasigao.' *Proceedings of the Royal Geographical Society*, 6, 10 (1884): 551–66.

Glassman, Jonathon. *Feasts and Riot: Revelry, Rebellion, and Popular Consciousness on the Swahili Coast, 1856–1888*. Portsmouth, NH: Heinemann, 1995.

Gleckler, P.J., K. AchutaRao, J.M. Gregory, B.D. Santer, K.E. Taylor, and T.M.L. Wigley. 'Krakatoa Lives: The Effect of Volcanic Eruptions on Ocean Heat Content and Thermal Expansion.' *Geophysical Research Letters*, 33, 17 (2006): 1–5.

Gleckler, P.J., T.M.L. Wigley, B.D. Santer, J.M. Gregory, K. AchutaRao, and K.E. Taylor. 'Krakatoa's Signature Persists in the Ocean.' *Nature*, 439 (9 Feb. 2006).

Hartwig, Gerald W. 'Social Consequences of Epidemic Diseases: The Nineteenth Century in Eastern Africa,' in *Disease in African History*, eds. Hartwig and Patterson: 25–45.

Hartwig, Gerald W. and K. David Patterson, eds. *Disease in African History: An Introductory Survey and Case Studies*. Durham, NC: Duke University Press, 1978.

Hastenrath, Stefan. 'Variations of East African Climate During the Past Two Centuries.' *Climatic Change*, 50 (2001): 209–17.

Hastenrath, Stefan, Dierk Polzin, and Pierre Camberlin. 'Exploring the Predictability of the "short rains" at the Coast of East Africa.' *International Journal of Climatology*, 24 (2004): 1333–43.

Hobley, C.W. 'Upon a Visit to Tsavo and the Taita Highlands.' *Geographical Journal*, 5, 6, (1895): 545–61.

Hodges, Geoffrey. 'Military Labour in East Africa and Its Impact on Kenya,' in *Africa and the First World War*, ed. Melvin E. Page. New York: Palgrave Macmillan, 1987: 137–51.

Hoppe, Kirk Arden. *Lords of the Fly: Sleeping Sickness Control in British East Africa, 1900–1960*. Westport, CN: Praeger, 2003.

Hore, Annie. *To Lake Tanganyika in a Bath Chair*. London: Sampson Low, Marston, Searle & Rivington, 1886.

Huijzendveld, Frans D. 'Changes in Political Economy and Ecology in West-Usambara, Tanzania: ca. 1850-1950.' *International Journal of African Historical Studies*, 41, 3 (2008): 383–509.

Johnson, Douglas and David Anderson, eds. *The Ecology of Survival: Case Studies from Northeast African History*. London: Lester Crook, 1988.

Johnston, H.H. *The Kilima-Njaro Expedition*. London: Kegan Paul, Trench & Company, 1886.

Kimambo, Isaria N. 'The Political History of the Pare People to 1900.' Unpublished Ph.D. Dissertation: Northwestern University, 1967.

Kimambo, Isaria N. 'Environmental Control and Hunger in the Mountains and Plains of Northeastern Tanzania,' in *Custodians of the Land*, eds. Maddox, Giblin, and Kimambo: 71–95.

Kjekshus, Helge. *Ecology Control and Economic Development in East African History*. London: Heinemann, 1977.

Kollman, Paul V. *The Evangelization of Slaves and Catholic Origins in Eastern Africa*. New York: Orbis Books, 2005.
Koponen, Juhani. *People and Production in Late Precolonial Tanzania: History and Structures*. Jyväskylä: Finnish Society for Development Studies, Finnish Historical Society, Finnish Anthropological Society, 1988.
Koponen, Juhani. 'War, Famine, and Pestilence in Late Precolonial Tanzania: A Case for a Heightened Mortality.' *International Journal of African Historical Studies*, 21, 4 (1988): 637–76.
Kopytoff, Igor, ed. *The African Frontier: The Reproduction of Traditional African Societies*. Bloomington: Indiana University Press, 1987.
Kremer, Eduard. *Die unperiodischen Schwankungen der Niederschläge und die Hungersnöte in Deutsch-Ost-Afrika*. Hamburg: Archiv der Deutschen Seewarte, 1910.
Larson, Lorne. 2010. 'The Ngindo: Exploring the Center of the Maji Maji Rebellion,' in *Maji Maji*, eds. Giblin and Monson: 71–114.
Last, James T. 'A Visit to the Wa-itumba Iron-Workers and the Mangaheri, Near Mamboia, in East Central Africa.' *Proceedings of the Royal Geographical Society*, New Series, 5, 10 (1883): 581–92.
Ledoulx, M. Ch. 'Explorateurs et missionaires dans l'est de l'Afrique.' *Comptes rendus de la société de géographie de Paris* (1885): 104–6.
Maddox, Gregory. 'Mtunya: Famine in Central Tanzania, 1917–20.' *Journal of African History*, 31, 2 (1990): 181–97.
Maddox, Gregory, James L. Giblin, and Isaria N. Kimambo, eds. *Custodians of the Land: Ecology and Culture in the History of Tanzania*. London: James Currey, 1996.
McCann, James C. 'Climate and Causation in African History.' *International Journal of African Historical Studies*, 32, 2/3 (1999): 261–79.
McGregor, Shayne, Miriam Khodri, Nicola Maher, Masamichi Ohba, Francesco S. R. Pausata, and Samantha Stevenson. 'The Effect of Strong Volcanic Eruptions on ENSO,' in *El Niño Southern Oscillation*, eds. McPhaden, Santoso and Cai: 267–87.
McPhaden, Michael J., Agus Santoso, and Wenju Cai, eds. *El Niño Southern Oscillation in a Changing* Climate. Hoboken, NJ: John Wiley Sons, 2021.
Merrit, E. Hollis. 'A History of the Taita of Kenya to 1900.' Unpublished Ph.D. Dissertation: Indiana University, 1975.
Mhajida, Samwel Shanga. *The Collapse of a Pastoral Economy: The Datoga of Central and Northern Tanzania from the 1830s to the 2000s*. Göttingen University Press, 2019.
Moloney, J.A.. *With Captain Stairs to Katanga*. London: Sampson Low, Marston & Co., 1893.
Morton, Fred. *Children of Ham: Freed Slaves and Fugitive Slaves on the Kenya Coast, 1873–1907*. Boulder: Westview Press, 1990.

Mtoro bin Mwinyi Bakari. *The Customs of the Swahili People*, trans. & ed. J.W.T. Allen. Berkeley, CA: University of California Press, 1981.
Musere, Jonathan. *African Sleeping Sickness: Political Ecology, Colonialism and Control in Uganda*. Lewiston, NY: Edwin Mellen Press, 1990.
Nicholson, Sharon E. 'A Semi-Quantitative, Regional, Precipitation Dataset for Studying African Climates of the Nineteenth-Century. Part I Overview of the Data Set.' *Climatic Change*, 50 (2001): 317–53.
Nicholson, Sharon E. 'Climate and Climatic Variability of Rainfall Over Eastern Africa.' *Reviews of Geophysics*, 55, (2017): 590–635.
Nicholson, Sharon E., Douglas Klotter, and Amin K. Dezfuli. 'Spatial Reconstruction of Semi-Quantitative Precipitation Fields Over Africa During the Nineteenth Century from Documentary Evidence and Gauge Data.' *Quaternary Research*, 78 (2012): 13–23.
Ó Gráda, Cormac. *Famine: A Short History*. Princeton, NJ: Princeton University Press, 2009.
Osmani, S.R. 'Comments on Alex de Waal's Re-Assessment of Entitlement Theory in the Light of Recent Famines in Africa.' *Development and Change*, 22, 3 (1991): 587–96.
Pawelczak, Marek. *The State and the Stateless. The Sultanate of Zanzibar and the East African Mainland: Politics, Economy and Society, 1837–1888*. Warszawa: Instytut Historyczny Uniwersytetu Warsawskiego, 2010.
Rigby, Peter. *Cattle and Kinship Among the Gogo: A Semi-Pastoral Society of Central Tanzania*. Ithaca and London: Cornell University Press, 1969.
Rockel, Stephen J. 'Caravan Porters of the *Nyika*: Labour, Culture and Society in Nineteenth Century Tanzania.' Unpublished PhD Dissertation: University of Toronto, 1997.
Rockel, Stephen J. *Carriers of Culture: Labor on the Road in Nineteenth-Century East Africa*. Portsmouth, NH: Heinemann, 2006.
Rockel, Stephen J. 'Forgotten Caravan Towns in Nineteenth Century Tanzania: Mbwamaji and Mpwapwa.' *Azania*, 41, (2006): 1–25.
Rubin, Olivier. 'The Entitlement Approach: A Case for Framework Development Rather Than Demolition.' *Journal of Development Studies*, 45, 4 (2009): 621–40.
Schmied, Doris. 'Managing Food Shortages in Central Tanzania.' *GeoJournal*, 30, 2 (1993): 152–58.
Schröder, Wilfried. 'The Krakatoa Event and Associated Phenomena: A Historical Review.' *Earth Sciences History*, 21, 2 (2002): 199–206.
Sen, Amartya. *Poverty and Famines: An Essay on Entitlement and Deprivation*. Oxford: Oxford University Press, 1983.
Sheriff, Abdul. *Slaves, Spices and Ivory in Zanzibar: Integration of an East African Commercial Empire into the World Economy 1770–1873*. London: James Currey, 1987.

Sheriff, Abdul and Engseng Ho, eds. *The Indian Ocean: Oceanic Connections and the Creation of New Societies*. London: Hurst and Co., 2014.
Shetler, Jan Bender. 'Interpreting Rupture in Oral Memory: The Regional Context for Changes in Western Serengeti Age Organization (1850–1895).' *Journal of African History*, 44, 3, (2003): 385–412.
Simpkin, Tom and Richard S. Fiske, eds. *Krakatoa 1883: The Volcanic Eruption and Its Effects*. Washington, DC: Smithsonian Institution Press, 1983.
Simpson, George L. and Peter Waweru. 'Becoming Samburu: The Ethnogenesis of a Pastoral People in Nineteenth-Century Northern Kenya.' *Journal of the Middle East and Africa*, 3 (2012): 175–97.
Sissons, Carol Jane. 'Economic Prosperity in Ugogo, East Africa, 1860–1890.' Unpublished Ph.D. Dissertation: University of Toronto, 1984.
Smith, C.S. 'Explorations in Zanzibar Dominions.' *Royal Geographical Society, Supplementary Papers* (1889): 101–25.
Speke, John Hanning. *Journal of the Discovery of the Source of the Nile*. New York: Harper, 1864.
Spinage, Clive Alfred. *African Ecology: Benchmarks and Historical Perspectives*. Berlin: Springer, 2012.
Stanley, R. and A. Neame, eds. *The Exploration Diaries of H. M. Stanley*. London: W. Kimber, 1961.
Sunseri, Thaddeus. 'Famine and Wild Pigs: Gender Struggles and the Outbreak of the Majimaji War in Uzaramo (Tanzania).' *Journal of African History*, 38, 2 (1997): 235–59.
Sunseri, Thaddeus. 'The Entangled History of *sadoka* (Rinderpest) and Veterinary Science in Tanzania and the Wider World, 1891–1901.' *Bulletin of the History of Medicine*, 89, 1 (2015): 92–121.
Symons, G.J. ed. *The Eruption of Krakatoa and Subsequent Phenomena: Report of the Krakatoa Committee of the Royal Society*. London: Harrisons & Sons, 1888.
Tagseth, Mattias. 'The Expansion of Traditional Irrigation in Kilimanjaro, Tanzania,' *International Journal of African Historical Studies*, 41, 3 (2008): 461–90.
Ten Raa, Eric. 'Bush Foraging and Agricultural Development: A History of Sandawe Famines.' *Tanzania Notes and Records*, 69 (1968): 33–40.
Thomson, Joseph. *Through Masai Land*. London: Sampson Low, Marston, Searle and Rivington, new and revised edition, 1887.
Thomson, Joseph. *To the Central African Lakes and Back*. London: S. Low, Marston, Searle, & Rivington, 1881.
Vaughan, Megan. *The Story of an African Famine*. Cambridge: Cambridge University Press, 1987.
Verschuren, Dirk. 'Lake-Based Climate Reconstruction in Africa: Progress and Challenges.' *Hydrobiologia*, 500, 1 (2003): 315–30.

Verschuren, Dirk, John Tibby, Peter R. Leavitt, and C. Neil Roberts. 'The Environmental History of a Climate-Sensitive Lake in the Former "White Highlands" of Central Kenya.' *Ambio*, 28, 6 (1999): 494–501.

Verschuren, Dirk, Kathleen R. Laird, and Brian F. Cumming. 'Rainfall and Drought in Equatorial East Africa during the Past 1,100 Years.' *Nature*, 403, 6768 (2000): 410–14.

Wakefield, E.S. *Thomas Wakefield: Missionary and Geographical Pioneer in East Equatorian Africa*. London: The Religious Tract Society, 1904.

Waller, Richard. 'Emutai: Crisis and Response in Maasailand 1883–1902,' in *The Ecology of Survival*, eds. Johnson and Anderson: 73–112.

Waller, Richard. 'Tsetse Fly in Western Narok, Kenya.' *Journal of African History*, 31, 1 (1990): 81–101.

Watt, Rachel Stuart. *In the Heart of Savagedom*. London: n.d. [1912].

Webel, Mari K. 'Ziba Politics and the German Sleeping Sickness Camp at Kigarama, Tanzania, 1907–14.' *International Journal of African Historical Studies*, 47, 3 (2014): 399–423.

Webel, Mari K. *The Politics of Disease Control: Sleeping Sickness in Eastern Africa, 1890-1920*. Athens, OH: Ohio University Press, 2019.

Webster, J.B., ed. *Chronology, Migration and Drought in Interlacustrine Africa*. Halifax, Nova Scotia: Dalhousie University Press, 1979.

Willis, Justin. 'The Nature of a Mission Community: The Universities Mission to Central Africa in Bonde.' *Past and Present*, 140 (1993): 127–54.

Willis, Justin. *Mombasa, The Swahili, and the Making of the Mijikenda*. Oxford: Oxford University Press, 1993.

Willoughby, John. *East Africa and Its Big Game: Narrative of a Sporting Trip from Zanzibar to Kilimanjaro and the Borders of the Masai*. London: Longmans, Green & Co., 1889.

Wolff, Christian, Iris Kristen-Jenny, Georg Schletter, Birgit Plessen, Hano Meyer, Peter Dulski, Rudolph Naumann, Achim Brauer, Dirk Verschuren, and Gerald H. Haug. 'Modern Seasonality in Lake Challa (Kenya/Tanzania) and Its Sedimentary Documentation in Recent Lake Sediments.' *Limnology and Oceanography*, 59, 5 (2014): 1621–36.

CHAPTER 11

'A Drought so Extraordinary': The 1911 ENSO and Disaster Nationalism in the American Colonial Philippines

Theresa Ventura

For Filipinos, the year 1911 started with a bang and ended with a whimper. Taal Volcano, located on *Isla ng Bulkan* (Volcano Island) in Luzon's southern Batangas Province, began a powerful week-long eruption on January 27. The volcano's short and squat stature produced a pyroclastic flow that, rather than send lava down the sides of the mountain, spewed molten ash, toxic fumes, and water vapor at speeds of 80 km an hour. Gas, steam, and ash reaching temperatures of 400 degrees Fahrenheit scalded and suffocated those in the way while a volcanic tsunami on Lake Taal's western shore claimed entire communities. An estimated 1200–2000 people lost their lives by the week's end. With at least 702 carabao dead and thick ash covering 2000-square kilometers of

T. Ventura (✉)
Department of History, Concordia University, Montreal, QC, Canada
e-mail: theresa.ventura@concordia.ca

P. Gooding (ed.), *Droughts, Floods, and Global Climatic Anomalies in the Indian Ocean World*, Palgrave Series in Indian Ocean World Studies, https://doi.org/10.1007/978-3-030-98198-3_11

farmland, countless more faced lost livelihoods.[1] One silver lining, wrote the Jesuit meteorologist Miguel Saderra Masó, was that fields were not yet planted in rice. Masó looked forward to the mid-year rainy season in which those fields, once cleansed of ash, 'will probably yield even more crops than before.'[2] But the rains did not come. Instead, a 'moderate' El Niño Southern Oscillation (ENSO) contributed to what another Jesuit with the Weather Bureau characterized as 'a drought so extraordinary that there is ground for the assertion that it was the severest ever observed in the Archipelago.'[3] This statement is borne out by raw rain gauge data in the European Climate and Assessment Dataset. In (from North to South) Aparri, Tuguegarao, Baler, Manilla, and Legazpi, the 1911 drought is either the deepest or one of the deepest on record.[4] The eruption that rained ash and fire in January had, by October, given way to the slow anxiety of waiting for rain (Fig. 11.1).

Whether Taal's eruption and the ENSO-related drought constitute two events in a single global climatic anomaly or were two distinct phenomena is unclear. Several climatological studies have investigated the possibility that sulfur-rich volcanic eruptions trigger positive ENSO anomalies, which then contribute to rainfall anomalies in diverse Indian Ocean World regions. Krakatau's 1883 eruption is often held as an example of a large eruption preceding an ENSO.[5] But correlation in timing is not the same as causation. More recent eruptions of comparable strength, such as El

[1] This description is drawn from: Miguel Saderra Masó, *The Eruption of Taal Volcano, January 30, 1911* (Manila: Bureau of Printing, 1911), 31; Charles Martin, 'Observations on the Recent Eruption of Taal Volcano,' *Philippine Journal of Science* (1911), 88; PJ Wester, 'The Situation in the Citrus District of Batangas,' *Philippine Agricultural Review* 6, 3 (1913), 127.

[2] Saderra Masó, *The Eruption of Taal Volcano*, 20.

[3] José Coronas, S.J., *The Extraordinary Drought in the Philippines, October 1911 to May 1912* (Manila: Bureau of Printing, 1912), 3. 'Moderate' according to: Joëlle L. Gergis and Anthony Fowler, 'A History of ENSO Events Since A.D. 1525: Implications for Future Climate Change,' *Climatic Change*, 92, 3 (2009), 368.

[4] Data taken from the European Climate and Assessment Dataset, which is searchable and freely available at: https://climexp.knmi.nl/selectstation.cgi?id=someone@somewhere [Accessed 8 Jan. 2021]. The Manila dataset starts in the 1860s, though is patchy for parts of the twentieth century, and stops in 1975. Most of the other datasets start in the 1900s and are largely continuous until the early twenty-first century.

[5] For the possible effects of volcanic eruption and ENSO on rainfall in the IOW 1883–1884, see: Rockel, this volume; Matthew S. Hopper, 'Cyclones, Drought, and Slavery: Environment and Enslavement in the Western Indian Ocean, 1870s to 1920s,' in *Natural*

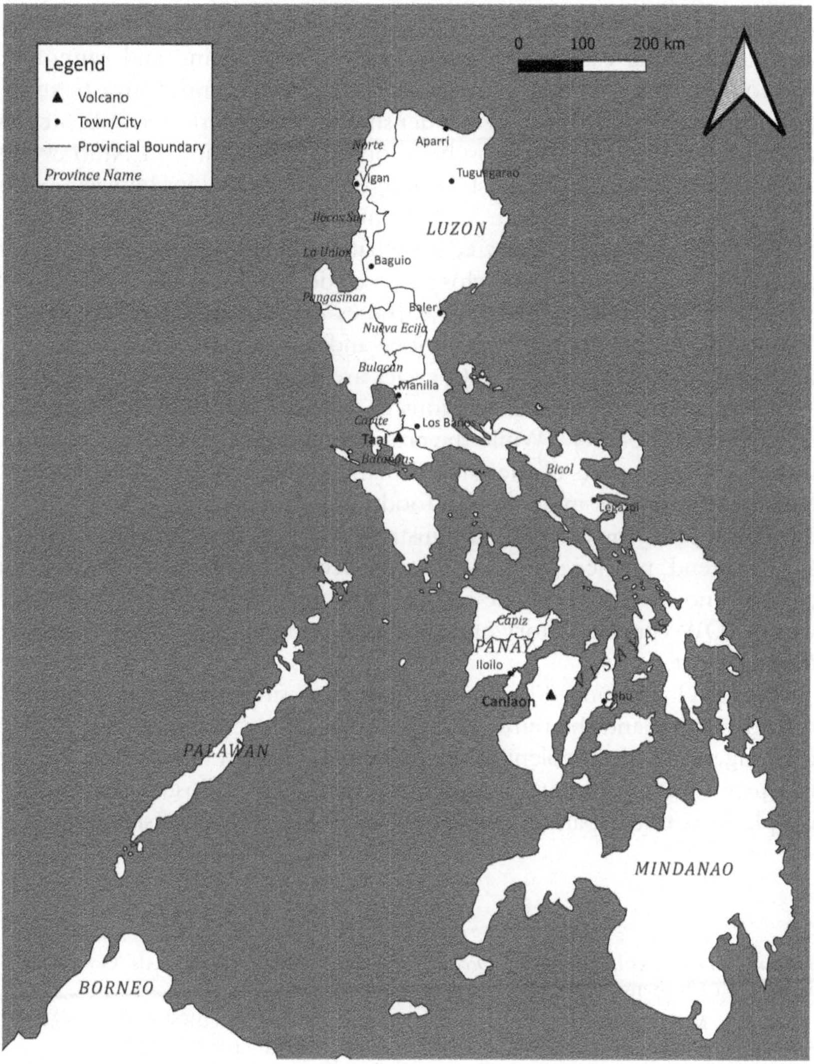

Fig. 11.1 Map of the Philippines showing places and regions mentioned in text. Drawn by Philip Gooding

Chichón, 1982 and Pinatubo, 1991 occurred as sea surface temperatures in the equatorial east-central Pacific were already rising and cannot be said to have triggered their corresponding El Niño events.[6] Investigations into the relationship between the density of atmospheric aerosols such as volcanic ash, known as 'stratospheric optical depth,' and El Niño events are also inconclusive. A 1997 study judged the correlations between peaks in stratospheric optical depth and El Niño events over the last 150 years to be 'what would be expected by chance,' while others using longer term proxy records suggest a closer correlation between high amounts of aerosols, SSTs, and the El Niño.[7] Studies accounting for different variables such as the timing, magnitude, and latitude of the volcano and ocean pre-conditions and ENSO cycles are ongoing, though none so far comment specifically on the relationship between the 1911 Taal eruption and that year's ENSO.[8] While the eruption and drought under investigation may be two parts of the same climatic event, here I treat each as two contributing factors in an overall food yield decline that, in turn, fueled a political crisis for colonial administrators and large landowners. Doing so, I contend, privileges the lived experiences of historical actors over our current knowledge of climate science, thus furthering the 'cultural turn' within IOW climate history and may also bring climate into histories of development.

The 1911 eruption—Taal's twentieth-sixth since the dawn of Spanish record keeping and the third since the American occupation—was neither the longest nor most violent in the volcano's recorded history. But the late nineteenth-century capitalist agriculture and land enclosure that pushed people into once sparsely populated areas like Volcano Island made it

Hazards and Peoples in the Indian Ocean World: Bordering on Danger, eds. Greg Bankoff and Joseph Christensen (New York: Palgrave, 2016), 268.

[6] S. Self, M.R. Rampino, J. Zhao, and M.G. Katz, 'Volcanic Aerosol Perturbations and Strong El Niño Events,' *Geophysical Research Letters*, 24, 10 (1997), 1247–50.

[7] Ibid., 1247; Masamichi Ohba, Hideo Shiogama, Tokuta Yokohata, and Masahiro Watanabe, 'Impact of Strong Tropical Volcanic Eruptions on ENSO Simulated in a Coupled GCM,' *Journal of Climate*, 26, 14 (2013), 5169–70.

[8] See, for example: Evgeniya Predybaylo, Georgiy Stenchikov, Andrew T. Wittenberg, and Sergey Ospisov, 'El Niño/Southern Oscillation Response to Low-Latitude Volcanic Eruptions Depends on Ocean Pre-conditions and Eruption Timing,' *Communications Earth and Environment* (2020), 1–12.

Taal's deadliest.[9] Survivors were hungry, and their immediate need placed a burden on already food short neighboring municipalities and Cavite Province. The eruption, then, exposed the thin line separating food sufficiency and scarcity in the political ecology of the American colonial Philippines. The drought pushed Batangas and portions of Luzon and the Visayas across that line. The positive ENSO anomaly that is associated with this drought is the last in a series of strong and highly destructive El Niño's that transcended the final third of the nineteenth- and the first quarter of the twentieth centuries. The other most significant El Niños in this context occurred during the 1860s and in 1877–1878, 1883–1884, 1888–1890, 1896, 1902, 1918–1919, and 1925.[10] Apart from in the Philippines, the 1911 ENSO event is also associated with drought in western India and in mainland southeast Asia, as well as with floods in the Yangzi River basin in China and with above-average rainfall in equatorial eastern Africa.[11] In subsequent years, the anomaly became stronger, reaching 'very severe' levels in 1912–1915, contributing to, for example, the worst drought of the twentieth century in Indonesia and Papua New Guinea.[12]

Although its effects were significant and widespread in the IOW, the 1911 ENSO event is not nearly as well known as others from around this period. Mike Davis' highly influential study on the links between El Niño events, European imperial policies in monsoon Asia, and the deaths of 50 million colonial subjects in 1877, for instance, is limited to the nineteenth century.[13] Studies approaching ENSO in the twentieth century

[9] Taal raged for nearly seven months in 1754 but, with no direct settlement on *Pulong Bulkan,* only four towns near Lake Bombón/Taal were destroyed. Saderra Masó, *The Eruption of Taal Volcano*, 18, 26.

[10] See, for example: Mike Davis, *Late Victorian Holocausts: El Niño Famines and the Making of the Third World* (London: Verso, 2001); Richard Grove and John Chappell (eds.), *El Niño: History and Crisis* (Cambridge: White Horse Press, 2000); Richard Grove and George Adamson, *El Niño in World History* (London: Palgrave Macmillan, 2018), 93–104; chapters by Clarence-Smith, Gooding, Rockel, Warren, and Williamson in this volume. For a Middle Eastern case study, see: Zozan Pehlivan, 'El Niño and the Nomads: Global Climate, Local Environment, and the Crisis of Pastoralism in Late-Ottoman Kurdistan,' *Journal of the Economic and Social History of the Orient*, 63, 3 (2020), 316–56.

[11] For mainland southeast Asia, see: Williamson, this volume.

[12] Grove and Adamson, *El Niño in World History*, 183.

[13] Davis, *Late Victorian Holocausts.*

have focused on India and other core regions of European empires to the exclusion of the Philippines and, hence, of the United States as a transpacific imperial formation.[14] A more holistic approach to ENSO and analysis of the 1911 event necessitates a re-alignment of regional focus toward the Philippines and Southeast Asia.[15] In the Philippines, accounts gleaned from the Spanish-language press (translated and archived by the colonial government) along with figures from the Bureaus of Agriculture and Weather, reveal widespread crop failures in central and northern regions caused by drought. Mindanao and the southern Philippines were largely spared. The central and northern crop failures threw the consequences of American tariff and public works policies favoring the export of sugar and abaca and the import of rice into relief. Furthermore, the macro-regional effects of the El Niño limited American officials' capacities to secure food relief through importation. Drought in Indochina and floods in the Yangzi River Valley created widespread rice crop shortages in the eastern IOW. As the press documented the movement of food short refugees from country to city, landed and educated elites read mass mobility, abandonment of labor contracts, and rising thefts as evidence of imminent social decay. Both press and elites held Americans responsible. Americans, they charged, were indifferent to Philippine lives, ignorant of tropical nature, and too incompetent to plan for weather anomalies.

Assembly and press criticisms of the US response to drought formed the outlines of a Philippine disaster nationalism that made an ability to plan for environmental crises a criterion of rule and the absence of hunger into a measure of good governance.[16] Disaster nationalism folded longer-standing frustrations with environmental management institutions like the Agriculture Bureau into anger over the lackluster US response to Taal's eruption but also reflected new ways of thinking about national space and future planning opened by meteorology. Whereas Americans measured Philippine progress in terms of long-distance trade, nationalist editors equated sovereignty with archipelago-wide sufficiency in rice. Advances in

[14] For a summary, see: Grove and Adamson, *El Nino in World History*, 183–34.

[15] See also: Williamson, this volume.

[16] On a parallel recasting of hunger in South Asia, see: James Vernon, *Hunger: A Modern History* (Cambridge: Harvard University Press, 2007) esp. Ch. 3; Sunil Amrith, *Unruly Waters: How Rains, Rivers, and Coasts Have Shaped Asia's History* (New York: Basic Books, 2018); Janam Mukherjee, *Hungry Bengal: War, Famine, and the End of Empire* (New York: Oxford University Press, 2015).

meteorology, meanwhile, suggested that state experts had an obligation to plan for weather anomalies. Planning for weather and food sufficiency came together in the nationalist movement for rice colonies on the southernmost island of Mindanao, which had a different experience of the positive ENSO. Americans responded to charges of mismanagement and neglect by both denying the trade imbalances that exacerbated rice shortages and, over the longer term, turning to technical solutions for food production. Philippine disaster nationalism and the American technological interventions in food production it facilitated re-merged in the 1960s to produce the high yield variety rice (HYV) strain, IR8, the aggressive spread of which constituted the 'Green Revolution' in the Philippines and India.

Most often understood as an American effort to combat communism and the poverty and food insecurity on which it fed, the Green Revolution saw the non-profit Rockefeller Foundation align with the US State Department and governments in the Philippines, India, Pakistan, Malaysia, and Indonesia to bring high yield variety (HYV) rice seeds to impoverished farmers. State actors in Asia, in turn, embraced HYV seeds as symbols of modernity. Indeed, the Philippines' own Benigno Aquino pointed to a paddy planted with the HYV called IR8 and told an American reporter in 1966, 'Here is the jumbo jet! The twentieth century!'[17] But this framing, as Prakash Kumar writes, positions agricultural modernization as an 'externally inspired attack on the problem of poverty' administered by a network of development experts and postcolonial elites at the expense of more indigenous and local forms of knowledge.[18] Wedding IOW climate histories to histories of development, I suggest, draws our attention from the 'external' toward how different social classes, empires, and postcolonial states have historically responded to weather anomalies. The Green Revolution, from this perspective, is not simply a grand project born of the global Cold War but instead the culmination of a particular kind of techno-nationalism that

[17] Joseph Lelyveld, 'Philippines Tries New Rice Strain,' *New York Times* (18 Dec. 1966), quoted in: Nick Cullather, 'Miracles of Modernization,' 227. Cullather makes a very strong case that the success of the modernization project rested on the power of seeds as symbols of modernity.

[18] Prakash Kumar's contribution to the forum: 'Roundtable: New Narratives of the Green Revolution,' *Agricultural History*, 91, 3 (2017), 401. See also: Raj Patel, 'The Long Green Revolution,' *Journal of Peasant Studies*, 40 (2013), 1–63.

emerged in response to early twentieth-century struggles with drought, food shortages, social unrest, and Euro-American colonialism. This attention to the *longue durée* of climate is one way to answer Raj Patel's call for histories of the Green Revolution that question its accepted origin story and claims to success. Patel's call is even more urgent as planners, policymakers, and philanthropists issue their calls for a second Green Revolution to tackle a host of ills stemming from the current climate crisis.[19]

To unpack these environmental, political, and intellectual connections and transformations, this essay begins not with drought but with the tensions of collaborative colonialism and environmental management. Bureaus of Weather and Agriculture encouraged Filipinos to lay down their arms and work with American administrators toward mutually beneficial profit. Though these institutions failed on their own terms, they incorporated Filipino students and 'barefoot' meteorologists into knowledge production and in the process opened new ways of imagining national space and the obligations of government toward subjects and citizens. The essay next turns to the 1911–1912 drought to assess its Philippine and regional impact before considering the political struggles over its meaning in the Philippines.

Collaborative Colonialism and the Environmental Management State

The American navy invaded Manila Bay in the spring of 1898 on the premise that it was liberating Filipinos from an especially corrupt Spanish empire. What became 'the black legend' in the United States—fantastic tales of Spanish venality and barbarism—included the failure to subdue nature among Spain's many crimes.[20] Americans were encouraged in this view by a host of authors for whom the War of 1898 was a publishing

[19] On this point, see: Patel, 'The Long Green Revolution'; Amrith, *Unruly Waters*, Ch. 8.

[20] Many journalists and authors emphasized this theme. See, for example: Murat Halstead, *Pictorial History of America's New Possessions* (Chicago: HL Barber, 1899); Benjamin Kidd, *The Control of the Tropics* (New York: Macmillan, 1898); Trumbull White, *Our New Possessions* (Chicago: National Education Union, 1899). A good discussion of the impact of these works with regard to the racialization of Filipinos is: Warwick Anderson, 'The Natures of Culture: Environment and Race in the Colonial Tropics,' in *Nature in the Global South: Environmental Projects in South and Southeast Asia*, eds. Paul Greenough and Anna Lowenhaupt Tsing (Durham: Duke University Press, 2003), 29–46.

opportunity. The Manila-born naturalized US citizen Ramon Reyes Lala accused Spaniards of 'look[ing] upon nature with a lazy eye, troubling himself little about anything that cannot be put to some immediate use.' His Philippines was 'in a large measure unexplored, waiting for the botanist to discover their treasures, the poet to sing their beauties, the practical man to develop their resources.'[21] Once explored and exploited, predicted the British arm-chair naturalist Benjamin Kidd, Americans would find 'that it is in the tropics, and not in the temperate zones, that we have the greatest food-producing and material producing regions of the earth.'[22] In the imperial imagination, then, mastery over nature was part of the civilizing mission and the tropics, a vast reserve of untapped materials to turn into commodities.[23] To this end, botanists, scientific foresters, and agronomists crossed the Pacific with the occupying army. Their expectations of virgin land unmet, these would-be experts came to depend on Spanish and Philippine scientific knowledge and institutions. Among the most indispensable was the *Observatorio de Manila*.

The Manila Observatory (f. 1865 as the Observatorio de Ateneo) was part of a global network of Jesuit observatories located outside of Europe. Jesuit work and education in meteorology, seismology, geography, and botany helped the order rebuild its international prestige upon the end of its worldwide suppression in 1814. Simultaneous advances in telegraphy raised the possibility that Jesuit observatories in areas as far-flung as Guatemala City (1851), Calcutta, India (1866), Zikawei, China (1872), and Tananarive, Madagascar (1890) could be linked to forecast the path

21 Ramon Reyes Lala, *The Philippine Islands* (New York: Continental Publishing Co., 1898), 151. Theresa Ventura, '"I Am Already Annexed": Ramon Reyes Lala and the Crafting of "Philippine" Advocacy for American Empire,' *Journal of the Gilded Age and Progressive Era*, 19, 3 (2020).

22 Benjamin Kidd, *The Control of the Tropics* (New York and London: The Macmillan Company, 1898), 84.

23 I am working with David Arnold's definition of 'tropicality'—the imperial socio-spatial discourse of human development that held the heat, humidity, and natural fertility of the tropics produced indolent 'natives' incapable of forming advanced civilizations. See: David Arnold, *The Problem of Nature: Environment, Culture and European Expansion* (Cambridge, MA: Blackwell, 1996). On tropicality and Americans in the Philippines, see also: Warwick Anderson, *Colonial Pathologies: American Tropical Medicine, Race, and Hygiene in the Philippines* (Durham: Duke University Press, 2006); Anderson, 'The Natures of Culture.'

of Atlantic- and Pacific-originated storms on the islands of the Caribbean and South China Seas. In Manila, the potential benefits that daily weather reports and storm warnings offered to overseas shipping earned the Observatory's director, Federico Faura, the financial support of private merchants, the Dutch Consul, and the Hong Kong and Shanghai Bank. Faura used the donated funds to acquire what one historian has called 'the finest array of astronomical and geophysical instruments in the colonized world.'[24] Merchants earned dividends on their investment in the form of fifty-three typhoon warnings between 1879 and 1882.[25] Jesuit climatologists also trained staff on the model farms established by the *Comisión Agronomica de Filipinas*' in the use of meteorological instruments.[26] The science of meteorology and the economic practices of capitalism, in this sense, were mutually constitutive institutions arising in 'an imperial context of global exchange.'[27]

The networks that linked meteorologists to bankers, shippers, and large planters ultimately helped Faura's successor, the Spanish-born and Georgetown-trained José Algué, weather the tumultuous transition from Spanish to American rule. Merchants and British officials protested loudly when the US Navy cut the undersea cable linking Manila to Hong Kong and banned Algué from issuing storm advisories. Properly chastened, Commodore George Dewey hosted Algué on the flagship *USS Olympia* where he pledged his 'trust that the United States government will make the necessary provisions for the continuance of the institution which you

[24] Gregory T. Cushman, 'The Imperial Politics of Hurricane Prediction: From Calcutta and Havana to Manila and Galveston, 1839–1900,' in *Nation-States and the Global Environment: New Approaches to International Environmental History*, eds. Erika Marie Bsumek, David Kinkela, and Mark Atwood Lawrence (New York: Oxford University Press, 2013), 146. On the relationship between meteorological instruments and Jesuit institutional authority in the Philippines, see: Kerby C. Alvarez, 'Instrumentation and Institutionalization: Colonial Science and the Obervatorio Meteorológico de Manila, 1865–1899,' *Philippine Studies*, 64, 3–4 (2016).

[25] Augustín Udías, *Searching the Heavens and the Earth: The History of Jesuit observatories* (Dordecht: Kluwer Academic, 2003); John Schumacher, 'One Hundred Years of Jesuit Scientists: The Manila Observatory, 1865–1965,' *Philippine Studies*, 13, 2 (1965).

[26] Alvarez, 'Instrumentation and Institutionalization,' 403.

[27] Lukas Rieppel, Eugenia Lean, and William Deringer, 'Introduction: The Entangled Histories of Science and Capitalism,' *Osiris*, 33, 1 (2018), 1.

conduct in such an able manner, and which has proved itself to be so great a benefit to maritime interests in this part of the world.'[28]

Dewey's pledge marked the beginning of a collaborative colonialism centered around environmental management institutions. Jesuits supplied Americans with an intimate knowledge of Philippine weather in exchange for financial support and the further expansion and institutionalization of their networks. Immediately after the *Olympia* meeting, the fathers at the *Observatorio* shared their notes on weather variations across the islands. The US federal government then published the notes as *Climatología de Filipinas*, which was translated into English and republished in the first report of the Philippine Commission to the US President in 1901.[29] Charged with investigating conditions in the islands and making recommendations for its governance, the seven-member presidentially appointed Philippine Commission also recommended that Aglué oversee an archipelago-wide Weather Bureau. Like the colony's new Bureaus of Science, Agriculture, Forestry, and Land, the Weather Bureau was modeled on the federal institutions that advanced US continental expansion, settler colonialism, and commercial agriculture. But unlike the others, the Weather Bureau was the only Philippine institution to remain under Jesuit control with a majority Filipino workforce. The Weather Bureau thus became an important training ground for Philippine meteorologists and scientific workers.[30]

The colonial environmental management institutions served important functions in both the United States and the Philippines. Within the United States, displays by each Bureau at the 1904 St. Louis World's Fair explained and legitimized the occupation to an otherwise wary public. The Weather Bureau's relief maps, intricate forecasting instruments, and thirty-meter-high galvanized towers emphasized the technological improvements the colonial state brought to the Philippines otherwise

[28] George Dewey (2 Feb. 1899), quoted in: Cymbeline R. Villamin, ed., *Biographies of Early Scientists in the Philippines, Vol. 1* (National Science Development Board: Manila, 1976 and 1978).

[29] José Coronas, *The Climate and Weather of the Philippines, 1903 to 1918* (Manila: Bureau of Printing, 1920), 14.

[30] James F. Warren, 'Scientific Superman: Father José Algué, Jesuit Meteorology, and the Philippines Under American Rule, 1897–1924,' in *Colonial Crucible: Empire in the Making of the Modern American state*, eds. Alfred W. McCoy and Francisco A. Scarano (Madison, WI: University of Wisconsin Press, 2009), 509.

cast as backward.[31] The Agriculture Bureau's display of economic botany offered visitors a chance to taste the fruits of empire. There they learned that Filipinos were an 'essentially agricultural people working in a very primitive fashion.' That the land produced at all was 'proof of the favorable character of the climate and the natural richness of the soil, which in many places seems to be practically inexhaustible.'[32] Or, as W. Cameron Forbes, a visitor to St. Louis before becoming a Philippine Commissioner, that 'American methods' combined with the 'wonderful fertility of the soil' would turn the islands into a 'veritable garden spot.'[33] Reference works of Philippine forest products, meanwhile, invited botanical entrepreneurs to invest in the archipelago's nascent timber industry, coconut plantations, and copra refining mills.[34]

The environmental management institutions also furthered the Pacific and Indian Ocean reach of the United States. In terms of network building, agronomists, botanists, and crop specialists seized on the Philippine occupation as a chance to visit European botanical outposts in Java and Sri Lanka. 'If the American agricultural and forestry administration in these islands are to succeed,' urged the botanist Elmer D. Merrill, 'both must be established on practically the same lines followed by the English and Dutch in their colonial possessions.'[35] The information exchanged included advice on carving forest reserves out of jungle and settling migratory people into disciplined forest users. While botanical exchanges furthered the personal ambitions of experts, collaboration with Jesuit meteorologists provided the outline of what would become an American-controlled Pacific-wide telecommunications infrastructure after the Second World War. The Philippine Commission funded

[31] Ibid., 509. Algué also contributed maps of the distribution of rainfall to the Panama Pacific International Exposition in 1915. See: Coronas, 'The Climate and Weather of the Philippines' (1920), 15.

[32] Dean C. Worcester, 'Present State of Agriculture,' *Report of the Philippine Commission*, Part 1 (1902), 297.

[33] W. Cameron Forbes to Hamilton Fish (7 Dec. 1907), W. Cameron Forbes Papers, MS 1366, vol. 6, letter 392, Houghton Library, Harvard University, Cambridge, MA.

[34] Brendan Luyt, 'Reading the Minor Forest Bulletins of the Philippine Bureau of Forestry: A Case Study on the Role of Reference Works in the American Empire of the Early Twentieth Century,' *Information and Culture*, 53, 1 (2018), 43.

[35] Elmer D. Merrill, 'Report of the Botanist' and 'Report of the Botanist on the Royal Botanical Gardens of Ceylon and the Botanical Gardens at Singapore,' *Philippine Commission Reports*, 600.

Algue's request for fourteen new observatory stations throughout the archipelago, each linked by telegraph and wireless cable to Manila, Hong Kong, and Japan. By the inter-war period, this networked infrastructure enhanced the work of the Pan Pacific Science Congress and the Institute of Pacific Relations. These institutions, writes Tomoko Akami, placed the 'regulation and management of the Pacific' under the auspices of American experts, and contributed to the carving of an American-led 'Asia Pacific Region.' By the postwar period, the networked infrastructure of meteorology contributed to the production of the world as a unit of knowledge.[36]

Within the Philippines, the environmental management institutions attracted landowners toward the colonial state by promising stability and increased profitability. The Land Bureau's registration laws encouraged large landowners and small farmers to lay down their arms by granting legal title to their land claims.[37] The Science Bureau's mobilization against rinderpest, the cattle epizootic that devastated 85% of the water buffalo population in three waves between the 1880s and 1910s, promised to decrease the high price of the draft animal necessary for wet rice agriculture.[38] The Agriculture Bureau appealed to planters by sending 'improved' seeds gathered from the United States, Buitenzorg, and Peradeniya through the mail.[39] Many planters already saw themselves as scientific agriculturalists and readily assisted with information gathering. The Visayan sugar planter Juan Araneta is a case in point. Perhaps best known for directing US forces to suppress rebellious sugar

[36] Tomoko Akami, 'The Open Ocean for Interimperial Collaboration: Scientists' Networks Across and in the Pacific Ocean in the 1920s,' in *Ocean Archives, Indigenous Epistemology, and Transpacific American Studies*, eds. Yuen Shu, Otto Heim, and Kendall Johnson (Hong Kong: Hong Kong University Press, 2019), 153; Paul N. Edwards, 'Meteorology as Infrastructural Globalism,' *Osiris*, 21, 1 (2006).

[37] Theresa Ventura, 'From Small Farms to Progressive Plantation: The Trajectory of Land Reform in the American Colonial Philippines, 1900–1916,' *Agricultural History*, 90, 4 (2016), 460.

[38] Daniel F. Doeppers, 'Fighting Rinderpest in the Philippines, 1886–1941,' in *Healing the Herds: Disease, Livestock Economies, and the Globalization of Veterinary Medicine*, eds. Karen Brown and Daniel Gilfoyle (Athens, OH: Ohio University Press, 2010).

[39] Richard A. Overfield, 'Science Follows the Flag: The Office of Experiment Stations and American Expansion,' *Agricultural History*, 64, 2 (1990); Elmer D. Merrill, 'Report of the Botanist,' 600; James H. Shipley, 'Report of the Farm Machine Expert. Report of the Chief of the Bureau of Agriculture for the Year Ending August 31, 1902. Exhibit C,' *Philippine Commission Reports*, 605.

workers led by the indigenous spirit medium, Dionisio Sigobela, Araneta cemented his relationship with Americans through rituals of collaborative environmental management.[40] He ceded his oversight of Negros' *La Granja Modelo* to the Agriculture Bureau before the official end of the war, donated his seed library to US census takers in 1903, hosted a visitor from the United States Department of Agriculture, and treated a small group of US geologists to an expedition on Mount Canalon's sulfur springs. Seventy assistants were on hand to serve five Americans. Araneta also renamed two haciendas 'Louisiana' and 'California'—the American agricultural export states he most admired.[41] Again, profit through the application scientific agriculture was the main attraction. Or, as one Bureau official simply stated to an audience of the Capiz Agriculture Society, 'One of the most important questions that is before the people of Capiz Province is, "How can we make more money?"'[42]

The environmental management bureaus ultimately failed to deliver on their promise of mutual profit through scientific mastery, and instead became places in and about which Americans and both urban and rural Filipinos engaged in what Paul Kramer calls a 'multivalent discourse on capacity.'[43] As American administrators shifted the benchmarks for an eventual independence further into the future, Filipino nationalists asserted their immediate right to rule through a criticism of these institutions. The Agriculture Bureau's inability to eradicate rinderpest led nationalist press editors to rechristen it 'the Bureau of Failure.' *La Democracia*, once an avowedly pro-American paper, concluded that the 'plague

[40] Alfred McCoy, 'Sugar Barons: Formation of a Native Planter Class in the Colonial Philippines,' in *Plantations, Proletarians and Peasants in Colonial Asia*, eds. E.V.B. Daniel, Henry Bernstein, and Tom Brass (London: Frank Cass, 1992), 121–23.

[41] On Araneta's census assistance, see: *Census of the Philippine Islands*, vol. IV (1905), 168; On the expedition, see: Warren D. Smith, *Geology and Mineral Resources of the Philippine Islands* (Manila: Bureau of Printing, 1924), 167. On his assistance to the USDA and oversight of the Negros model farm, see: Alonzo Stewart, 'Agricultural Conditions in the Philippine Islands,' presented to USDA Secretary James Wilson in 1903. Published in serial set of the 60th Congress, 1st Session, Document 535, 23.

[42] R.L. Chute, 'The Importance of Agricultural Education for Capiz Province. Delivered April 13, 1909,' *Philippine Agricultural Review*, II, 7 (July 1909), 405.

[43] Paul Kramer, *Blood of Government: Race, Empire, the United States, and the Philippines* (Chapel Hill: University of North Carolina Press, 2006), 288. See also: Ruby Paredes, ed., *Philippine Colonial Democracy* (Quezon City: Ateneo de Manila University Press, 1989).

of "experts" imported from the United States proves more disastrous for the country than the calamities which they seek to exterminate.'[44] Landowners and the well-to-do challenged US priorities and projects in the Philippine Assembly—a representative body elected by a limited all-male franchise beginning in 1907. Though the Commission retained full veto power over Assembly legislation, symbolic votes funding large-scale rice irrigation suggested that some were uncomfortable with the American contention that rice, to quote one British merchant living in the islands, was a 'low civilization' crop best imported rather than grown locally.[45] In early January 1911, just a week before Taal's eruption and months before the drought, a 'League of all Patriots without Distinction of Party, Race, or Sect' formed in Manila. Its goal, according to *El Comercio*, was to 'defend the political rights and material interests of the Filipinos,' the most basic of which was the right to 'properly prepare the budgets without killing the people with hunger.'[46]

It is harder to gauge the impact of collaborative colonialism and environmental management institutions on the intellectual and cultural life of Filipinos outside of the landholding and merchant classes. Just as Bureaus

[44] 'Serum and the Experts,' *La Democracia* (15 Apr. 1911), in Elliott Papers, Box 3, Bound Volume: 'Translations from Spanish and Filipino Newspapers.' All newspaper excerpts come from Elliott's bound volumes. On American and Philippine press relations, see: Carson Taylor, *History of the Philippine Press* (No publisher: Manila, 1927); Sheila S. Coronel, 'The Media, the Market, and Democracy: The Case of the Philippines,' *The Public*, 28, 2 (2001); Glòria Cano, 'Filipino Press Between Two Empires: *El Renacimiento*, a Newspaper with Too Much *alma Filipina*,' *Southeast Asian Studies*, 49, 3 (Dec. 2011).

[45] Frederic Sawyer, *The Inhabitants of the Philippines* (Charles Scribner's Sons: New York, 1900), 130.

Bataan Representative JM Lerma presented a petition from Cerferino Tiangco, Julian Calimbos, Pedro Paquio, and Felipe de los Reyes for irrigation in Pilar, Bataan, Committee Report, No. 18. Governor General James Smith returned the petition with a request for more information, *Journal of the Philippine Commission (JPC)*, 2, 1, 32. The Commission unanimously approved AB No. 233, 'An Act to Authorize the appropriation of P750,000 annually for the promotion, establishment, and maintenance of irrigation systems in the Philippine Islands,' on 10 June 1908 but rejected AB No. 241, 'An Act Authorizing municipal and provincial governments to grant, under certain conditions, privileges, and concessions for the utilization of public waters for agricultural irrigation and for other purposes,' on the basis that it did not have the power to decree water use and water rights. *JPC*, 2, 1, 302, 305, and 320.

[46] 'League of All Patriots Without Distinction of Party, Race, or Sect, to Defend the Political Rights and Material Interests of the Filipinos Grievously Neglected by Those Most Called to Support Them,' *El Comercio* (10 Jan. 1911).

of Agriculture, Lands, and Forestry drew in and alienated landowners toward and from the colonial state, the institutions undoubtedly shaped material life as new titles and proof of land ownership reclassified the poor as squatters. Yet while many were pushed to the margins of the state, others were incorporated into the state through the practice of what we would today call 'crowd-sourced sciences.' Children attending the colony's extensive English-language public schools filled out questionnaires about crop and soil conditions in their communities, thereby learning how to identify information deemed important to agronomists. Beginning in 1908, student answers composed the new *Philippine Agricultural Review's* monthly crop reports. The Weather Bureau also trained scores of rural Filipinos to monitor local conditions at more than seventy field stations across the archipelago. 'Barefoot meteorologists,' writes James F. Warren, completed decades of service and often remained on duty for up to fourteen hours a day over periods of three to four days during storms.[47] The work of reporting on local crops and weather to a central authority in Manila may have affirmed that those local spaces belonged to a larger national unit. The 'numbers- and instruments-based' reading of meteorology, argues Kerby C. Alvarez, suggested that the causes of inclement weather were not moral but instead rational, cyclical, and predictable.[48] Taken together, the environmental management state and the calculative imagination of meteorology may have created an expectation of state action during weather anomalies. The American failure to prepare for and act within the face of disaster gave rise to a Philippine disaster nationalism.

The crises precipitated by Taal's eruption and the subsequent drought made 1911 a pivotal year in the articulation of disaster nationalism. After the eruption, Assembly fights moved to the politics of relief. Assembly

[47] Warren, 'Scientific Superman,' 513.

[48] Alvarez, 'Instrumentation and Institutionalization,' 395; Ibid., 517; James F. Warren, 'Philippine Typhoons, Sources and the Historian,' *Water History*, 7, 2 (2015), 225–26. Mark Elvin calls the attribution of weather to human behavior and/or spirits 'moral meteorology.' See: Mark Elvin, 'Who Was Responsible for the Weather? Moral Meteorology in Late Imperial China," *Osiris*, 13, 1 (1998). See also: Chapter by Schottenhammer, this volume. At the same time, it is important not to overestimate the providentialism of past interpretations of the weather and the secularism of present. See: Gregg Bankoff, 'In the Eye of the Storm: The Social Construction of the Forces of Nature and the Climatic and Seismic Construction of God in the Philippines,' *Journal of Southeast Asian Studies*, 35, 1 (2004).

members proposed appropriating P100,000 for disaster relief; commissioners refused to authorize more than P25,000. The nationalist press condemned the paltry amount. According to *La Democracia*, when 'in the presence of a scene of misery and much tears which urgently require the sacrifice of the most rebellious self-love,' American Commissioners instead chose 'to establish supremacy, to show power.'[49] *El Ideal* charged that the Commission's 'passivity in the disaster' was due, 'in part, to the fact that the dead and wounded are only Filipinos. If the tragedy had occurred elsewhere in the United States and the Government had showed the same attitude the roar of indignation of the people would have made the White House tremble on its foundations.'[50] Another editorial proposed that volcanic eruptions, earthquakes, and typhoons constituted 'an invisible thread, which united common interests and makes one those who were born under the same sky.'[51] From this perspective, the islands were one despite the American contention that multiple languages, ethnic groups, and so-called tribal units spread out across a diverse terrain foreclosed on the possibility of Philippine nationhood. Far from a land inhabited by warring tribes kept at peace by a benevolent but firm American presence, Philippine-led responses to the natural disaster were proof that there was a 'public spirit [that] in private and social matters watches over its own.'[52]

Rice Famine and Disaster Nationalism in the ENSO of 1911

The Commission's tight-fisted response to Taal was a mere foreshadowing of its response to the drought. Neither commissioners nor experts in the environmental management bureaus foresaw that the year's monsoon rains would be weak. This is understandable. While nineteenth and early twentieth century meteorologists pioneered the study of cyclonic storms, they did not scientifically identify and name the ENSO until the 1920s. However, Observatory meteorologists and journalists did recognize signs of drought in the summer months as meteorological practice at the time

49 'About the Failure of a Bill,' *La Democracia* (7 Feb. 1911).

50 'They... and We,' *El Ideal* (7 Feb. 1911).

51 'There Is a Public Spirit,' *El Ideal* (11 Feb. 1911).

52 Ibid.

entailed recording and comparing daily temperatures and atmospheric pressure to previous years. The Jesuit José Coronas noted that, except for several strong typhoons in the eastern Philippines, the 1911 rainy season was weaker and shorter, and its temperatures on average higher. Eastern storms, he warned, masked what appeared to be drought.[53] An April heatwave in the island province of Cebu had already caused the maize crop to fail and, as reported by the Iloilo paper *El Tiempo*, people were hungry.

> The intense heat of the time, the dry-season and an infinity of causes and joint causes that we all know, have destroyed the little crops and have made unable to those who own small parcels of land to cultivate them. The maize which is the main food of the poor is very scarce and has its highest price, and in such circumstances fatally combined for the misfortunes of the countrymen, urge an implacable and terrific sinister image of the famine.[54]

The Cebuano paper *El Revolucion* reported that 'the huts of the mountains are going to be abandoned' by people who would make their way to the port city 'with their pale faces and empty stomachs.'[55] 'Famine,' another paper warned that same month, 'is imminent.'[56]

The April food scarcities in the Visaysas spread to Luzon when rain practically ceased in October 1911. It did not return until December 1912. Between that time, Manila's Central Observatory collected a scant 94.6 mm of rain—an amount far smaller than what Coronas deemed a more 'normal' 580.6 mm . This average, of course, obscures variation by locality. Vigan, in Ilocos Sur, experienced an unprecedented 165 consecutive days without rain. Temperatures in Manila hovered around 36 °C for sixteen days in April and May of 1912. The last comparable stretch of dry and hot weather occurred in 1889—a year we now recognize as an El Niño year.[57] The 1911–1912 drought, Coronas wrote several years later, was 'the worst ever experienced since the foundation of the Manila Observatory in 1865.'[58]

[53] Coronas, *Extraordinary Drought*, 4.

[54] 'The Famine in Cebu,' *El Ideal* (12 Apr. 1911).

[55] *La Revolucion*, quoted in 'The Famine in Cebu,' *El Tiempo* (12 Apr. 1911).

[56] 'Fratricidal Silhouette,' *El Ideal* (18 Apr. 1911).

[57] Coronas, *Extraordinary Drought*, rainfall averages, 3; Vigan, 9; temperatures, 5.

[58] Coronas, *The Climate and Weather of the Philippines*, 115.

The Agriculture Bureau's monthly crop reports documented the local impact of the drought. The central Visayan Island of Panay lost an estimated 50% of that season's rice crop. Ambos Camarines (now the provinces of Camarines Norte and Sur) saw a 60% drop in its annual yield. In southeastern Luzon's Bicol region, the province of Albay lost its entire crop of mountain rice, which would have met the needs of the upland population 'without the necessity of buying imported grain.' Just northwest of Bicol, Batangas was still recovering from Taal's eruption when drought struck. According to the *Philippine Agricultural Review* (*PAR*), 'many people in different sections are living on root and forage.' In neighboring Cavite, the rice crop was 'ruined by the continuous drought, even those lands under irrigation having suffered for lack of water.' The food crops of Laguna's San Pedro Tunasan municipality, the *PAR* also reported, was 'completely killed by the drought.'[59] Further north, yields in Central Luzon's Pangasinan and Nueva Ecija Provinces—Manila's 'rice bowl'—were down by one-third and two-thirds, respectively. To the north and west, Ilocos Sur and Norte, along with La Union, also lost significant acreage to drought.[60]

Competing demand for Indochinese rice meant that neither merchants nor colonial state officials could secure imports as local crops failed. Heavy rainfall from May through July in central and eastern China led to flooding along the Yangzi River. The valley's entire rice crop was destroyed, bringing the grim reality of famine to hundreds of thousands of people.[61] The floods spurred the Japanese government to eliminate duties on rice imports, allowing merchants to create stores of the grain. Faced with their rice shortages, authorities in Indochina contemplated

[59] 'Monthly Crop Conditions—November 1911,' *Philippine Agricultural Review* (hereafter: *PAR*) V, 2 (1912), 106, 107.

[60] Daniel F. Doeppers, *Feeding Manila in Peace and War* (Madison: University of Wisconsin Press, 2016), 91.

[61] Kathryn Edgerton-Tarpley, 'Tough Choices: Grappling with Famine in Qing China, the British Empire, and Beyond,' *Journal of World History*, 24, 1 (2013).

a ban on imports by September of 1911.[62] Philippine prices skyrocketed as a result. In Bulacan Province, a *cavan* of *palay* (approximately 130 pounds of un-husked rice) rose from P3.10 to P4.90. The Spanish language daily *La Vanguardia* noted that such a 'jump… had never been seen in this locality.'[63] Cavan prices in drought-gripped Vigan reached P7 in November 1911.[64]

The drought, local shortages, and high rice prices raised the specter of famine and, with that, an elite panic. Already in October, *La Vanguardia* reported, 'the house of the wealthiest man in [Gapang, Nueva Ecija], Mr. Simeón Linsañgan, was assaulted and his storehouse plundered by persons, perhaps famished, who are still unknown and have not yet been captured.'[65] The paper foresaw more 'robberies, forays, assaults, and other violations of the law.' But perhaps most ominously for the landowning planter class, the paper also predicted an increase in the already many pending cases before the courts for 'non-performance of contract, which are attributed to this extraordinary crisis in which the price of rice is rising, not from hour to hour, but from minute to minute.'[66] Adequate food supplies was the basis on which social stability, including landlord-tenant relations, depended.

As the crisis posed by Taal gave way to drought, the Philippine Assembly and nationalist press called on Philippine Commissioners to act. Weather forecasting had bestowed on the state a responsibility to plan for dearth. 'Why,' an editorial in *La Vanguardia* asked, did Commissioners 'not foresee that the rice crop would one day not be sufficient equal to consumption?'[67] *La Democracia* praised Japan for its accumulation of rice

62 This timeline of events comes from the diary of Philippine Commissioner Charles B. Elliott (8 Oct. 1911), see Box 1, Folder 2, Charles B. Elliott Papers, Library of Congress, Washington, DC. More research in French sources is needed to substantiate whether or not Indochinese authorities banned exports. The diaries of US officials and editorials in the Philippine press indicate the ban was imposed. US Consular reports are less clear. For claims that the ban was contemplated but not imposed, see: The 1912 Consular Report of Great Britain's Foreign Office, 10. For a report on rice crop conditions in Indochina in 1911 see: The Daily Consular and Trade Reports produced by the United States Bureau of Foreign and Domestic Commerce, Vol. 1, 1–75, 903.

63 'The Present Scarcity of Rice,' *La Vanguardia* (8 Sept. 1911).

64 Coronas, *Extraordinary Drought*, 5.

65 'The Present Scarcity of Rice,' *La Vanguardia* (18 Sept. 1911).

66 Ibid.

67 'I Told You So!' *La Vanguardia* (20 Sept. 1911).

stores, calling its government 'more foresighted than ours.'[68] Seeking to follow Japan's example, the Assembly passed legislation authorizing the government to purchase and distribute rice. Commissioner Charles B. Elliott found the bill 'poorly drawn' and its proposed state intervention in private markets 'vicious in principle.' He 'objected to it going through without amendment' and the bill was vetoed.[69]

Nationalist editors identified American understandings of tropical nature as a cause of the subsistence crisis, talking back to the assumptions of imperial tropicality clearly and loudly. They charged Americans, especially Forbes, with the inability to imagine the tropics as anything but an Eden. According to *El Comercio*, the Governor-General's Philippines were 'a happy Acadia which has sprung up at the invention of the magic rod of the Executive. Who can resist believing in this felicity, which is offered us? Let us believe, let us believe, let us enjoy and read once more the message of our governor! "Nature has been lavish in her gifts to these islands." "Exports and imports have increased more than 90%."' If Forbes' Philippines were indeed a garden, the paper asked, 'Is this not a veritable paradise without a serpent?'[70] The serpent, in this case, were policies that favored export agriculture over food crops for domestic consumption. In the words of a *La Democracia* editorial, it was American colonialism that had brought a 'falling-off in the production of rice, a falling-off which has been increasing since the American occupation.' By holding US tariffs, public works priorities, and an overall misunderstanding of tropical nature responsible for the crisis, La *Democracia* concluded, 'we cannot but recognize that [the rice shortage] is not due to indolence and slothfulness but to the change which has turned our way of living, which has turned out activities into other channels and towards other sources of production.'[71] Disaster nationalists claimed foresight and technical knowledge for Filipino elites and rescued peasant cultivators from myths of the lazy native. In this reversal, it was Americans who had descended on the tropics in search of an ease enabled by the exploitation of others. As *La Vanguardia* assailed, 'the clamor of the people has not reached the yacht on which the gentlemen of the payroll

[68] 'Face to Face,' *La Democracia* (18 Sept. 1911).

[69] Elliott, Diary (8 Oct. 1911).

[70] *El Comercio* (16 Oct. 1912).

[71] 'A Serious Question,' *La Democracia* (28 Sept. 1911).

have just made a trip for pleasure and investigation. The music on board drowns the imprecations of the multitude which supports the state.'[72]

Nationalist criticism ultimately gave rise to the sentiment that Philippine sufficiency in rice and independence were entwined along with a wholesale rejection of the colonial economic policies shrouded in the language of civilization. So many assurances 'that the importation of rice was a sure sign of reigning prosperity, an incontrovertible demonstration of the purchasing power of the country,' *La Vanguardia* charged, had turned the staple into 'an article of luxury.' 'Official prosperity,' to paraphrase the editorial's title, meant 'popular misery.'[73] *La Democracia* accused US tariff and public works policies of tethering the Philippines to the American market for its exports and Indochina for imports. Instead of rising 'up to Java and other rice countries which are provided with irrigation systems,' Americans forced Filipinos to pay 'tribute to Saigon and other rice countries.'[74] In the paper's rendering, agricultural modernization was not the stick by which Americans should measure Philippine readiness for independence. The continued reliance on draft animals rather than machinery was a sign that Americans had failed Filipinos.

The Philippine Commission finally secured rice imports in September, though the archival record is silent on where the rice originated. The cost of the imports—just over 11,000 tons cost a staggering P900,000—may offer some insight into the regional impact of the 1911 ENSO-related drought. As Fiona Williamson's contribution to this volume demonstrates, the drought persisted in parts of mainland Southeast Asia. Coupled with the floods in the Yangzi River basin, the 1911 ENSO event likely contributed to rice crop shortages and corresponding price rises across a wide expanse. The cost of the rice, ongoing revenue shortages in the Insular Treasury, and an American reluctance to extend charity meant that the colonial government was intent on selling the relief rice rather than distributing it for free. Rice entering the ports of Manila, Iloilo, and Cebu was sent to municipal councils with the instruction that it should be distributed through a ticketing system. Consumers could purchase no more than a *ganta* (approximately three quarts of dry rice/just under one kilo) at a time. The system capped the price of a

72 'Rice and Meat,' *La Vanguardia* (14 Sept. 1911).

73 'Official Prosperity and Popular Misery,' *La Vanguardia* (20 Sept. 1911).

74 'Let Us Reflect a Little,' *La Democracia* (20 Sept. 1911).

ganta at 36 centavos, just slightly above the estimated average cost of rice during normal rainfall. According to Forbes, the relief rice solved the crisis: 'The people began purchasing freely, panic was averted, and those who had tried to take advantage of the food shortage to enrich themselves were blocked.'[75] But for many in the nationalist press, the Commission's measures were 'tardy, too tardy, perhaps.' Sheltered on the decks of their yachts or in the summer capital, Philippine Commissioners had waited until the predictable 'fatal, imposing, crushing crisis arrived.'[76] In October of 1911, Assembly delegates passed AB 601, 'An Act to prevent distress which may be caused by the excessively high price and probable shortage of rice.' The measure once again asserted that foresight was necessary to assure an adequate food supply in the future by granting the authorities to increase rice imports at the first sign of shortages. It also asserted that either provincial or municipal governments—the local state rather than the private market or the Commission—were best suited to distribute rice. Though not stated explicitly, the act insinuated that many also held private rice wholesalers, in particular Chinese wholesalers, responsible for the shortages—an association that would increase as Filipinos replaced departing American bureaucrats in the 1920s and 1930s.[77] The Assembly's Bill 1039, proposed in February of 1912, obligated the insular government to import 15,000 tons of rice that year as a preventative measure.[78] Rice, each of the bills asserted, was too precious to leave to the free market. Scholars of development generally identify the North China Famine of 1920–1921 as the moment in which international relief efforts laid the basis for famine prevention. But these Assembly bills indicate that proposals for addressing the systemic causes of food shortages began at least a decade earlier, and not at the impetus of Americans.[79]

Accounts of American colonialism in the Philippines rarely take the environment and linkages to other regions in the IOW Indian Ocean

[75] Forbes, Journal IV, 425–26 (23 July 1911).

[76] 'The Situation,' *El Ideal* (22 Sept. 1911).

[77] *Journal of the Philippine Commission*, V, 1 (16 Oct. 1911), 28, 32.

[78] AB 1039, 'An Act to Prevent Distress Among the People of the Philippine Islands from Failure of Food Staples,' 1 Feb. 1912, *Journal of the Philippine Commission*, 656.

[79] David Ekbladh, *The Great American Mission: Modernization and the Construction of an American World Order* (Princeton: Princeton University Press 2010), 27–30.

world into consideration. This is due, in part, to the American reaction to the drought and rice crisis. Neither commissioners nor Governor-General Forbes acknowledged the severity of rice crop failures in 1911 and 1912. Forbes approached the press coverage of the 'rice famine' as a personal attack. He sought to reign in what he considered 'typical Filipino newspaper abuse of the government.'[80] After an alleged 'firm talk' with a co-owner of *El Ideal*, he noted in his journal that the paper's next editorial, 'discussed the rice situation... and remarked that the government had done its parts in remedying the situation.'[81] Forbes also barred *La Vanguardia's* journalists from daily press briefings for its 'long series of abusive and unjust attacks culminating in insult[.]'[82] In his view, the crisis ended once his administration secured relief rice. That the insular treasury incurred an additional $900,000 of debt to purchase and distribute over 11,000 tons of rice was simply proof of his munificence.[83] American officials at the Bureau of Agriculture viewed the crisis as a chance to reassert their expertise. 'The present shortage of rice,' the new Director of Agriculture Frederick Taylor lectured in the *Philippine Agricultural Review*, 'ought to teach the Philippine people a lesson.' It had been foolish to depend on rice as a staple grain in the first place. As a crop, rice was fickle and weak—'more or less subject to damage by heavy rains and the consequent floods' or indeed 'any cause that changes the peculiar requirements of lowland rice.' Corn, Taylor continued, was the ideal supplement and the only thing standing in the way of its widespread consumption was 'custom.' Beginning in 1911, his Bureau of Agriculture combined forces with the Bureau of Education to wage the 'Corn Campaign'—'a campaign,' as he told a group of teachers assembled in Baguio, 'for a better food supply of the masses.'[84] When these campaigns failed to change tastes, BA administrators turned to improve the yields of rice seeds. The changed focus moved agricultural 'improvement' from the

[80] Forbes, *The Philippine Islands*, vol. II, 487. Forbes included examples of this abuse in his appendix.

[81] Ibid., 73, 75.

[82] Ibid., 74, 77. See also *Journal*, IV. 433 (7 Aug. 1911).

[83] Forbes, *Philippine Islands*, vol. II, 73.

[84] Taylor, 'Speech to the 1912 Teachers' Conference in Baguio,' *Philippine Craftsman* 1, 1 (1912), 54–55. On the corn campaign, see: Ventura, 'Medicalizing *Gutom*'; Glenn Anthony May, 'The Business of Education in the Colonial Philippines,' in *Colonial Crucible*, eds. McCoy and Scarano.

farm to the laboratory. The University of the Philippines College of Agriculture at Los Baños conducted much of this work, eventually partnering with Cornell University. The infrastructure that emerged out of the 1911 ENSO-drought led the Rockefeller Foundation, with prodding by the US Central Intelligence Agency, to choose Los Baños as the site of the International Rice Research Institute (IRRI) in the late 1950s. Officially opened in 1961, IRRI developed the HYV rice seeds that American Cold War planners hoped would fight communism by ending rural hunger and poverty—a Green Revolution to detract from one in red.[85]

Assembly members, landowners, and other elites also seized on the rice shortages to pursue longer term goals. Among the most prominent was the colonization of Mindanao by displaced people from Ilocano- and Tagalog-speaking provinces. Then part of 'Moro Province,' Mindanao had been under US military rule since 1898; the province sent no representatives to the Assembly. But despite this administrative segregation, US officials pursued the same commercial export policies as those in Luzon and the Visayas. Assembly members countered American control by proposing nine bills pertaining to the establishment of agricultural colonies on Mindanao between 1907 and 1911.[86] None received the support of American commissioners until 1911. The justification now depended on a new geopolitical imagination opened by the meteorological imagination. With Mindanao positioned just south of the typhoon belt, it did not have the same flash flood ecology taking shape in Luzon and the Visayas. As one BA administrator noted, 'a general crop failure throughout all three regions of the archipelago will very rarely occur during the same year.'[87] Commissioner Newton Gilbert urged Forbes to see, 'the two-fold purpose of increasing the food supply of the Islands and of encouraging immigration of people from the more populous and

[85] Nick Cullather, 'Miracles of Modernization: The Green Revolution and the Apotheosis of Technology,' *Diplomatic History*, 28, 2 (2004), 227–54. See also: Nick Cullather, *The Hungry World: America's Cold War Battle Against Poverty in Asia* (Cambridge, MA: Harvard University Press, 2010); Patel, 'The Long Green Revolution,' 14.

[86] Nobutaka Suzuki, 'Upholding Filipino Nationhood: The Debate Over Mindanao in the Philippine Legislature, 1907–1913,' *Journal of Southeast Asian Studies*, 44, 2 (2013), 275–77.

[87] *Philippine Commission Report* (1914), 376.

sterile districts into those parts of the Archipelago where nature has been more bounteous in her gifts.'[88] Mindanao, in short, offered liberation from Indochinese rice imports. Its place in Philippine nationalism grew in accord with the tightening supply of rice and a nationalism that equated sufficiency in food with political independence.

Conclusion

Taal's January 1911 eruption and that year's ENSO may be two parts of one related climatic phenomenon in which volcanic ash in the atmosphere changed sea surface temperatures resulting in rainfall anomalies in the IOW. In the Philippines, as in much of mainland Southeast Asia, drought prevailed even as overly abundant rainfall in the Yangzi River basin brought floods. While climate historians continue to debate the exact nature of the relationship between atmospheric volcanic ash, SSTs, and the ENSO, approaching the eruption and the drought as a political crisis mirrors the on-the-ground experience of historical actors and therefore allows us to consider the interactions between climate and culture. In the American colonial Philippines, the ENSO-related drought and rice crop shortages further delegitimized American claims to mastery over tropical nature and its right to rule. The wider IOW context is important here. As the Philippines experienced rice shortages, which themselves were exacerbated by disastrous American colonial grain policies and by the aftermath of the Taal eruption in Batangas, food relief was hard to come by, as adverse weather contributed to shortages elsewhere as well.

The drought and rice shortages of 1911 and 1912 resulted in novel conversations about tropical agriculture, food needs, and the relationship of the islands to one another. The nationalist press held Americans responsible for hunger and in the process articulated a disaster nationalism in which an intimate knowledge of nature became a criterion of government. Weathering the crucibles inflicted by volcanos, earthquakes, typhoons, and drought had made Filipinos one people despite distance and linguistic difference. Disaster nationalism also framed sufficiency in rice as a goal of independence. Drawing on the geopolitics of meteorology, disaster nationalists imagined Mindanao as a granary for Luzon and the Visayas. To this end, the Philippine Assembly pushed the Commission to support

[88] Gilbert to Forbes, *Journal of the Philippine Commission*, 1913, 23, cited in: Suzuki, 'Upholding Filipino Nationhood,' 283.

and coordinate the migration of landless Ilocano and Tagalog cultivators to the south. Americans, meanwhile, continued to deny the severity of the drought by casting corn as an alternative staple crop before conceding to the improvement of rice seeds. Disaster nationalism also raised possibilities not taken by either Americans or Filipino rice landlords. Though the Philippine Assembly and press editors proposed that state funding for large-scale irrigation could make Philippine rice production competitive with that of Indochina and Java, this simply did not happen. Instead, disaster nationalists reiterated the rhetorical template blaming 'outsiders' for rice shortages. This tendency grew in accord with the power of Central Luzon's rice landlords to extract high rents from tenants and high *palay* prices from millers and merchants. When the 1918–1919 ENSO resulted in region-wide rice crop failures, a ban on Indochinese exports, and skyrocketing Philippine prices, landlords and the nationalist press directed public anger toward Chinese merchants. Rice landlords, in turn, strengthened their influence over national policy. Their discouragement of rice imports, while continuing to extract high rents from cultivators, had the effect of keeping prices high. Rather than bringing the country toward self-sufficiency, the rhetoric that began after 1911 maintained starvation. The pattern, writes Yoshihiro Chiba, which was repeated in 1935, saw rice landlords once again blame Chinese merchants for high prices.[89]

The nationalist push for sufficiency in rice and the technology of high yield variety seeds re-merged in the postwar Philippines during what became known as the 'Green Revolution.' Here, the Philippine government, the Rockefeller Foundation and American Cold War planners built upon ideas and the research infrastructure developed in the wake of the 1911–1912 food shortages. Rockefeller officials chose the University of the Philippines' College of Agriculture at Los Baños to host the International Rice Research Institute because agronomists at Los Baños had been working on increasing the yield of annual rice crops. Techno-nationalist governments in India, Pakistan, Indonesia, Malaysia and elsewhere embraced the seeds developed by IRRI as part of their own quest for food sufficiency after colonial deprivation and in response to the specter of shortages during the drought of 1965–1966. The global Green Revolution raised the total output of grains but increased the

[89] Yoshihiro Chiba, 'The 1919 and 1935 Rice Crises in the Philippines: The Rice Market and Starvation in American Colonial Times,' *Philippine Studies*, 58, 4 (2010), 535.

need for fertilizer, pesticide, and water and, according to some, increased rural poverty and urban migration. While advocates of a second 'Green Revolution' point to the first as a way to combat hunger and climate change, wedding the *longue durée* of climate history to a *longue durée* analysis of the idea of a Green Revolution, helps, as Raj Patel writes, 'in understanding the transformations at hand as a decades-long complex of discourse, technology, state power, class politics, national and international relations, private investment, cultural intervention, education and ecological change.'[90] Attention to environmental teleconnections and the origins of techno-nationalist responses may better inform weather mitigation and famine relief policies moving forward.

Bibliography

Manuscripts

Charles B. Elliott Papers, MS 19687, Library of Congress, Washington DC.

W. Cameron Forbes Papers, MS Am 1365, Houghton Library, Harvard University, Cambridge, MA.

Published and Unpublished Sources

Akami, Tomoko. 'The Open Ocean for Interimperial Collaboration: Scientists' Networks Across and in the Pacific Ocean in the 1920s,' in *Ocean Archives*, eds. Y. Shu, O. Heim, and K. Johnson. Hong Kong: HKU Press, 2019: 149–71.

Alvarez, Kerby C. 'Instrumentation and Institutionalization: Colonial Science and the Obervatorio Meteorológico De Manila, 1865–1899.' *Philippine Studies: Historical and Ethnographic Viewpoints*, 64, 3–4 (2016): 385–416.

Amrith, Sunil. *Unruly Waters: How Rains, Rivers, Coasts and Seas Have Shaped Asia's History*. New York: Basic Books, 2018.

Anderson, Warwick. 'The Natures of Culture: Environment and Race in the Colonial Tropics,' in *Nature in the Global South: Environmental Projects in South and Southeast Asia*, eds. Paul Greenough and Anna Lowenhaupt Tsing. Durham: Duke University Press, 2003: 29–46.

Anderson, Warwick. *Colonial Pathologies: American Tropical Medicine, Race, and Hygiene in the Philippines*. Durham: Duke University Press, 2006.

[90] Patel, 'The Long Green Revolution,' 2.

Arnold, David. *The Problem of Nature: Environment, Culture and European Expansion*. Cambridge, MA: Blackwell, 1996.
Bankoff, Greg. 'In the Eye of the Storm: The Social Construction of the Forces of Nature and the Climatic and Seismic Construction of God in the Philippines.' *Journal of Southeast Asian Studies*, 35, 1 (2004): 91–111.
Brown, Karen, and Daniel Gilfoyle. *Healing the Herds: Disease, Livestock, and the Globalization of Veterinary Medicine*. Athens: Ohio University Press, 2010.
Bsumek, Erika Maria, David Kinkela, and Mark Atwood Lawrence, eds. *Nation-States and the Global Environment*. New York: Oxford University Press, 2013.
Cano, Gloria. 'Filipino Press Between Two Empires: *El Renacimiento*, a Newspaper with Too Much *alma Filipina*.' *Southeast Asian Studies*, 49, 3 (2011): 395–430.
Chiba, Yoshihiro. 'The 1919 and 1935 Rice Crises in the Philippines: The Rice Market and Starvation in American Colonial Times.' *Philippine Studies: Historical and Ethnographic Viewpoints*, 58, 4 (2010): 523–55.
Chute, R.L. 'The Importance of Agricultural Education for Capiz Province: Delivered April 13, 1909.' *Philippine Agricultural Review*, II, 7 (1909): 405–7.
Coronas, José S.J. *The Extraordinary Drought in the Philippines, October 1911 to May 1912*. Manila: Bureua of Printing, 1912.
Coronas, José S.J. *The Climate and Weather of the Philippines, 1903 to 1908*. Manila: Bureau of Printing, 1920.
Cullather, Nick. 'Miracles of Modernization: The Green Revolution and the Apotheosis of Technology.' *Diplomatic History*, 28, 2 (2004): 227–54.
Cullather, Nick. *The Hungry World: America's Cold War Battle Against Poverty in Asia*. Cambridge, MA: Harvard University Press, 2010.
Cushman, Gregory T. 'The Imperial Politics of Hurricane Prediction: From Calcutta and Havana to Manila and Galveston, 1839–1900,' in *Nation-States and the Global Environment*, eds. Erika Marie Bsumek, David Kinkela, and Mark Atwood Lawrence. New York: Oxford University Press, 2013: 137–62.
Daniel, E.V.B., Henry Bernstein, and Tom Brass, eds. *Plantations, Proletarians and Peasants in Colonial Asia*. London: Frank Cass, 1992.
Davis, Mike. *Late Victorian Holocausts: El Niño Famines and the Making of the Third World*. London: Verso, 2001.
Doeppers, Daniel F. 'Fighting Rinderpest in the Philippines, 1886–1941,' in *Healing the Herds*, eds. Karen Brown and Daniel Gilfoyle. Athens, OH: Ohio University Press, 2010.
Edgerton-Tarpley, Kathryn. 'Tough Choices: Grappling with Famine in Qing China, the British Empire, and Beyond.' *Journal of World History*, 24, 1 (2013): 135–76.
Edwards, Paul N. 'Meteorology as Infrastructural Globalism.' *Osiris*, 21, 1 (2006): 229–50.

Ekbladh, David. *The Great American Mission: Modernization and the Construction of an American World Order*. Princeton: Princeton University Press, 2010.

Elvin, Mark. 'Who Was Responsible for the Weather: Moral Meteorology in Late Imperial China.' *Osiris*, 13, 1 (1998): 213–37.

Forbes, W. Cameron. *The Philippine Islands*, 2 Volumes. Boston: Houghton Mifflin and Sons, 1928.

Gergis, Joëlle L., and Anthony M. Fowler. 'A History of ENSO Events Since A.D. 1525: Implications for Future Climate Change.' *Climatic Change*, 92, 3 (2009): 343–87.

Grove, Richard, and John Chappell, eds. *El Niño: History and Crisis*. Cambridge: White Horse Press, 2000.

Journal of the Philippine Commission. AB No. 233, 'An Act to Authorize the Appropriation of P750,000 Annually for the Promotion, Establishment, and Maintenance of Irrigation Systems in the Philippine Islands.' 10 June 1908, *JPC*, 2, 1, (1908): 302–5.

Journal of the Philippine Commission. AB No. 241, 'An Act Authorizing Municipal and Provincial Governments to Grant, Under Certain Conditions, Privileges, and Concessions for the Utilization of Public Waters for Agricultural Irrigation and for Other Purposes.' *JPC*, 2, 1, (1908): 320.

Journal of the Philippine Commission. AB 1039, 'An Act to Prevent Distress Among the People of the Philippine Islands from Failure of Food Staples,' 1 Feb. 1912, *JPC*, 5, 1 (1912): 656, 699, 782–83, 809–10, 857–58, 1079.

Kidd, Benjamin. *The Control of the Tropics*. New York: The Macmillan Company, 1898.

Kramer, Paul. *The Blood of Government: Race, Empire, the United States, and the Philippines*. Chapel Hill: University of North Carolina Press, 2006.

Kumar, P., T. Lorek, T.C. Olsson, N. Sackley, S. Schmazler, and G. Soto Lavaga. 'Roundtable: New Narratives of the Green Revolution.' *Agricultural History*, 91, 3 (2017): 397–422.

Lala, Ramon Reyes. *The Philippine Islands*. New York: Continental Publishing Co., 1898.

Luyt, Brendan. 'Reading the Minor Forest Bulletins of the Philippine Bureau of Forestry: A Case Study on the Role of Reference Works in the American Empire of the Early Twentieth Century.' *Information and Culture*, 53, 1 (2018): 43–66.

Martin, Charles. 'Observations on the Recent Eruption of Taal Volcano.' *Philippine Journal of Science*, VI, II (1911): 87–91.

May, Glenn Anthony. 'The Business of Education in the American Colonial Philippines,' in *Colonial Crucible*, eds. A.W. McCoy and F.A. Scarano. Madison: The University of Wisconsin Press, 2009: 151–62.

McCoy, Alfred. 'Sugar Barons: Formation of a Native Planter Class in the Colonial Philippines,' in *Plantations, Proletarians and Peasants*, eds. E.V.B. Daniel, Henry Bernstein, and Tom Brass. London: Frank Cass, 1992: 106–41.

McCoy, Alfred, and Francisco Scarano, eds. *Colonial Crucible: Empire in the Making of the Modern American State*. Madison: University of Wisconsin Press, 2009.

Merrill, Elmer D. 'Report of the Botanist.' *Report of the Philippine Commission* Part 2. Manila: Bureau of Printing, 1902: 599–601.

Merrill, Elmer D. 'Report of the Botanist on the Royal Botanical Gardens of Ceylon and the Botanical Gardens at Singapore.' *Report of the Philippine Commission*. Part 2. Manila: Bureau of Printing, 1902: 627–30.

'Monthly Crop Conditions—November 1911.' *Philippine Agricultural Review*, V, 2 (1912): 105–10.

Overfield, Richard A. 'Science Follows the Flag: The Office of Experiment Stations and American Expansion.' *Agricultural History*, 64, 2 (1990): 31–40.

Paredes, Ruby, ed. *Philippine Colonial Democracy*. Quezon City: Ateneo de Manila University Press, 1989.

Patel, Raj. 'The Long Green Revolution.' *The Journal of Peasant Studies*, 40, 1 (2013): 1–63.

Rieppel, Lukas, Eugenia Lean, and William Deringer. 'Introduction: The Entangled Histories of Science and Capitalism.' *Osiris*, 33, 1 (2018): 1–24.

Saderra Masó, Miguel. *The Eruption of Taal Volcano, January 30, 1911*. Manila: Bureau of Printing, 1911.

Sawyer, Frederic. *The Inhabitants of the Philippines*. New York: Charles Scribner's Sons, 1900.

Schumacher, John. 'One Hundred Years of Jesuit Scientists: The Manila Observatory, 1865–1965.' *Philippine Studies: Historical and Ethnographic Viewpoints*, 13, 2 (1965): 258–86.

Shipley, James H. 'Report of the Farm Machine Expert. Report of the Chief of the Bureau of Agriculture for the Year Ending August 31, 1902. Exhibit C.' *Report of the Philippine Commission*. Manila: Bureau of Printing, 1902: 604–6.

Shu, Yeun, Otto Heim, and Kendall Johnson, eds. *Ocean Archives, Indigenous Epistemology, and Transpacific American Studies*. Hong Kong: Hong Kong University Press, 2019.

Stewart, Alonzo. 'Agricultural Conditions in the Philippine Islands, Presented to USDA Secretary James Wilson, 1903.' Serial Set of the 60th Congress, 1st Session, Document 535, 1908.

Suzuki, Nobutaka. 'Upholding Filipino Nationhood: The Debate Over Mindanao in the Philippine Legislature, 1907–1913.' *Journal of Southeast Asian Studies*, 44, 2 (2013): 266–91.

Taylor, Carson. *History of the Philippine Press*. Manila: No Publisher, 1927.

Taylor, Frederick. 'Speech to the 1912 Teachers' Conference in Baguio.' *Philippine Craftsman*, 1, 1 (1912): 54–55.

Udías, Augustin. *Searching the Heavens and the Earth: The History of Jesuit Observatories*. Dordecht: Kluwer Academic, 2003.

Ventura, Theresa. 'Medicalizing *Gutom*: Hunger, Diet, and Beriberi During the American Period.' *Philippine Studies: Historical and Ethnographic Viewpoints*, 63, 1 (2015): 39–69.

Ventura, Theresa. 'From Small Farms to Progressive Plantations: The Trajectory of Land Reform in the American Colonial Philippines.' *Agricultural History*, 90, 4 (2016): 459–83.

Ventura, Theresa. 'I Am Already Annexed: Ramon Reyes Lala and the Crafting of "Philippine" Advocacy for American Empire.' *Journal of the Gilded Age and Progressive Era*, 19, 3 (2020): 426–46.

Vernon, James. *Hunger: A Modern History*. Cambridge, MA: Harvard University Press, 2007.

Villamin, Cymbeline R., ed. *Biographies of Early Scientists in the Philippines, Vols. 1*. Manila: National Science Board, 1976 and 1978.

Warren, James F. 'Scientific Superman: Father José Algué, Jesuit Meteorology, and the Philippines Under American Rule,' in *Colonial Crucible*, eds. A.W. McCoy and F.A. Scarano. Madison: The University of Wisconsin Press, 2009: 508–22.

Warren, James F. 'Philippine Typhoons, Sources and the Historian.' *Water History*, 7, 2 (2015): 213–31.

Wester, P.J. 'The Situation in the Citrus District of Batangas.' *Philippine Agricultural Review*, 6, 3 (1913): 127.

Worcester, Dean C. 'Present State of Agriculture.' *Report of the Philippine Commission* Part 1. Manila: Bureau of Printing, 1902: 297.

Correction to: Droughts, Floods, and Global Climatic Anomalies in the Indian Ocean World

Philip Gooding

Correction to:
P. Gooding (ed.), *Droughts, Floods, and Global Climatic Anomalies in the Indian Ocean World*, Palgrave Series in Indian Ocean World Studies, https://doi.org/10.1007/978-3-030-98198-3

The original version of the book was previously published with incorrect text in Chapters 3 and 6, which has now been corrected. The book has been updated with the changes.

The updated version of these chapters can be found at
https://doi.org/10.1007/978-3-030-98198-3_3
https://doi.org/10.1007/978-3-030-98198-3_6

P. Gooding (ed.), *Droughts, Floods, and Global Climatic Anomalies in the Indian Ocean World*, Palgrave Series in Indian Ocean World Studies, https://doi.org/10.1007/978-3-030-98198-3_12

Index

P. Gooding (ed.), *Droughts, Floods, and Global Climatic Anomalies in the Indian Ocean World*, Palgrave Series in Indian Ocean World Studies, https://doi.org/10.1007/978-3-030-98198-3

The manufacturer's authorised representative in the EU is Springer Nature Customer Service Centre GmbH, Europaplatz 3, 69115 Heidelberg, Germany. If you have any concerns regarding our products, please contact ProductSafety@springernature.com

Printed and bound by CPI Group (UK) Ltd, Croydon, CR0 4YY
22/09/2026
02220984-0007